CALIFORNIA GEOLOGICAL SURVEY
JAMES F. DAVIS, STATE GEOLOGIST

STATE OF CALIFORNIA-GRAY DAVIS, GOVERNOR
THE RESOURCES AGENCY-MARY D. NICHOLS, SECRETARY
CALIFORNIA DEPARTMENT OF CONSERVATION-DARRYL YOUNG, DIREC

Compiled by C.W. Jennings , 1995

0	30	62 Mile
0	50	100 Kilometer

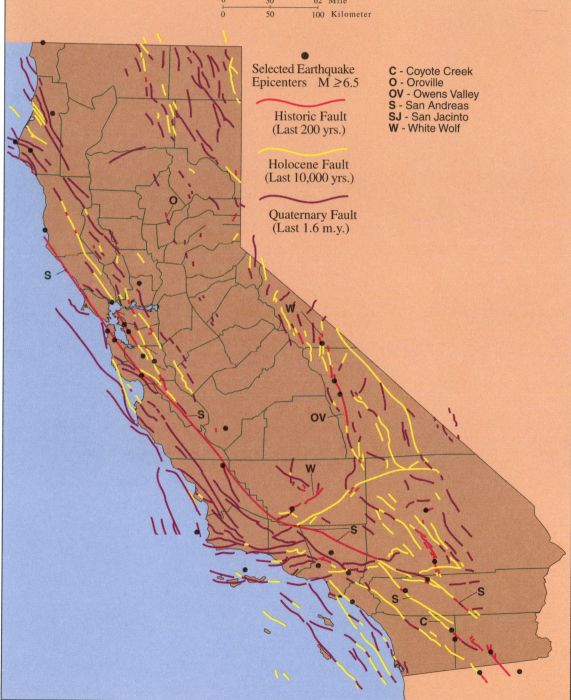

● Selected Earthquake
Epicenters   M ≥6.5

C - Coyote Creek
O - Oroville
OV - Owens Valley
S - San Andreas
SJ - San Jacinto
W - White Wolf

Historic Fault
(Last 200 yrs.)

Holocene Fault
(Last 10,000 yrs.)

Quaternary Fault
(Last 1.6 m.y.)

Simplified Fault Activity Map of California

California Geology

Second Edition

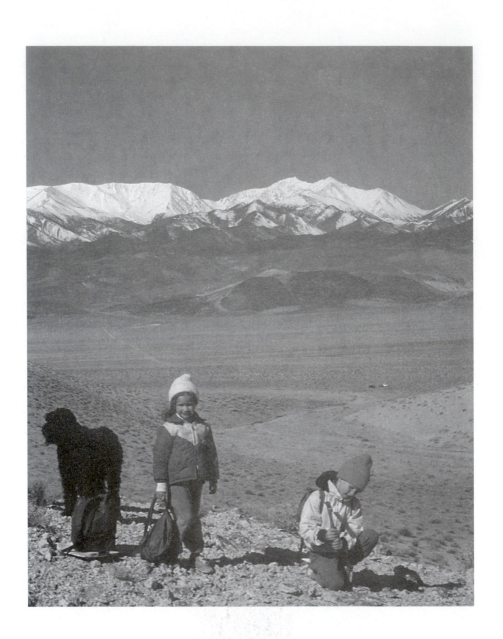

California Geology

Second Edition

Deborah R. Harden

San José State University

PEARSON

Prentice
Hall

Upper Saddle River, NJ 07458

Library of Congress Cataloging-in-Publication Data

Harden, Deborah Reid.
    California geology / Deborah R. Harden.— 2nd ed.
        p. cm.
    Includes bibliographical references.
    ISBN 0-13-100218-X
    1. Geology—California. I. Title.
QE89 .H37 2004
557.94—dc22

2003019497

Editor-in-Chief: John Challice
Executive Editor: Patrick Lynch
Assistant Editor: Melanie Van Benthuysen
Vice President and Director of Production and Manufacturing, ESM: David W. Riccardi
Executive Managing Editor: Kathleen Schiaparelli
Production Editor: Robert Runck
Director of Creative Services: Paul Belfanti
Art Director: Jayne Conte
Cover Designer: Bruce Kenselaar
Art Editor: Connie Long
Manufacturing Manager: Trudy Pisciotti
Manufacturing Buyer: Lynda Castillo
Cover Image: Tufa Formations at Sunrise, Mono Lake, California. © Catherine Gehm

© 2004, 1998 Pearson Education, Inc.
Pearson Education, Inc
Upper Saddle River, NJ 07458

Pearson Prentice Hall® is a trademark of Pearson Education, Inc.

Printed in the United States of America

10 9 8 7 6 5 4 3

ISBN: 0-13-100218-X

Pearson Education Ltd., *London*
Pearson Education Australia Pty. Ltd., *Sydney*
Pearson Education Singapore, Pte. Ltd.
Pearson Education North Asia Ltd., *Hong Kong*
Pearson Education Canada, Inc., *Toronto*
Pearson Educación de Mexico, S.A. de C.V.
Pearson Education—Japan, *Tokyo*
Pearson Education Malaysia, Pte. Ltd.
Pearson Education, Inc., *Upper Saddle River, New Jersey*

With love and thanks to Andrei for all his help and support.

Contents

17 The Peninsular Ranges   464

PART III   CALIFORNIA GEOLOGY: AN INTEGRATED VIEW

18 The Evolution of California Through Geologic Time   483

Preface

Almost 100 years ago, on April 18, 1906, Californians encountered one of the unwelcome consequences of living on an active tectonic plate boundary, when the 1906 earthquake destroyed many of northern California's communities. As we approach the centennial of the earthquake, we can appreciate our increased understanding of California geology and the greater preparedness for natural disaster that has come with that knowledge. Most Californians are aware that earthquakes, volcanic eruptions, floods, and landslides have geological causes, but they are less aware of the positive influences of plate tectonics on the state's history, economy, and quality of life. Resources such as gold, petroleum, and fertile soils are products of California's geologic history, as is the diverse and beautiful landscape of our state. The supply, distribution, and safety of water in California will continue to be a major force in the state's future, and they are partly controlled by geology.

These are some of the reasons that students are attracted to beginning geology courses. Many of these students are not particularly interested in detailed information about rocks, formations, geologic models, or ages of materials. Many come to these courses without prior background in geology and without fluency in the complex vocabulary of geology and science in general. This book is intended to acquaint those students with the basic geologic history of California and to cover those topics that do interest non-geologists. I have included examples of natural disasters, aspects of California history that are tied to geology, geological resources, and discussions of the impact of geology on California society. These topics bring the relevance of geology to the daily context of California life.

PURPOSE

This book is intended for a semester-long course in California geology, but covering all of the chapters in detail during that time is not a realistic goal. Instructors are encouraged to cover the introductory and final chapters and to select the chapters from Part II that focus on their own location and interest. A college student, pre-college teacher, or interested citizen should be able to comprehend the material without having had a prerequisite course in geology. However, the content is also informative to geology majors, graduate students, and practicing geologists not intimately acquainted with California geology.

Readers who complete this book should have a solid understanding of the following major concepts:

- Geology is a logical, non-intimidating, and enjoyable science.
- Geologists study geologic processes to better understand California's geologic history as it is preserved by rocks, sediments, and structures.
- Maps and cross sections are a fundamental tool of geology.
- Plate tectonics has been the dominant factor that shapes California's geology and landscape.
- California's natural resources and hazardous natural events result from our geologic and tectonic setting.
- Geologists work with government agencies to mitigate the impact of earthquakes, volcanic eruptions, floods, and landslides on California citizens, and to inform us about risks.
- California has many beautiful natural landscapes and geologic wonders that we can enjoy at hundreds of parks and preserves throughout the state.

ORGANIZATION

California Geology is divided into three sections. The first four chapters are intended to introduce readers to the basic principles of geology. Because the basic concepts are illustrated with California examples, those with previous geology background will also benefit from the material in Part I.

Part II, Chapters 5 through 17, covers the major features of California's provinces and important geologic processes that are not specific to a single province. For each province, I have selected the most important geologic features or events. Therefore, these chapters do not cover all rock types, events, or geologic features found in each part of California. These chapters begin with less geologically complex areas, and later chapters assume that the reader has acquired the fundamental knowledge given in earlier chapters. However, it is possible to cover these chapters in a variety of orders and to eliminate some entirely during a semester, as I do during my own course. For courses with emphasis on southern California, I would advise beginning with the desert provinces and the Sierra Nevada before moving to the Transverse or Peninsular ranges chapters.

In my opinion, it is important to cover the final two chapters (Part III) during any course on California geology because they unify the material presented in earlier chapters. A synthesis of the geologic development of California through time, presented in Chapter 18, is a good way to review the material in Part II. The final chapter of the book reviews the geologic processes that created hazards for California citizens, the geologic causes of those hazards, and the measures taken to mitigate and prepare for hazardous events. This is perhaps the most important "take-home lesson" for non-geologists who plan to live and work in California.

NEW FOR THE SECOND EDITION

Chapter Summaries: A list of major points or events appears at the end of each chapter. For chapters covering California's geomorphic provinces, major geologic events are listed in chronological order.

Province Highlights: Major geographic features, dominant rock types and tectonic setting, recent earthquakes, and major parks are summarized in a table at the beginning of chapters covering California's geomorphic provinces. Simplified index maps and an improved map of the geomorphic provinces allow students to better locate themselves within the state.

Web and Media Resources: An updated list of Web sites and videos is included in the reference section at the end of each chapter.

Highlights of National Parks: Chapters covering California's geomorphic provinces also highlight the geologic setting and major features of California's 12 national parks and monuments, as well as selected state parks.

Updated Maps and Diagrams: Newly available maps and diagrams, many from the U.S. Geological Survey and the California Geological Survey, help to clarify basic principles and California's geologic history.

More Coverage of Southern California Geology: Updates are included on tectonics, basement geology of southern California, and late Cenozoic tectonic rotation.

Paleogeographic Maps: Generalized maps, provided by Dr. Ron Blakey of Northern Arizona University, show the tectonic development of California in geologic time. These maps appear in Chapter 18.

New Topics: Additional topics covered in this edition include: Owens Lake and its impact on the environment, caves of the Sierra Nevada Province, Lake Tahoe, water supply controversies in southern California, groundwater contamination by perchlorate, vernal pools, tsunami hazards along the California coast, the Hector Mine earthquake, and the St. Francis dam disaster.

Slide Set: Also available with the new edition is a slide set of 50 slides from locations throughout California, which illustrate various geologic processes discussed in the book. Contact your local Prentice Hall representative to obtain a set.

ACKNOWLEDGMENTS

As was true for the making of the first edition, the driving force behind the second edition are the students taking the course "Geology of California" at San José State University. Using the first edition in the classroom during the past five years was the best way to discover points of confusion and to improve organization and clarity.

Colleagues, particularly those at the U.S. Geological Survey and the California Geological Survey, were most generous with their time and provided new and improved information and illustrations for this edition. I am particularly indebted again to Andrei Sarna-Wojcicki, of the U.S. Geological Survey and the Harden-Sarna family, for his extensive help with illustrations.

I am grateful to the peer reviewers who provided many suggestions for improving both the content and form of the first edition: Michael Forrest of Rio Honda College, Sally McGill of California State University-San Bernardino, Elizabeth Miller of Stanford University, Robert Wallen of Mendocino College, and Thomas Williams of U.C. Berkeley Extension. Dr. Forrest made the excellent suggestion of highlight tables, and both he and Dr. McGill provided helpful resources for improving the coverage of southern California topics. I would also like to thank several other readers of the book for sending me careful lists of errors and thoughtful suggestions: W.C. Bradley of the University of Colorado, James Court of City College of San Francisco, N.K. Huber of the U.S. Geological Survey, and Jeffrey Shaeffer of Napa, California.

Thanks are also due to the many people, including my students, who pointed out errors (most frequently the mis-captioned photograph of Tenaya Lake!). I appreciate the time and thoughtful consideration that went into these reviews.

Many people at Prentice Hall deserve much credit for bringing the second edition to fruition. Patrick Lynch supplied the encouragement to begin the task, and Melanie Van Benthuysen, the book's assistant editor, was the capable coordinator of the effort. Robert Runck and Erin Connaughton led the nSight team in production, and Connie Long coordinated the graphics program.

Deborah R. Harden

Please contact Dr. Harden with comments, corrections, and suggestions:

Dr. Deborah R. Harden
Dept. of Geology
San José State University
San José, CA 95192-0102
harden@geosun.sjsu.edu

About the Author

Deborah R. Harden is a Professor in the Department of Geology at San José State University, where she teaches undergraduate and graduate courses in California geology and Earth-surface processes. Her particular interests are in landscape-forming processes, rivers, the effects of faults on streams, human impacts on watersheds, and California geology.

Before coming to San José State University, Dr. Harden worked as a geologist for a private engineering consulting firm and as a hydrologist for the U.S. Geological Survey. She has a Ph.D. from the University of Colorado and B.S. and M.S. degrees from Stanford University, all in geology.

Dr. Harden has had the privilege and opportunity of visiting and studying geology and landscapes in many parts of the world, both professionally and with her family.

California Geology

Second Edition

Basic Principles of Geology

The first four chapters of this book are designed to introduce readers with little or no geology background to some basic principles of geology. These principles are discussed with particular application to California geology, and geologic maps and cross sections are also introduced. Readers with a limited background in geology are also encouraged to supplement this information with additional readings from introductory geology textbooks as needed.

1

Plate Tectonics and California

OVERVIEW

Most Californians know that earthquakes and the San Andreas fault have something to do with movement of the Earth's plates, but few residents realize the importance of *plate tectonics* in shaping California. According to the plate tectonics theory, which has been widely accepted for the past 25 years, the Earth's outer layers consist of a number of rigid plates (Fig. 1-1). These plates are moving, and their encounters with each other cause many of Earth's dynamic events. Today, plate tectonics is responsible for some obviously unwelcome aspects of our lives—earthquakes and volcanic eruptions—as well as some of the scenery that attracts people to California—rugged mountain ranges and beautiful, rocky coastlines. The region that we now call California has been at an active plate boundary for the past 230 million years, and this has been the dominant factor in California's geologic history.

THE LAYERED EARTH

Although the deepest holes drilled into the Earth's interior have only penetrated 10,000 meters, scientists have long known from indirect evidence that the Earth is a layered sphere. The outermost thin skin of the Earth is the *crust*, a rigid layer of rocks rich in silicon and oxygen (Fig. 1-2). The crust is only about 5 kilometers thick beneath the oceans and 50 or more kilometers thick beneath continental mountains like the Sierra Nevada (Fig. 1-3). Beneath the crust is the *mantle,* which extends inward about halfway to the center of the Earth and makes up more than 80 percent of the Earth's volume. Like the crust, the mantle is composed of rocks rich in silicon and oxygen, but its rocks also contain more iron and magnesium, making them denser than those found in the crust.

Rocks believed to originate in the mantle are found in several places in California. They were brought to the surface by different processes, including volcanic eruptions and

3

Fig. 1-1 Map showing the Earth's major plates. (Source: Modified from Coch, N., and Ludman, A. 1991. *Physical Geology.* New York: Macmillan Publishing Co., pp. 14-15.)

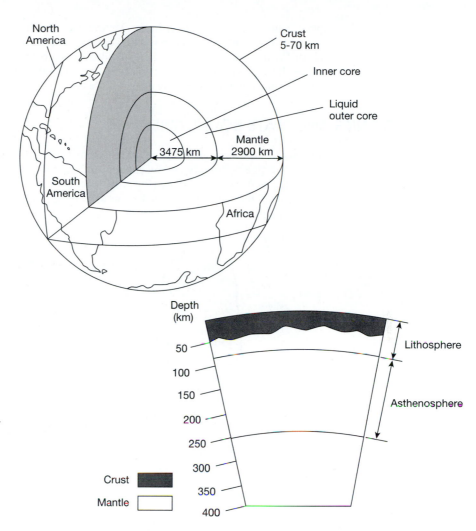

Fig. 1-2 Slice through the Earth, showing its layered interior (top). Enlargement of the outer 400 km (right). (Source: Modified from Coch, N., and Ludman, A. 1991. *Physical Geology.* New York: Macmillan Publishing Co., pp. 30, 32.)

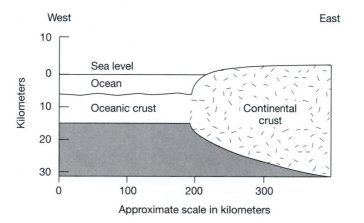

Fig. 1-3 A cross section beneath a continental margin showing the contrasts in the thickness of the crust. (Source: Modified from Skinner, B., and Porter, S. 1989. *The Dynamic Earth.* New York: John Wiley & Sons, p. 7.)

plate collisions. For example, in the Mojave Desert, bits of mantle rock have been brought to the Earth's surface in recently erupted lava (see Chapter 7). Indirect evidence about Earth's deeper interior is also provided by the behavior of seismic (earthquake) waves. Beneath the mantle is the *core*, which has a liquid outer part and a solid inner part. Both the inner and outer core are thought to be composed mostly of iron and nickel.

PLATE TECTONICS AND PLATE BOUNDARIES

Interactions between plates generate much of Earth's geologic activity. California owes its active seismicity and recent volcanic eruptions to the fact that both the Pacific Plate and the Gorda Plate (Fig. 1-4) meet the North American Plate in California. Most of the world's earthquakes occur at plate boundaries; in fact, the zones of concentrated seismicity are one of the main lines of evidence used to identify plate

Fig. 1-4 Major plate tectonic features of the western United States. (Source: Modified from the U.S. Geological Survey, *This Dynamic Earth.* *http://pubs.usgs.gov/ publications/text/ understanding.html.*)

Fig. 1-5 Characteristic features of divergent plate boundaries. (Source: Skinner, B., and Porter, S. 1989. *The Dynamic Earth.* New York: John Wiley & Sons, p. 388.)

boundaries. Along plate boundaries, the different motions of the plates create stress, and that stress is relieved during earthquakes (see Chapter 13).

Plates can interact in three ways. They may separate along *divergent* plate boundaries, where the crust is stretched and thinned (Fig. 1-5). At divergent boundaries, molten magma rises and erupts to form new oceanic crust. Today an active divergent boundary in the Gulf of California is causing Baja California, Mexico to break away from the Mexican mainland as new oceanic crust forms in the Gulf. This process is one example of *seafloor spreading,* the process that gave birth to the Atlantic Ocean as the Americas separated from Europe and Africa. Today satellite systems and laser-instrumented surveying devices allow scientists to measure the opening of the Gulf of California from space. Based on the results of repeated surveys with global positioning satellite (GPS) systems, it appears that the area is spreading at the rate of about 5 centimeters per year.

Tectonic plates collide at *convergent* boundaries. At these zones, crust is either piled up into mountain ranges by buckling, faulting, and folding along the plate boundary, or shoved into the mantle during *subduction*. Subduction occurs if one or both of the converging plates is an oceanic plate. At the point of convergence, one plate—always the oceanic crust if the other plate is a continent—is overridden and driven beneath the other plate along a subduction zone. Along most subduction zones, a deep ocean *trench* marks the point where the plates converge (Fig. 1-6). As

Fig. 1-6 Characteristic features of subduction zones. (Source: Modified from the U.S. Geological Survey, *This Dynamic Earth:* http://pubs.usgs.gov/publications/text.)

the plate descends into the mantle, earthquakes are generated. Subduction zones are the only tectonic settings where earthquakes originate at great depths, and they also give rise to the largest magnitude earthquakes. Earthquakes and their magnitudes are discussed in Chapters 13 and 14.

As subduction drives the dense, water-rich rocks of the oceanic plate deeper within the Earth's interior, water is driven out of the subducting rocks. This dewatering causes partial melting in the mantle above the subducted rocks, creating subduction-zone volcanoes. As we will see in Chapter 5, these volcanoes are quite different from those found at divergent boundaries. Today along the northern margin of California, the Gorda Plate is being subducted beneath the North American Plate (see Fig. 1-4). As a result, northwestern California faces the threat of a great subduction-zone earthquake, and areas farther east are at risk of major volcanic eruptions from Mt. Shasta and Lassen Peak.

During convergence and subduction, pieces of the subducted plate may also be scraped off along the plate margin, where they become attached to the upper plate. These pieces are referred to as ***accreted terranes***. Throughout California, examples of ancient collisions are found in the form of oceanic fragments attached to the North American Plate (see Chapters 8, 9, and 12). In many accreted terranes, fragments of the mantle have been scraped up along with the oceanic crust, providing geologists with another opportunity to study rocks formed at great depths (see Chapter 9).

Plate boundaries where two plates are moving past each other without diverging or converging are known as ***transform*** plate boundaries. Along transform boundaries, earthquakes record the plate motions as they do along the other boundary types. However, volcanic eruptions are generally absent. The most famous transform plate boundary in the world is California's San Andreas fault system, which separates the Pacific and North American plates (Fig. 1-4). The faults of the San Andreas system show that the Pacific Plate is moving to the northwest relative to the North American Plate. The relative motion, or the ***displacement,*** on these faults is in a right-lateral direction, as illustrated in Fig. 1-7.

The actual boundary between the Pacific and North American plates is best thought of as a wide zone rather than a line. The effects of plate motion, such as active

Fig. 1-7 Right-lateral displacement is characteristic of the transform plate boundary in California. In the San Francisco Bay area, right-lateral movement takes place on several faults of the San Andreas system. (Source: Modified from U.S. Geological Survey, *San Andreas Fault,* a general-interest publication.)

Fig. 1-8 Topography of the United States, illustrating the rugged landscape of the western continent. The effects of recent tectonics are seen far east of the San Andreas fault. (Source: Thelein, G., and Pike, R. 1991. U.S. Geological Survey, digital shaded-relief map of the conterminous U.S., Map I-2206.)

transform faults and recently uplifted mountain ranges, are seen across the width of California and even farther east (Fig. 1-8). The San Andreas fault is conventionally thought of as the plate boundary, but only about half to two-thirds of the plate motion takes place within the San Andreas fault zone. In the San Francisco Bay area and south of the Transverse Ranges, plate motion takes place on other faults of the San Andreas system east and west of the San Andreas itself (Fig. 1-7). Additional lateral motion takes place along the eastern edge of the Sierra Nevada, in the central Mojave desert (see Chapter 14), and even east of California. The complexities of California geology are revealed when one realizes that even the question "Where is the exact boundary between the Pacific and North American plates?" has no precise answer.

COASTLINES AND PLATE BOUNDARIES

As shown on the map of the Earth's tectonic plates (Fig. 1-1), not all plate boundaries are at the borders between continents and oceans. Places where the edge of a continent does coincide with a plate boundary are termed ***active continental margins,*** whereas continental margins far from active plate boundaries are known as ***passive continental margins.*** Today, the east coast of the United States is a passive margin, whereas the west coast, including California, is an active margin.

During its early history, in Paleozoic time, the area we now call California was a passive margin (see Chapter 7). It then became a subduction-zone active margin during the Mesozoic and early Cenozoic era (see Chapters 8 and 12). During the past 25 million years, a transform plate boundary has been present along most of California, although subduction continues along northern California.

CAUSES OF PLATE MOTIONS

Within the Earth's upper mantle is a layer known as the *asthenosphere,* a weak, semiplastic layer that underlies the moving plates, or *lithosphere.* This layer is important in understanding plate tectonics, because the slow flow of the asthenosphere is the "conveyor belt" that moves the plates. The boundary at the base of the plates, where relatively cool lithosphere meets hotter, more plastic asthenosphere, is about 100 to 250 kilometers beneath the Earth's surface. Scientists believe that they have located the top of the asthenosphere using seismic techniques, but they are less certain about the mechanisms that cause plates to move.

Until recently, geologists believed that *convection currents* in the asthenosphere caused the rigid plates above it to move. These convection currents cause hot mantle rock to rise, cool, and then sink again, creating circulation within the upper mantle (Fig. 1-9). More recent ideas about plate tectonics focus on density and gravity differences within the lithosphere as the driving forces. After molten material rises to the surface and forms new oceanic crust at divergent boundaries, the plate cools and becomes denser. As a result, the cooler part of the plate sinks back into the asthenosphere at convergent boundaries, pulling the rest of the plate with it (Fig. 1-9). Geologists hypothesize that this cooling and sinking of a plate could take as much as 60 to 70 million years.

TRACKING THE MOVEMENT OF PLATES

Because all of the Earth's plates are mobile, the plate boundaries provide clues only to the relative motions between plates. The actual movement directions and rates for individual plates—as they would be seen from space, for example—are

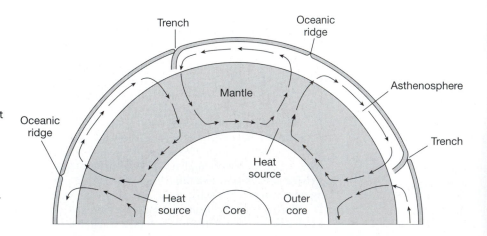

Fig. 1-9 Plate movement is thought to be related to convection in the Earth's mantle. (Source: Hamblin, W., and Christiansen, E. 1995. *Earth's Dynamic Systems*, 7th ed. Upper Saddle River, NJ: Prentice Hall, p. 490.)

Fig. 1-10 Station map showing the Southern California Integrated GPS Network. Repeated surveys of the stations show tectonic movements in the Los Angeles basin (Source: *http://www.scign.org*.)

measured in several ways. GPS systems are one method for precisely determining plate motions or the vertical growth of mountain ranges. GPS data are currently being used for many geologic applications, including the monitoring of plate motions. For example, the Southern California Integrated GPS Network (*http://www.scign.org*) is a network of approximately 250 GPS receivers in the Los Angeles Basin and surrounding region (Figs. 1-10 and 1-11). This network continuously records the tectonic movement of the Earth's crust and can measure changes in the ground surface on a scale of millimeters. The Southern California Earthquake Center, NASA's Joint Propulsion Laboratory, and the U.S. Geological Survey established the network following the 1994 Northridge earthquake (see Chapters 13 and 14).

Geologists can also determine the long-term movement of plates at *hot spots.* These are concentrated areas of very high heat flow deep within the mantle that cause rocks to melt above them. Because these hot spots are located below the

Fig. 1-11 The GPS site at the East Los Angeles Science Center. The survey point is located beneath the tripod. (Source: J. Galetzca, U.S. Geological Survey, Pasadena, CA.)

plates, scientists believe that the spots remain fixed while the plates move over them. Directly above the hot spots, lava rises through the plate to form volcanoes. As the plate moves, new volcanoes appear in the part of the plate newly positioned over the hot spot. After millions of years of plate motion over a fixed hot spot, a chain of volcanoes and volcanic rocks is left as a marker of the direction and the long-term rate of plate motion. In this way, the Hawaiian islands record the movement of the Pacific Plate (Fig. 1-12). Similarly, the volcanic rocks of eastern Oregon, Idaho, and Wyoming record the movement of the North American Plate over the Yellowstone hot spot.

The absolute direction of movement of the Pacific Plate is northwestward, whereas that of the North American Plate is southwestward. Where the two plates meet along the northwest-oriented boundary, the relative motion between

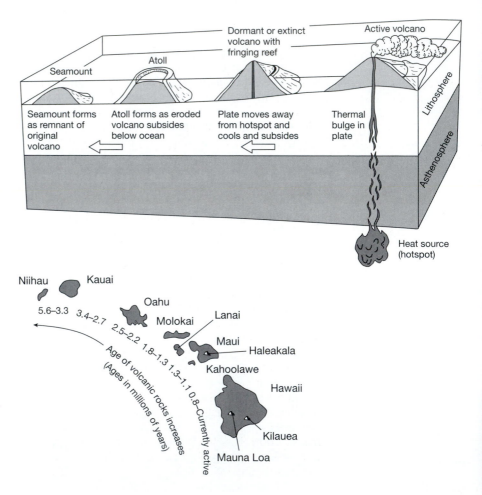

Fig. 1-12 The Hawaiian islands record the movement of the Pacific Plate during the past 5.6 million years. (Source: Modified from Coch, N., and Ludman, A. 1991. *Physical Geology.* New York: Macmillan Publishing Co., p. 444.)

them results in the right-lateral motion that characterizes the San Andreas fault system (Fig. 1-4; **see also** Chapter 14). A small amount of this motion, about 10 percent, is convergent, resulting in the formation of mountains, as discussed next.

DEFORMATION

Movement between plates creates forces that act on the rocks located near plate boundaries. In any given area, force applied to a particular body of rock, regardless of its size, creates *stress* on the rock. Tectonic stress causes rocks and landforms to be disrupted, creating features that geologists can use to interpret the forces that created them. By studying these patterns of *deformation* and analyzing their geometry, geologists can infer the modern stress conditions, as well as the ancient forces recorded in the geologic features of an area. However, because plate boundaries are not simple planes between homogeneous blocks of rock, the patterns of deformation are often complex and difficult to interpret.

In general, convergent motion causes rocks to be *compressed* (Fig. 1-13). Compression causes rocks to be crumpled and buckled by folding and faulting, all of which act to shorten the Earth's crust (Fig. 1-14). *Reverse and thrust faults* result from shortening (Fig. 1-15a), and these are the types of faults found in regions undergoing compression, such as the Transverse Ranges (see Chapter 16). Divergent motion causes the crust to be *extended.* As the crust is extended or stretched, blocks of rock can slide apart, creating depressions (Fig. 1-16). In these settings, including California's Basin and Range Province (see Chapter 7), *normal faults* result (Fig. 1-15b). Further complexity can be added to the picture. For example, a mountain range might be uplifted in response to compression, and the uplift itself might at the same time cause extension along the crest of the range.

An area affected by purely transform motion is subjected to *shear* stress (Fig. 1-13), creating *transform or strike-slip faults* (Fig. 1-15c) like the major faults of the San Andreas system (see Chapter 14). However, the forces operating along the San Andreas fault system are not created by pure transform motion. In many areas, convergent motion accompanies the transform motion along the plate boundary; geologists refer to this combination of forces as *transpression* (Fig. 1-17). As a result, vertical movements are recorded during earthquakes, and entire mountain ranges have been buckled and uplifted in the fault zone during the past few million years. California's Coast Ranges, Transverse Ranges, and even the Sierra Nevada are testimony to the significant recent compression along the plate boundary. During the geologically recent past, a significant component of extension accompanied transform motion along the San Andreas system, caused by *transtension* (Fig. 1-17) along the plate boundary. This past extension is recorded in a series of fault-related marine basins and volcanism, as reconstructed from rocks of that period.

Compression

Extension

Transform motion

Fig. 1-13 This cartoon illustrates compression, extension, and transform motion. (Source: Coch, N., and Ludman, A. 1991. *Physical Geology.* New York: Macmillan Publishing Co., p. 466.)

Fig. 1-14 Folded and faulted rocks record the shortening of the Earth's crust that results from compression. (Source: Creely, R.S. 1997.)

(a) Reverse Fault   (b) Normal Fault

(c) Right Lateral

Fig. 1-15 Block diagrams illustrating: **A,** Normal fault; **B,** Reverse and blind thrust faults; **C,** Transform or strike-slip faults showing right- and left-lateral offset. (Source: U.S. Geological Survey, Southern California Earthquake Center, *Putting Down Roots in Earthquake Country.* *http://www.scec.org.*)

Fig. 1-16 Depressions on the land surface may form as the crust is pulled apart. **A,** A fault-bounded valley, or **graben,** develops; **B,** a smaller and more localized depression contains a **sag pond**. (Source: Creely, R.S. 1997.)

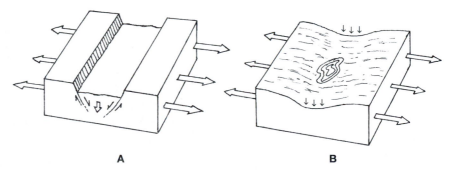

A   B

Fig. 1-17 Transpression or transtension can be created if compression or extension accompany transform motion. Bends in a fault (**B**) can create transtension (**A**) or transpression (**C**), depending on the type of fault motion (shown by arrows). (Source: Creely, R.S. 1997.)

A   B   C

SUMMARY

1. According to the theory of plate tectonics, the Earth's lithosphere is composed of a number of rigid plates that are in motion. The plates are driven by slow flow within the asthenosphere beneath them.

2. The crust is much thinner beneath oceans than under continents, making continental plates thicker and less dense than oceanic plates.

3. Plate motions can be measured from satellite instruments using the GPS system and from the long-term movement of the plates over hot spots.

4. Much of Earth's geologic activity occurs at the boundaries between plates. Geologists know where the boundaries between plates are by the worldwide pattern of earthquake epicenters. Shallow earthquakes occur at all active plate boundaries.

5. Plate boundaries are divergent, convergent, or transform. At divergent boundaries, plates move apart, and new lithosphere forms along the boundary. At oceanic spreading ridges, lava rises to create ocean-floor volcanoes and new ocean floor.

6. Convergent boundaries are of three types, and all result in the destruction of lithosphere. Convergence between two continental plates involves collision and formation of new mountain ranges. If the converging plates are both oceanic plates or if one of the plates is oceanic and one continental, then subduction occurs. Along an oceanic–continental subduction zone, the denser oceanic plate is carried beneath the continental lithosphere. At subduction zones, volcanoes are created on the upper plate, a deep ocean trench marks the point on Earth's surface where the plates meet, and deep earthquakes are generated as the subducting plate descends.

7. Transform boundaries involve the lateral motion of two plates without convergence or divergence. Today, the San Andreas fault is a transform boundary between the Pacific Plate and the North American Plate. It runs along most of California, with its northern end near Cape Mendocino and its southern end near the California–Mexico border. Along the transform boundary, the relative motion between the two plates is right-lateral, with the North American side of the fault moving southeast relative to the Pacific side, which is moving northwest.

8. North of the San Andreas fault, the plate boundary along the western edge of North America is a subduction zone, with the Gorda and Juan de Fuca plates being driven beneath the North American Plate. Offshore, the Pacific Plate and the Gorda/Juan de Fuca Plate form a divergent boundary at a spreading ridge.
9. South of the San Andreas fault, a divergent boundary is separating the mainland of Mexico from the Baja Peninsula, forming the Gulf of California.
10. Rocks provide information about the tectonic forces acting on them, even if those forces are no longer present. Folds and faults of different types record compression, extension, and shearing.

REFERENCES

HOWELL, D.G., ed. 1985. *Tectonostratigraphic Terranes of the Circum-Pacific Region.* Houston: Circum-Pacific Council for Energy and Mineral Resources, Earth Science Series, No. 1, 581 pp.

MARSHAK, S. 2001. *Earth: Portrait of a Planet.* New York: Norton, 735 pp.

MOORES, E.M., and TWISS, R.J. 1995. *Tectonics.* New York: W.H. Freeman & Co., 415 pp.

TARBUCK, E.J., and LUTGENS, F.K. 2002. *The Earth,* 7th ed. Upper Saddle River, NJ: Prentice Hall, 670 pp.

Web Sites and Media Resources

This Dynamic Earth: The Story of Plate Tectonics: *http://pubs.usgs.gov/publications/text*

Plate Tectonics for Beginners: Geology in the Parks: *http://www2.nature.nps.gov/grd/usgsnps/pltec/pltec1.html*

TASA Graphic Arts. 2002. *The Theory of Plate Tectonics,* version 2.0: interactive CD-ROM (E.J. Tarbuck, D. Tasa, and L. McNeill, authors): *http://www. tasa-graphicarts.com*

National Earthquake Information Center and World Data Center for Seismology: *http://neic.usgs.gov/*

GPS Network, Southern California Integrated GPS Network: *http://www.scign.org/*

Animations of plate motions, U.S. Geological Survey: *http://www.aqd.nps.gov/grd/usgsnps/animate*

Animations of plate motions, including California: Dr. Tanya Atwater, University of California, Santa Barbara, including free download of viewing software: Navigate to "Freeware" from the Outreach site: *http://www.geol.ucsb.edu/L2/Outreach-FR.html*

Animations from the University of California, Berkeley, Museum of Paleontology: *http://www.ucmp.berkeley.edu/geology/tectonics.html*

California's Rocks

Because of its long history as an active plate margin, California has an unusual assortment of rocks. Rocks that are common in many parts of the world—such as limestone—are uncommon in much of California, and rocks that are very rare in most places—such as *serpentinite*—are fairly abundant. Although it is beyond the scope of this book to discuss the formation and classification of rocks in detail, this chapter presents a brief overview of California's rocks. Later chapters include more detailed discussions of some rock types, and many excellent sources are available for more comprehensive reading.

Even rocks that appear to be completely homogeneous are, in most cases, composed of a multitude of individual grains. These grains can be mineral crystals, pieces of rock or minerals, fragments of animal or plant remains, or a combination of two or more of these. The types of grains and their proportionate amounts define a rock's *composition*. The size and shape of the grains and the way the grains fit together define its *texture*. Both the composition and texture provide important clues about the origin of each rock, which is the reason that identification and classification of rocks form an important part of any geologist's work. Field study of rocks as they appear in the context of surrounding rocks and the landscape is an important part of the identification process (Fig. 2-1). Closer examination of hand samples with a magnifying lens can aid in the identification of minerals or fossils. Rocks composed of grains too small to be seen with a hand lens require examination with a microscope (Figs. 2-2 and 2-3).

The minerals and textures of a given rock reflect its source, providing clues about either the preexisting rocks or the fluids from which it formed. For a particular mineral to form, the correct chemical "building blocks" must be present. For example, iron oxide cannot form in an environment where oxygen is absent, and quartz cannot crystallize from pure water. Even if the correct ingredients are present, some minerals only form if certain temperatures and pressures exist. After decades of detailed laboratory studies, geologists also know the temperature and pressure conditions favorable for the formation of most minerals. The presence of certain minerals in a rock reveals whether the rock formed at or near Earth's surface or thousands of meters beneath it.

Fig. 2-1 A geologist examines volcanic rocks and sediments near Mono Lake. (Source: Sarna-Wojcicki, A. 1995. U.S. Geological Survey.)

Fig. 2-2 Rock samples can be cut into *thin sections.* Geologists view the thin sections under a polarizing microscope to identify the component minerals and study the microscopic texture of the rock. (Source: Coch, N., and Ludman, A. 1991. *Physical Geology.* New York: Macmillan, p. 77.)

Fig. 2-3 Thin section of granite showing an interlocking network of silicate minerals.

MINERALS: THE BUILDING BLOCKS OF ROCKS

When confronted with the variety and complexity of California's rocks, a beginning student of geology has a difficult time believing that rocks are constructed in any systematic order. In fact, rocks are made of well-ordered building blocks that are themselves composed of orderly materials. Rocks are composed of **minerals**, minerals are composed of one or more **chemical elements**, and the chemical elements are made up of **atoms**, the basic building blocks of all matter. It is the variety of possible combinations of elements and minerals and the variety of physical and chemical conditions of formation that create the variety in rock composition. On a larger scale, the hierarchy of Earth's building blocks continues: bodies of similar rock are grouped together in geologic **formations**, as discussed in Chapter 4. Rock formations can in turn constitute a distinctive geologic terrane. On a still grander scale, lithospheric plates—some of them aggregates of terranes—are the building blocks of the Earth's crust.

A mineral is a naturally occurring, inorganic, crystalline solid. A **crystalline solid** is a material with atoms that are arranged in an orderly, regularly repeating arrangement. Minerals can be composed of atoms of only one element, as is the case with gold nuggets. However, most of the nearly 3,000 known minerals are made up of more than one element combined in fixed proportions in a regular crystalline arrangement. Even minerals that appear to be dirty, lumpy specks

are in fact constructed with beautifully ordered atoms of their component elements. If minerals grow under ideal circumstances—with adequate space, time, and chemical conditions—the internal order may be reflected in the formation of crystals (Fig. 2-4A, 2-4B).

Many people are familiar with the term *mineral* as it is sometimes used for nutritional supplements, such as iron, calcium, or magnesium. Strictly speaking, these are chemical elements rather than minerals. Only 10 of the 105 known chemical elements make up 99.9 percent of the Earth's crust, and two—oxygen and silicon—account for almost 75 percent. It is therefore not surprising that the most common minerals contain oxygen and silicon. These minerals are known as the **silicate minerals**, and they are the building blocks of most rocks (see Table 2-1). Although more than 600 minerals are found in California, and about 50 of these are found nowhere else, most of California's rocks are made up predominantly of the common rock-forming minerals.

Some of the rare elements in the Earth's crust are extremely valuable to society, and rare minerals and rocks that contain these elements have long been sought after by miners. Geologic events have enriched California rocks with pockets of valuable commodities such as gold, mercury, lithium, chromium, boron, and other elements. As we will see in subsequent chapters, mineral resources have always played an important role in California's history.

Sodium ion

Chloride ion

Fig. 2-4 A, The internal atomic structure of halite. **B**, Halite (salt) crystal. (Photo by author.)

A

B

The texture of a rock also reflects its conditions of formation. In general, all rocks display one of two general textures. Rocks with a ***crystalline*** texture display a network of interlocking mineral grains, indicating that crystals grew together (Fig. 2-5). In contrast, rocks with ***clastic*** texture are composed of mineral or rock

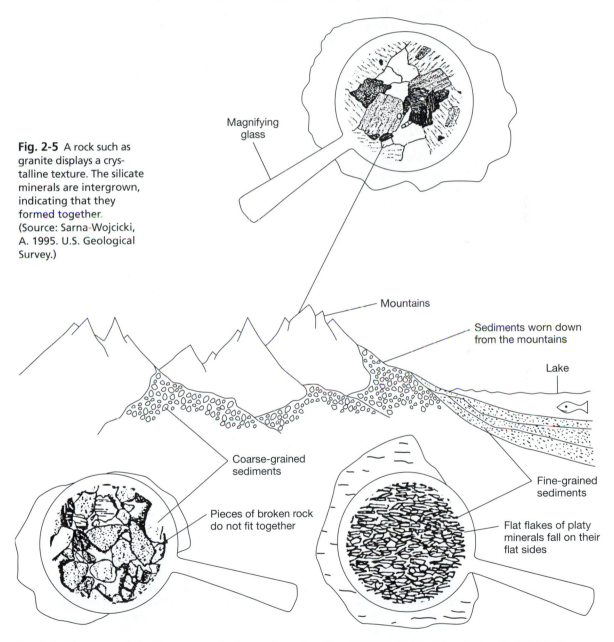

Magnifying glass

Fig. 2-5 A rock such as granite displays a crystalline texture. The silicate minerals are intergrown, indicating that they formed together. (Source: Sarna-Wojcicki, A. 1995. U.S. Geological Survey.)

Mountains

Sediments worn down from the mountains

Lake

Coarse-grained sediments

Pieces of broken rock do not fit together

Fine-grained sediments

Flat flakes of platy minerals fall on their flat sides

Fig. 2-6 Rocks composed of fragments of other rocks, such as the sediments shown here, display clastic texture. (Source: Sarna-Wojcicki, A. 1995. U.S. Geological Survey.)

TABLE 2-1   EXAMPLES OF COMMON MINERALS IN CALIFORNIA ROCKS

Mineral Group	Minerals	Component Elements	Occurrence
Silicates	Quartz	Silicon (Si), oxygen (O)	All rock and sediment types
	Feldspars	Si, O, aluminum (Al), plus:	All rock and sediment types
	Potassium	Potassium (K)	
	Plagioclase	Calcium (Ca), sodium(Na)	
	Clays	Si, O, Al, hydrogen (H), many others	Sediments, soils
	Olivine	Si, O, magnesium (Mg), iron (Fe)	Igneous rocks
	Serpentine	Si, O, Mg, Fe, O, H	Serpentinite
	Pyroxenes Augite	Si, O, Al, Ca, Mg, Fe	Igneous, metamorphic rocks
	Amphiboles Hornblende	Si, O, Al, Na, Ca, Mg, Fe, and others	Igneous, metamorphic Rocks
	Biotite mica	Si, O, K, Mg, Fe, O, H	All rock types
	Garnet	Si, O, Al, Ca, Mg, Fe	Metamorphic rocks
Carbonates	Calcite	Carbon (C), O, Ca	Sedimentary rocks, sediments, marble
	Dolomite	C, O, Ca, Mg	Sedimentary rocks, Sediments, marble
Oxides	Hematite (iron oxides)	Fe, O	Sediments, soils, sedimentary rocks, all weathered rocks
Sulfides	Pyrite (fool's gold)	Fe, sulfur (S)	All types of rocks
	Galena	Lead (Pb), S	Metamorphic rocks
	Cinnabar	Mercury (Hg), S	Metamorphic rocks

fragments held together by a cementing mineral (Fig. 2-6). Rocks with a clastic texture are formed at least in part from pieces of older, preexisting rocks or from pieces of once-living organisms.

IGNEOUS, SEDIMENTARY, AND METAMORPHIC ROCKS

Igneous Rocks

Rocks are classified into three major groups, based on the way in which they formed. *Igneous* rocks form by the solidification of *magma*, molten silicate material formed by melting or partially melting rocks. Most magma is created at active plate boundaries or at hot spots (see Chapter 1). Igneous rocks are further broken down into two major subgroups: *volcanic* rocks that crystallize rapidly from *lava*, which is magma

SILICON: FROM BEACH SAND TO COMPUTER CHIP

In 1958, scientists produced the first integrated circuit by putting two transistors onto a silicon crystal. This led to the invention of today's microprocessors, devices that can execute hundreds of millions of instructions per second on a single chip. Microprocessors are used in automobiles, medical instruments, televisions, and computers, and modern life wouldn't be possible without them. The dominance of the microprocessor led to the economic rise and dominance of Silicon Valley, northern California's Santa Clara Valley, in the 1990s.

Many materials, including most metals, allow electrical current to flow through them. These are known as **conductors**. Materials that do not allow electrical current to flow through them are called **insulators**. Because of its atomic structure, silicon is considered a semiconductor. It is the building material of most transistors (devices that allow the flow of binary information) because its conductivity can be modulated by the introduction of impurities.

Single crystals of pure silicon—so pure that they contain less than a part per billion of impurities—are necessary for manufacturing microprocessors. The process of making chips begins with commercially processed silicon, known as brown silicon. Brown silicon is made from $SiO2_2$, the common mineral quartz, by reducing it with carbon in an electric furnace. This process produces silicon that is about 97 percent pure. For use in microprocessors, this silicon is treated with hydrogen chloride and then cast into a rod of nearly pure silicon. The rod is used as a sort of seed crystal to grow large cylindrical ingots of ultrapure silicon (Fig. 2-7). The crystals are sliced into thin silicon wafers. A microprocessor is composed of extremely thin layers of silicon, some "doped" with tiny amounts of chemical impurities arranged in a very specific pattern. The introduced impurities, mainly the elements boron or phosphorus, enhance and control the flow of electricity within the silicon crystal.

The quartz used for manufacturing silicon is mined from sedimentary deposits of relatively pure quartz sand. These deposits most likely originally formed in beach or dune deposits. Silica sand for semiconductor use is mined from Washington State and from North Carolina, as well as from other localities worldwide. Silica mined from quartz sand and sandstone is also used for making glass and ceramics, making steel, for abrasives, and for making silicon carbide.

Silicon is the second most common chemical element in the Earth's crust (after oxygen), and quartz is one of the most common minerals. From this humble and common material comes a cornerstone of today's culture and economy, the microprocessor.

6"

Fig. 2-7 A crystal of pure silicon, manufactured for use in the semiconductor industry. (Source: Texas Instruments.)

that reaches the surface; and *plutonic* rocks that crystallize slowly beneath the surface. Beneath the surface, molten magma rises into solid rocks and intrudes into them: for this reason, plutonic rocks are also commonly termed *intrusive* rocks. As we will see in Chapter 8, not all of the mechanisms of plutonic rock formation are completely understood.

Plutonic rocks crystallize slowly underground because the overlying rocks are good heat insulators. A body of magma might require many thousands of years to completely solidify. In contrast, magma that reaches the Earth's surface as lava cools very rapidly because heat dissipates rapidly into the atmosphere or ocean. Because plutonic rocks have so much more time to cool, minerals grow to larger sizes, producing a rock with characteristic coarse-grained texture (see Figs. 2-3 and 2-6; Table 2-2). Volcanic rocks are characteristically finer grained, and mineral grains are commonly so small that they can be seen only by using a microscope. Igneous rocks displaying both visible and microscopic minerals—a *porphyritic* texture—have undergone slow cooling followed by rapid crystallization (Fig. 2-8). In all three of these cases, the igneous rocks display crystalline texture, indicating that crystals grew together. Some volcanic rocks produced during violent eruptions display clastic texture, produced by broken fragments. Lava might also cool so rapidly that the atoms in the magma have no time to form orderly crystals. When this occurs, volcanic glass forms. Obsidian is the most common type of volcanic glass. Volcanic eruptions and the rocks they produce are discussed more completely in Chapter 5, which includes a classification table for volcanic rocks (Table 5-1).

Plutonic and volcanic rocks are further classified according to their mineral composition, which in turn reflects the composition of the magmas that generate them (Table 2-2). The relative abundance of silicon and oxygen (silica) and iron and magnesium are the key elements of igneous rock classification. Although there are dozens of volcanic and plutonic rock types, some of which are discussed in subsequent chapters, they can be grouped into general categories. *Mafic* (sometimes called *basaltic*) rocks are relatively low in silicon and oxygen (silica) and high in iron and magnesium. It should be noted that mafic magma

Fig. 2-8 A rock with porphyritic texture is evidence that slow, early cooling was followed by rapid cooling.

TABLE 2-2   COMMON IGNEOUS ROCKS

Origins and Texture	Description of Texture	Compositional Category	Rock Color	Minerals Present	Rock Name
Plutonic: Coarse-grained	Visible, interlocking crystals	Silicic	Light; salt-pepper	Quartz, potassium feldspar, sodium feldspar, biotite and muscovite mica, amphibole	Granite
		Intermediate	Intermediate; pepper-salt	Sodium-calcium feldspar, amphibole, biotite mica	Diorite
		Mafic	Dark	Calcium feldspar, pyroxene	Gabbro
Volcanic: Fine-grained or porphyritic	Microscopic crystals or visible crystals in microscopic crystalline or glassy matrix	Silicic	Light; white or pink	Quartz, potassium feldspar, sodium feldspar, biotite and muscovite mica, amphibole; glass possibly also present	Rhyolite, rhyolite porphyry
		Intermediate	Intermediate gray or spotted-looking	Sodium-calcium feldspar, amphibole, biotite mica	Andesite, andesite porphyry
Volcanic: Fine-grained	Microscopic crystals	Mafic	Dark	Calcium feldspar, pyroxene	Basalt
Volcanic: Glassy or vesicular	No crystals present	Silicic	Black, red, or light	No minerals present	Obsidian, pumice

See Chapter 5 for further information about volcanic rock classification and Chapter 8 for further information about plutonic rocks.

still contains about 50 percent silica. The term *mafic* denotes the high content of magnesium and iron, whose chemical symbol is Fe. About 90 percent of all erupted lava is mafic lava. Volcanic rock formed from mafic lava is called **basalt**, and its plutonic equivalent is **gabbro**. Both are dark colored because of the high content of dark, mafic silicate minerals. ***Ultramafic*** rocks contain only about 40 percent silica. ***Silicic*** or ***felsic*** magma has a high silica content (about 70 percent) and is low in iron and magnesium. Light-colored minerals dominate in **granite**, the plutonic rock formed from silicic magma, and **rhyolite**, its volcanic equivalent.

Sedimentary Rocks

Sedimentary rocks form at or near the Earth's surface as accumulations of mineral, rock, and (or) plant and animal fragments. These sedimentary rock types display clastic (from the Greek word meaning *broken*) texture (Fig. 2-6). Other sedimentary rocks form by the crystallization of minerals from groundwater or from evaporating sea or lake water. These rocks display crystalline texture, but, in contrast to igneous rocks with crystalline texture, they contain minerals that form at near-surface temperatures and pressures. Other sedimentary rocks are organic, formed from the remains of living organisms. It is easier to conceptualize the formation of sedimentary rocks than the other rock types, because the processes take place in surface environments, where they can be observed. In their daily lives people deal with many of the agents that transport sediment, including rivers, waves, and wind. The processes that form marine sedimentary rocks (those formed in the ocean) are less familiar, but they can be directly observed by oceanographers.

All rocks and minerals at the Earth's surface are affected by **weathering**. Chemical reactions between minerals and water can decompose rocks or cause new minerals to form. Minerals can also be chemically changed by interactions with the atmosphere and water. One example of **chemical weathering** is the formation of iron oxide on the surface of rocks. The outside of a freshly broken piece of rock almost always appears more red or brown than the inside, because iron oxide has formed on the exposed surface (Fig. 2-9). Acting together with chemical weathering, physical processes cause rocks to break into smaller pieces. Freezing and thawing, crystallization of salt, and the growth of roots are examples of **physical weathering** processes that disintegrate rocks. Over geologic time, weathering weakens even the most solid rocks. Massive rock disintegrates into small pieces available for transport by water, wind, or glaciers. After decomposition and disintegration by weathering, rock materials are removed by **erosion**.

Sedimentary rocks are usually layered (Fig. 2-10). The layering, or **bedding**, results from the successive accumulation of sediments over time. Sedimentary beds, each of which generally represents a single depositional event such as a windstorm or a flood, can be millimeters or meters thick. The bedding is visible because of differences between layers. An interval of no deposition or weathering between successive deposits also produces layering. Volcanic rocks can also be layered, but these are easily distinguished from sedimentary layers because they are composed of volcanic materials.

When classifying sedimentary rocks, geologists are most interested in determining the environment where the sediments were originally deposited. In the case of clastic rocks, the nature of the sediment reflects the energy conditions in the environment where it accumulated. One important diagnostic characteristic of

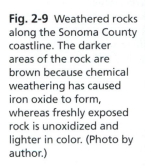

Fig. 2-9 Weathered rocks along the Sonoma County coastline. The darker areas of the rock are brown because chemical weathering has caused iron oxide to form, whereas freshly exposed rock is unoxidized and lighter in color. (Photo by author.)

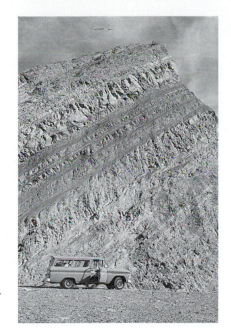

Fig. 2-10 Tilted sedimentary rocks along Highway 99 near Fort Tejon. (Source: Sarna-Wojcicki, A. 1995. U.S. Geological Survey.)

sediment and sedimentary rocks is the *grain size* of the particles in the rock (Table 2-3). For example, in high-energy environments like steep mountain streams, huge boulders might be part of the river channel. In contrast, in the almost flat channels of a delta, only the finest mud is found. Other important features of sediments and clastic sedimentary rocks are their *sorting* and *rounding*. Sediments that are well sorted contain particles of about the same size and density, whereas those that are poorly sorted contain particles of many different sizes. The rounding of grains is another measure of how well-traveled and tumbled the

TABLE 2-3 COMMON SEDIMENTS AND SEDIMENTARY ROCKS

CLASTIC SEDIMENTARY ROCKS

Group/Texture	Sediment Type and Size	Characteristics	Rock Name
Clastic	Gravel, >2 mm	Rounded fragments	Conglomerate
		Angular fragments	Breccia
	Sand, 1/16–2 mm	Mostly quartz	Quartz sandstone
		Quartz and feldspar	Arkose sandstone
		Rock fragments, quartz, feldspar; brown or gray	Greywacke sandstone
	Mud (silt and clay), <2 mm		Shale or mudstone

CHEMICAL/BIOLOGICAL SEDIMENTARY ROCKS

Group	Texture	Composition	Rock Name
Inorganic	Clastic or crystalline	Calcite	Limestone
	Clastic or crystalline	Dolomite	Dolomite
	Crystalline	Halite (NaCl)	Rock salt
	Crystalline	Gypsum ($CaSO_4H_2O$)	Gypsum
Biologic	Clastic	Calcite	Limestone
	Clastic and crystalline	Microcrystalline quartz (radiolaria)	Chert
	Clastic	Silica (diatoms)	Diatomite Peat, coal
		Altered plant remains	

grains are. Sand dunes and beaches are the two environments that produce highly rounded and well-sorted sediments (Fig. 2-11).

Using all of these characteristics, together with the composition of the grains and the type of fossils present, geologists are able to determine with some certainty the environments where sedimentary rocks originally formed (Fig. 2-12). For example, a shale might originally have been deposited in a coastal lagoon. Even though that lagoon and its surrounding landscape have long since vanished, the rock preserves key evidence about the size, location, water chemistry, and surroundings of that environment.

Fig. 2-11 Examples of well-sorted beach sand and poorly sorted sediment from a landslide.

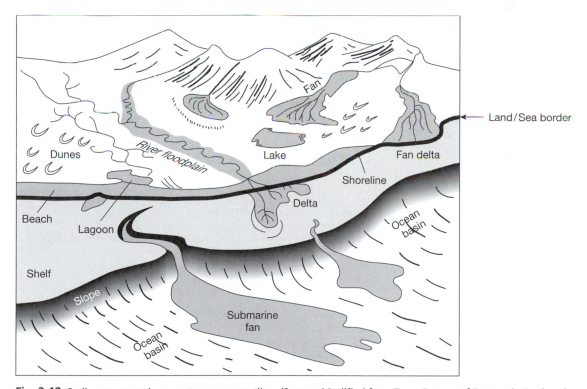

Fig. 2-12 Sedimentary environments near a coastline. (Source: Modified from Texas Bureau of Economic Geology.)

Fig. 2-13 Mud cracks in Death Valley. The cracks form when the mud dries and shrinks (Photo by author.)

Other distinctive features could indicate a sedimentary rock's environment of deposition very precisely. For example, mud cracks (Fig. 2-13) are evidence that the surface of the sediment experienced episodic wetting and drying, indicating an environment such as a tidal flat or intermittently flooded lake bed. Other sedimentary features that aid interpretation of depositional environments include ripple marks, channels, cross-beds, mud lumps, and, of course, fossils. California fossils will be discussed further in Chapter 3.

Sedimentary rocks can also provide important information about the tectonic setting of a region. The sudden appearance of coarse-grained sediments in a section

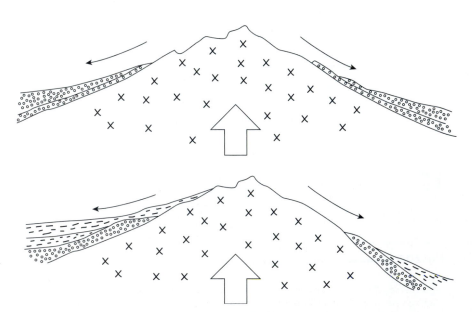

Fig. 2-14 As tectonic forces uplift mountains, erosion removes rock materials from high areas and deposits them in lower areas.

of rocks might indicate that uplift has occurred in nearby areas, causing an influx of sediment to be shed into nearby lowlands (Fig. 2-14). If the sediments contain distinctive rock or mineral types, it might be possible to identify their geologic source with some precision, even if the sediments have been displaced from their sources by faulting. Indicators of current directions in the sediments might identify the direction of sediment transport. Through careful study of sedimentary rocks, geologists are thus able to identify the precise location of newly uplifted mountain ranges long after those mountains have been eroded away. Reconstruction of the ancient geography or *paleogeography* of an area can, in turn, provide clues about past tectonic events.

Metamorphic Rocks

Metamorphic rocks have been subjected to enough heat and pressure to cause grains to grow together, but not enough to melt them. During metamorphism, hot fluids containing reactive chemicals can also circulate through rocks and substantially change their composition. In classifying metamorphic rocks, geologists are particularly interested in interpreting the degree of metamorphism and the evidence of stresses during metamorphism. By reconstructing the temperature and pressure conditions that existed during metamorphism, geologists can estimate how deep beneath the Earth's surface the rocks were when they were metamorphosed (Fig. 2-15).

One type of metamorphism occurs when magma intrudes into a body of rock. The unmelted rocks surrounding the magma are heated. This type of metamorphism

Fig. 2-15 Generalized diagram showing the increase in pressure and temperature with depth beneath the Earth's surface. (Source: Tarbuck, E., and Lutgens, R. 1993. *The Earth*, 4th ed. New York: Macmillan, p. 176.)

is known as ***contact metamorphism*** because it affects rocks that are in contact with igneous intrusions. Hot fluids can also be released into the rocks, reacting with them to form new minerals; this process is known as ***hydrothermal alteration***. One important result of contact metamorphism is the concentration of valuable mineral deposits during hydrothermal alteration.

Most metamorphic rocks form during ***regional metamorphism***. Large areas are subjected to high temperature and pressure when rocks are intensely deformed. The most obvious setting for regional metamorphism is along a convergent plate boundary, where rocks are buried to great depths in a subduction zone. An important characteristic of regional metamorphism is that rocks are subjected to stresses that are not equal in all directions. Rocks that are compressed by directed stress during metamorphism develop a ***foliated*** texture. Foliation is a distinctive layered texture that develops during metamorphism. If rocks are subjected to stress that is

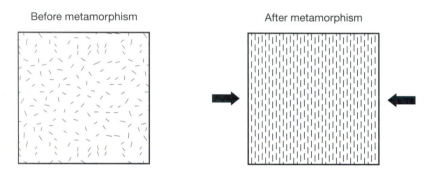

Before metamorphism   After metamorphism

Fig. 2-16 When rocks are subjected to directed pressure during metamorphism, the mineral grains typically grow preferentially in the direction where stress is minimal. (Source: Tarbuck, E., and Lutgens, R. 1993. *The Earth*, 4th ed. New York: Macmillan.)

Fig. 2-17 Foliated schist from the Sierra Nevada foothills. (Source: Harwood, D., U.S. Geological Survey.)

Fig. 2-18 Gneiss, Sierra Nevada. Note geologist's Brunton compass, shown for scale. (Source: Lockwood, J., U.S. Geological Survey.)

stronger in one direction, then new metamorphic minerals will crystallize in a preferred orientation (Fig. 2-16). It is the alignment of metamorphic minerals that produces foliation.

As the degree of metamorphism increases, rocks are subjected to higher temperatures and directed stresses. As a result, more highly metamorphosed rocks show more obvious foliation. The obvious layering of *schist* results from the growth of platy silicate minerals like mica, *chlorite*, or *talc* during metamorphism (Fig. 2-17). Less metamorphosed *slate* is also foliated, but the new metamorphic minerals are very small. Foliated metamorphic rocks can also contain other silicate minerals formed only by metamorphism, such as garnet, kyanite, or staurolite. *Gneiss* is a banded metamorphic rock composed mostly of mica, feldspar, and quartz, indicating a very high degree of metamorphism (Fig. 2-18).

Nonfoliated metamorphic rocks commonly display a grainy texture caused by the intergrowth of minerals with no preferred alignment. Metamorphism of limestone and dolomite produces larger crystals of calcite and dolomite with a characteristically sugary texture—the metamorphic rock *marble*. Metamorphism of shale without directed stresses produces a hard, dark-colored, nonfoliated rock called *hornfels*. California's metamorphic rocks are discussed in more detail in Chapter 8 and Chapter 12.

THE ROCK CYCLE

The rocks that we find at any given location have often experienced a complex history of transformations as they have been subjected to the forces that shape the Earth. Throughout all but its earliest history, when major extraterrestrial impacts were frequent, the mass of the planet has changed very little. Only a tiny amount of new material has been added to the Earth by collision with meteors. The endless creation of

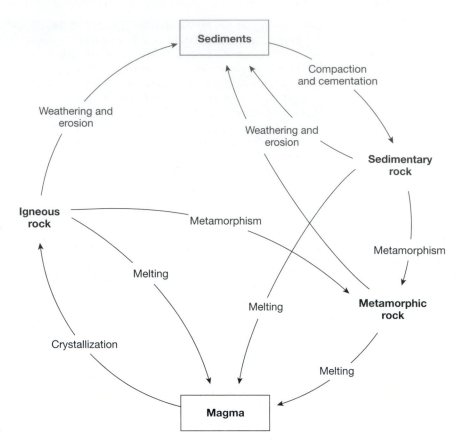

Fig. 2-19 The rock cycle.

new rocks and the destruction of old rocks result from continuous recycling of rock material because of plate tectonics. The transformation from one rock type to another is referred to as the ***rock cycle***. The concept of the rock cycle conveys the idea that a rock can be changed over geologic time. Although it is useful to use a diagram to visualize the rock cycle (Fig. 2-19), the diagram fails to convey the almost endless possibilities created as rocks are subjected to geologic forces over millions, or even billions, of years.

Rocks are melted when they are subducted to depths where temperatures are sufficiently high. Also because of plate tectonics or at isolated hot spots, magma rises and cools to form new igneous rocks. Rocks that are buried to great depths without melting can be changed to metamorphic rocks. As mountains are uplifted, they are worn down as rocks are weathered, eroded, transported, and deposited to form sediments. These sediments can then be gradually buried by younger sediments, eventually becoming sufficiently compacted and cemented to form sedimentary rocks (the process of ***lithification***).

By identifying the sediments that make up a sedimentary rock such as a sandstone, geologists might be able to decipher the sandstone's ancestor rock. For example,

140 million-year-old sandstone found today in the Great Valley contains abundant fragments of volcanic particles, suggesting that the sources of the sediments were volcanoes. The type of volcanic particles in the sandstone might even enable the geologist to infer what type of rocks were melted to generate the magma that in turn created the volcanoes. Using the principles of the rock cycle to trace the history of a rock back through geologic time is not unlike tracing a family's genealogy. In many cases, the "parents" of rocks can be identified, but their lineage becomes very sketchy beyond that—not surprising considering the time spans involved!

It is important to realize that rocks can be transformed by many different pathways (Fig. 2-19). For example, a volcanic rock can be transformed directly into a metamorphic rock—specifically a *metavolcanic* rock in this case—by subjecting it to sufficient heat and pressure. Belts of slate and schist in the western foothills of the Sierra Nevada and the Klamath Mountains were originally layers of volcanic rock formed from lava erupted from submarine volcanic chains. Any type of rock can be melted to form magma. As we will see in Chapter 5, lavas formed by melting oceanic rocks are recognizably different from lavas formed by melting continental rocks.

All of the rocks in California's landscape are subjected to erosion and weathering. In Death Valley, sand dunes contain quartz grains that were eroded from sedimentary rocks that formed hundreds of millions of years ago. Some of those ancestral rocks were themselves deposited in ancient sand dunes several hundred million years ago. The sediments carried by modern rivers from the high Sierra to San Francisco Bay might someday become sandstone. That sandstone will carry compositional clues that it was eroded from the Sierra Nevada— mountains of mostly granitic rock—even if the mountains themselves have been completely eroded away. Any rock has almost endless possibilities of becoming a very different type of rock in its geologic future. The only certainty is that, if a rock melts to become magma, the magma can only be transformed to an igneous rock.

COMMON ROCKS IN CALIFORNIA

Igneous Rocks

Igneous rock types common in California include the following:

- *Granitic plutonic rocks* (shown in red on Endpaper 1, inside the front cover). Granitic rocks make up much of the Sierra Nevada Province. Smaller granitic bodies are found in the Klamath Mountains, in the Coast Ranges west of the San Andreas fault, in the Mojave Desert, and in the Peninsular Ranges of southern California. Types of granitic rocks and their origins are discussed in Chapter 8.
- *Mafic and ultramafic plutonic rocks* (shown in purple on Endpaper 1). Serpentinite, which is actually a metamorphosed ultramafic rock, is included in this unit. Geologists believe that these rocks formed in the lower crust and upper mantle beneath oceanic plates. These rocks are discussed further in Chapter 9.

- *Intermediate and silicic volcanic rocks* (shown in pink on Endpaper 1). Young volcanic rocks cover most of the Cascades Province, are abundant east of the Sierra Nevada, and are also found in the northern Coast Ranges. In these three areas, young volcanic rocks include large deposits of rocks with intermediate and high silica contents. California's volcanoes and volcanic rocks are discussed further in Chapter 5.
- *Mafic volcanic rocks* (shown in pink on Endpaper 1). Recently erupted mafic volcanic rocks, mostly basalt, are most common in California in the Modoc Plateau, the Mojave Desert, and in the Basin and Ranges of southeastern California.

    Basalt covers more of the Earth's surface than any other type of rock—it forms the floor of all the ocean basins. In California, older volcanic rocks (included in the green and blue areas on Endpaper 1) occur in the Coast Ranges, where they have commonly been slightly metamorphosed, and in the Klamath Mountains and western Sierra Nevada, where they have been more highly metamorphosed. Even these metavolcanic rocks can show well-developed rounded forms called **pillows**, formed when lava entered the ocean and congealed into blobs. Studies have confirmed that many of California's metavolcanic rocks formed in ancient midocean ridges, seamounts, and island arcs and were then accreted to the North American continent. Mafic volcanic rocks are discussed further in Chapter 5 and Chapter 12.

Sedimentary Rocks

On the California geologic map, sedimentary rocks are grouped according to their age. From youngest to oldest, the sedimentary rocks are colored tan, green, and blue. California's metamorphic rocks are grouped with the metasedimentary and metavolcanic rocks of Mesozoic (green) and Paleozoic (blue) age. Serpentinite is included with the ultramafic rocks (purple), and highly metamorphosed rocks of unknown or pre-Paleozoic age are shown in dark brown.

Sedimentary rocks common in California include the following:

- *Sandstone*. Sandstone is found in all provinces of California. The majority of California's sandstone was originally deposited in marine environments. Much of the sandstone in California is gray or brown and contains chips of shale and other rocks. Because of its location along a steep continental margin, California has produced these types of "dirty" sandstones rather than quartz-rich sandstones. The deposition of these sandstones, called *greywacke*, is discussed further in Chapter 12.
- *Shale*. Dark gray or black marine shale is also very common in California and usually is found together with sandstone (see Chapter 12). Metamorphosed sandstone and shale, examples of **metasedimentary** rocks, are common in the Sierra Nevada, Klamath Mountains, and other provinces.

- *Chert*. Chert is abundant in the Coast Ranges and can be black, red, green, brown, or white. Much of California's chert formed by the slow settling of microorganisms called **radiolaria** onto the floors of the deep ocean basins. It is commonly found together with submarine volcanic rocks. Chert is also found in the Klamath Mountains, the Transverse Ranges, and the metamorphic rocks of the Sierra Nevada. Chert is discussed further in Chapter 3 and Chapter 12.
- *Limestone and dolomite (carbonate rocks)*. Carbonate rocks are common only in the mountain ranges of the Basin and Ranges and Mojave Desert provinces. Metamorphosed carbonate rocks (marble) are also common in the eastern Klamath Mountains and parts of the Sierra Nevada. Limestone also occurs as small blocks within the Coast Ranges and in the mountains of southern California. The formation of these rocks is discussed further in Chapter 7.

Metamorphic Rocks

Metamorphic rocks common in California include the following:

- *Serpentinite*. This is California's state rock, commonly found along faults in the Coast Ranges, Klamath Mountains, and Sierra Nevada. Serpentinite forms when rocks originally formed as ultramafic rocks in the mantle are hydrated (see Chapter 9).
- *Slate*. Slate is abundant in the western Sierra Nevada foothills. It was originally mafic volcanic rock, shale, or sandstone that changed during metamorphism (see Chapter 8).
- *Schist*. Schist occurs in the Santa Monica, San Bernardino, and San Gabriel Mountains in southern California, in the Sierra Foothills, and in the Klamath Mountains. Like slate, many of the schists in California were originally sedimentary or volcanic rocks. A distinctive type of schist, blueschist, is found in the Coast Ranges. Metamorphic rocks and their significance are further discussed in Chapter 8 and Chapter 12.
- *Marble*. Marble, metamorphosed limestone or dolomite, is relatively rare in California and occurs in blocks within the high Sierra Nevada, in the northwestern Sierra Foothills, and in the Klamath Mountains.

Unconsolidated sediments are recent deposits that have not been buried, compacted, and cemented to form sedimentary rocks. These sediments cover virtually all of the low-lying areas of California and are shown in yellow on the California geologic map. Most of these sediments are **alluvium**, sediments deposited by rivers. The processes that deposit nonmarine sediments are discussed further in Chapters 6, 10, and 11. The processes of marine and beach deposition along the California coast are discussed in Chapter 15 and Chapter 16.

SUMMARY

1. Minerals are naturally occurring, inorganic, crystalline solids composed of one or more chemical elements. Silicate minerals make up most of the rocks in the Earth's lithosphere.
2. Rocks are composed of one or more minerals. The composition and the texture of a rock are used to identify its origin and to classify it.
3. Igneous rocks form by the crystallization of silicate minerals from magma. Magma can cool slowly under the Earth's surface to form plutonic rocks, or it can cool rapidly at the surface to form volcanic rocks. A porphyritic rock contains both large crystals that cooled slowly and microscopic crystals or glass that cooled rapidly when the magma erupted at the surface.
4. Sedimentary rocks form at the Earth's surface by the accumulation of rock and mineral fragments or the crystallization of minerals from water. Clastic sedimentary rocks display a fragmental texture and form when sediments are lithified. Sediments form when rocks are chemically and physically weathered and eroded by forces operating at the Earth's surface.
5. Metamorphic rocks form beneath the Earth's surface when rocks are changed without melting by the action of heat, pressure, and chemical fluids. If pressures within the Earth are directional, then new minerals will grow in the direction of least resistance, creating foliation.
6. Rocks are continually transformed by the processes operating within the Earth. Because Earth's mass is not increasing, new rocks form only by the destruction of old rocks. The continuous transformation of rocks is described by the rock cycle.
7. The distribution of California's rocks and sediments is portrayed on Endpaper 1, the geologic map of California. All of the major types of rocks are found in California, including most common rocks as well as unusual rocks that owe their origins to California's unique tectonic history.

REFERENCES

BARKER, R. 1990. *Collecting Rocks*. U.S. Geological Survey. General interest pamphlet.

DIETRICH, R.V., and SKINNER, B.J. 1979. *Rocks and Rock Minerals*. New York: Wiley, 319 pp.

ESTAVILLO, W. 1992. *Gems and Minerals of California: A Guide to Localities*. Frederick, CO: Renaissance House, 48 pp.

Special California Issue. 1994. Rocks and Minerals, 69(6), 360–430.

SYMES, R.F. 1988. *Rocks and Minerals*. New York: Knopf, Eyewitness Books, 64 pp.

TOOKER, E.W., and BEEBY, D.J. 1990. *Industrial Minerals in California: Economic Importance, Present Availability, and Future Development*. U.S. Geological Survey, Bulletin No. 1958.

Web Sites and Media Resources

California Geological Survey: *http://www.consrv.ca.gov/CGS/*

Intel Museum: Virtual Tour of chip manufacturing process: *http://www.intel.com*

TASA Graphic Arts, *The Wonders of Rocks and Minerals; Introduction to Minerals; Rock Cycle I* (interactive CD-ROMs): *http://www.tasagraphicarts.com*

TASA Graphic Arts. 2001. *Earth's Dynamic Surface*: interactive CD-ROM (F. Pazzaglia, author): *http://www.tasagraphicarts.com*

U.S. Geological Survey, Minerals Information: *http://minerals.usgs.gov/minerals/*

U.S. Geological Survey, *Rocks and Geology in the San Francisco Bay Region* (Phil Stoffer, author): *http://geopubs.wr.usgs.gov/bulletin/b2195*

3

Geologic Time, Dating Earth Materials, and California Fossils

Time is one factor that sets geology apart from other sciences. Beginning students of geology often have difficulty comprehending the vastness of geologic time. Many of the processes that shape the Earth have little immediate effect on the overall landscape or distribution of materials. However, these processes operate over a seemingly endless time span when compared to the centuries or decades easily imagined by humans. Over tens of millions of years, entire oceans and continents can be created and destroyed. An understanding of how geologists determine rates and sequences of geologic events is important for deciphering California's complex geologic history.

RELATIVE DATING

Until the 1950s, it was not possible to determine the actual numerical age of a rock or sediment. Geologists relied on a scheme of *relative age dating*, which enabled them to order rock formations from oldest to youngest. Relative dating is still an important tool in many geologic investigations. Geologists use four fairly simple principles to determine the relative order of geologic events in an exposed section of rocks or sediments:

1. In an undisturbed sequence of layered materials, the layers at the bottom are oldest, and those at the top are youngest (Fig. 3-1A): the principle of *superposition*. Anyone who wades through a stack of papers or dirty laundry is familiar with this principle.
2. Sediments are originally deposited in horizontal, or nearly horizontal layers (Fig. 3-1A): the principle of *original horizontality*. Sediments accumulate in low-energy, gently sloping, or level environments.

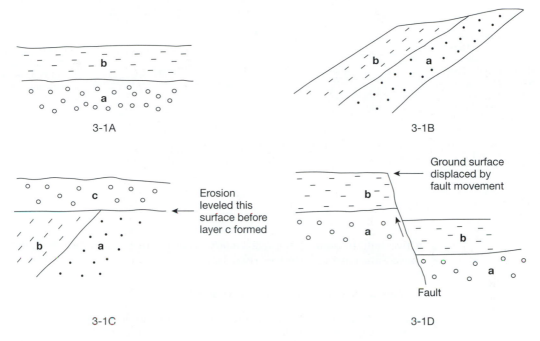

3-1A

3-1B

Erosion
leveled this
surface before
layer c formed

Ground surface
displaced by
fault movement

Fault

3-1C

3-1D

Fig. 3-1 The principles of relative age dating are used to determine the sequencing of geologic events. **A,** Layer a was deposited, followed by layer b. **B,** Layers a and b are tilted following their deposition. **C,** During a period of erosion a level surface is formed on the tilted layers, and then layer c was deposited on the eroded surface. **D,** A fault offsets the layers that were previously deposited.

One can apply the first two principles to an exposure of folded or tilted rocks (Fig. 3-1B) and conclude that tilting and folding must have occurred after the rocks were deposited. Sediments deposited horizontally on tilted rocks (Fig. 3-1C) give us two important pieces of information: A period of erosion occurred after the tilting, forming a relatively horizontal surface, and the sediments deposited on that surface are younger than the erosional period.

3. A rock unit or feature such as a fault that cuts through a layer of rock must have formed after that rock layer was formed: the principle of ***cross-cutting relations***. A vein of quartz that cuts through a granite body was intruded after the granite cooled. A fault that displaces rock layers moved after those layers formed (Fig. 3-1D).

4. A rock or sediment containing fragments of other rocks must be younger than the rocks it contains. For example, lava might incorporate pieces of rock from the sides of a fissure as it erupts. The resulting volcanic rock will contain ***inclusions*** of the older, incorporated rock.

One important aspect of these principles is that the landscape is younger than any of the rock or sediment beneath it. Hills, valleys, and plains have been sculpted on the materials beneath them, and the forces of erosion and deposition that shape the landscape are constantly reshaping the landscape. If a fault causes displacement of a landscape feature such as a stream or a level field (Fig. 3-1D), it is most likely an

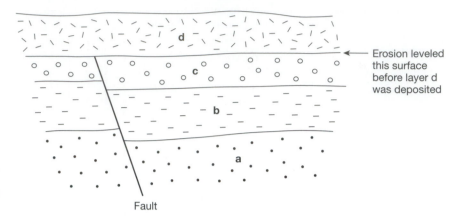

Fig. 3-2 Relative age dating is used to establish the time of movement along a fault. The fault moved after layers a, b, and c were deposited, because these layers are broken (or offset) by the fault. The fault has not moved since layer d was deposited and the landscape developed, because layer d continues unbroken across the fault and the landscape.

Erosion leveled this surface before layer d was deposited

Fault

active fault. In contrast, if a sedimentary deposit buries a fault, then the fault has not moved since that layer formed (Fig. 3-2).

Fossils and the Geologic Time Scale

Many sedimentary and metasedimentary rocks contain *fossils*—remains of plants and animals—that can aid geologists in assigning ages to the rocks. Fossils have been found from rocks formed in every conceivable environment that supports life, but the vast majority of fossils are the remains of marine (ocean-dwelling) life. This is not surprising, considering that oceans make up over 70 percent of Earth's surface and that they are low-lying areas where sediments accumulate, making marine environments favorable for fossil preservation. Rocks formed in ancient, low-energy terrestrial environments such as lake beds, peat bogs, and marshes also commonly contain fossils. Species with hard parts like shell, bone, and wood are also more likely to be preserved than more delicate organisms. Environments conducive to rapid burial are particularly favorable for preservation of fossils and indirect evidence of life such as worm trails, leaf impressions, and footprints.

The principles of *paleontology*, the study of ancient life, were developed during the 19th century, when paleontologists began to systematically collect and classify fossils. Evolutionary sequences of species were identified as paleontologists pieced together the successions of life forms represented in the fossil record. Although some species (e.g., ferns) have persisted through long stretches of geologic time without much change, others made only brief appearances on Earth. The presence of these time-specific, characteristic **index fossils** is particularly useful. Their presence enables paleontologists to identify rocks formed during the same restricted time interval, even if the rocks are found in widely separated areas.

Long before methods of numerical dating of rock materials were developed, paleontologists established a geologic time scale based on the evolutionary sequence of species through time. Based on characteristic assemblages of species, geologic time was broken into major **eras** that encompassed several **periods**; periods were subdivided into **epochs** (Fig. 3-3). The boundaries between eras are marked by first appearances of groups of species and sudden, catastrophic **extinctions** of many life forms from the geologic record. Less dramatic changes in life forms mark the boundaries between periods and epochs. The seemingly odd

Age	Eon	Era	Period	Epoch	Animals				Plants			
		Cenozoic	Quatemary	Holocene Pleistocene								
			Tertiary	Pliocene Miocene Oligocene Eocene Paleocene				Humans				
65 MA			Cretaceous	Late Early								Flowering plants
		Mesozoic	Jurassic	Late Middle Early			Birds					
			Triassic	Late Early		Mammals					Gingkos	
250 MA	Phanerozoic		Permian	Late Early								
			Pennsylvanian	Late Middle Early		Reptiles				Pines		
			Mississippian	Late Early	Amphibians							
		Paleozoic	Devonian	Late Middle Early				Horsetail rushes	Ferns			
			Silurian	Late Middle Early			Club mosses					
			Ordovician	Late Middle Early								
			Cambrian	Late Middle Early	Fishes							
540 MA	Proterozoic	Late Proterozoic Middle Proterozoic Early Proterozoic			Animals with shells							
	Archean	Late Archean Middle Archean Early Archean										
		Pre-Archean										

Fig. 3-3 The geologic time scale and the history of life on Earth (MA = millions of years ago).

names for some geologic periods arise from localities where the best examples of the fauna of that period are found, such as the Jura Mountains in Europe for the Jurassic period. Other periods are named for ancient peoples who inhabited parts of Britain where characteristic fauna occur, such as the Ordovician period, named for the Ordovice Celtic people of Wales.

ABSOLUTE DATING

Since the 1950s, geologists have relied on the Earth's natural radioactive "clocks" to determine the numerical ages of geologic materials. These are sometimes referred to as "absolute" ages. Some of the naturally occurring chemical elements have forms or *isotopes* that are unstable. These unstable forms account for a tiny fraction of the total amount of that element. They break down spontaneously to more stable atoms, and in the breakdown process they give off energy in the form of radiation. Unstable forms of elements are referred to as *radioactive isotopes*, and the breakdown process is known as *radioactive decay*. The decay of radioactive isotopes takes place at a specific and constant rate, as does the buildup of the products of decay. The time it takes for half of the initial amount of radioactive material to decay is the *half-life* of that particular isotope. Physicists have measured the half-lives of radioactive isotopes with great precision. They range from seconds to billions of years, but the half-life of each isotope is unique and unchanging.

The constant rate of decay of radioactive elements provides us with naturally occurring clocks that can be used to date Earth materials. Knowing the original amount of a radioactive isotope in a rock, sediment, or fossil, and knowing its half-life, one can calculate the amount of time since the radioactive "clock" began ticking. The use of isotopes to date Earth materials is known as *radiometric dating*.

A radiometric dating method commonly used for dating materials that are geologically young uses carbon-14, a naturally occurring radioactive isotope of carbon. Carbon-14 is produced when cosmic rays bombard nitrogen in the upper atmosphere, and a fixed amount of carbon-14 is present in all living tissues, including our own. When organisms die, they stop taking in carbon-14, and the carbon-14 in their bodies begins to decay (Fig. 3-4). Knowing that the half-life of carbon-14 is 5,730 years, one can measure the amount of carbon-14 in formerly living material and calculate the time that has passed since it died. After about 50,000 years, the amount of carbon-14 remaining is so small that it can't be measured accurately. For this reason, carbon-14 dating can be used to date only recent materials and events. Fossils of wood, charcoal, bone, shell, and soil materials are all potentially useful for carbon-14 dating. Geologists rely heavily on this technique to determine the ages of sediments formed by floods, landslides, earthquakes, and other potentially hazardous processes.

The "workhorse" of radiometric dating techniques for most rocks and geologic events is potassium-argon (K-Ar) dating. This method works best on igneous rocks and measures the time that has passed since the rock crystallized from magma or lava. If igneous or other rocks are metamorphosed, the K-Ar method measures the time that has passed since the rocks cooled after metamorphism. Because the decay of radioactive potassium-40 to argon-40 has a half-life of 1.3 billion years, the K-Ar technique can be used on all but the youngest geologic materials. The most commonly used technique for K-Ar dating involves measurement of the buildup of argon-40, a gas, in rocks.

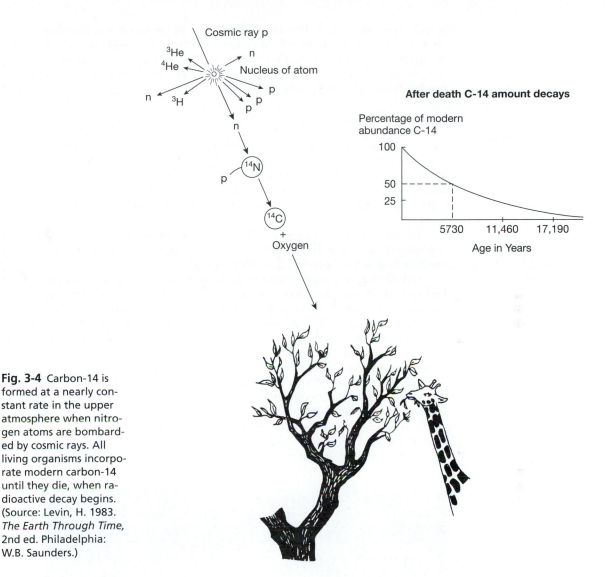

Fig. 3-4 Carbon-14 is formed at a nearly constant rate in the upper atmosphere when nitrogen atoms are bombarded by cosmic rays. All living organisms incorporate modern carbon-14 until they die, when radioactive decay begins. (Source: Levin, H. 1983. *The Earth Through Time,* 2nd ed. Philadelphia: W.B. Saunders.)

By finding localities where sedimentary rocks containing key fossil assemblages are in contact with ancient lava flows, geologists have used K-Ar dating to add the numbers to the geologic time scale (see Fig. 3-3), as well as to date rocks from the moon. Other commonly used radiometric dating techniques use uranium-lead and rubidium-strontium isotopes.

CALIFORNIA'S FOSSILS

Because rocks of pre-Mesozoic age are relatively rare in California, fossils of the Paleozoic era are also restricted. Mesozoic rocks contain a wide variety of mostly marine fossils, because most of California's Mesozoic sedimentary and metasedimentary rocks were deposited in marine environments. The Cenozoic fossil record in California is rich and diverse. Marine fossils of many types are abundant, and California is also home to some of North America's richest late Cenozoic terrestrial fossil finds. Although it is not possible to list all of California's fossil localities, it is worth describing some of the typical fossils of each era. Subsequent chapters include discussion of additional fossil localities in specific provinces.

Paleozoic Fossils

Paleontologists have identified at least 95 species of trilobites from early Paleozoic rocks in southeastern California. Trilobites were ocean-dwelling arthropods, **invertebrate** animals related to arachnids, insects, and crustaceans (Fig. 3-5). Because many distinctive species developed during Paleozoic time, and because trilobites became extinct at the end of the Paleozoic era, they are excellent index fossils for Paleozoic rocks.

Trilobite fossils are present in the Cadiz Formation of Cambrian age in the Marble Mountains of San Bernardino County, and in similar carbonate rocks in the southern Nopah Mountains near Death Valley.

Metasedimentary rocks found in the high parts of the Sierra Nevada west of Bishop contain graptolite fossils that place the formation of the original marine sedimentary rocks in Ordovician time. Graptolites, which persisted until Mississippian time, are very important index fossils for Ordovician and Silurian rocks. They were colonial organisms that secreted small skeletons; in rocks, graptolites look like small dashes or rods (Fig. 3-6). Fossil brachiopods are found in Silurian and Devonian rocks in the eastern Klamath Mountains.

Permian rocks are more widespread in California than rocks from earlier Paleozoic periods. Corals, including many forms that are extinct today, are found in Permian rocks near Death Valley and in the eastern Klamath Mountains. Fusilinids, large, single-celled foraminifera that are extinct today, are also found in Permian sedimentary rocks. Crinoids, particularly their cylindrical remnants (Fig. 3-7), and brachiopods are also well represented in fossiliferous rocks of Permian age.

Fig. 3-5 Cambrian trilobite fossil from the Marble Mountains, San Bernardino County.

Fig. 3-6 A, Ordovician graptolite fossils from metasedimentary rocks in the Sierra Nevada. **B,** Pennsylvanian tabulate corals, Death Valley. (Source, McMackin, M. San José State University.)

A   B

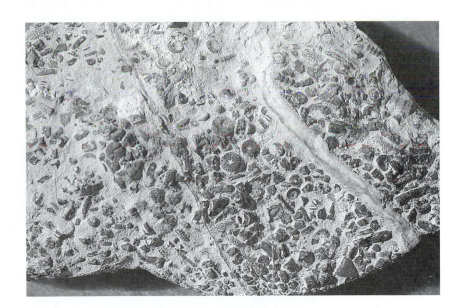

Fig. 3-7 Ordovician crinoid fossils from the Inyo Mountains, shown at actual size.

Mesozoic Fossils

Several groups of marine invertebrate organisms are important fossils in California's Mesozoic rocks. Most of these are shells of mollusks, such as clams and snails. Ammonites, which are cephalopods that resemble the modern chambered nautilus, are

Fig. 3-8 Cretaceous Scaphites ammonite, 3.175 cm long. (Source: Courtesy of Dr. Rick Miller, San Diego State University.)

found in Mesozoic rocks throughout California. In the Peninsular Ranges, Triassic ammonites are found in the Santa Ana Mountains. The age of the widespread Franciscan Assemblage in the Coast Ranges was originally determined from the presence of Jurassic and Cretaceous ammonites, and excellent Cretaceous specimens have been found along the western Great Valley (Fig. 3-8). Many other mollusks are abundant in California's Cretaceous rocks, including *Turritella,* a gastropod, and *Buchia,* a distinctive pelecypod found in the Great Valley Sequence (Fig. 3-9).

The Mesozoic era is famous for its dinosaurs, and in the early 1990s almost all Californians were introduced to the Jurassic period by the movie *Jurassic Park.* Little of today's California was dry land during Mesozoic time, and erosion has removed the Mesozoic mountain chain that existed where the Sierra Nevada is today.

Fig. 3-9 Cretaceous marine fossils from California: *Buchia,* about 2 to 3 cm long, from the western edge of the Sacramento Valley.

As a result, one must travel east from California to find abundant fossils of the famed dinosaurs. However, in 1991, Richard Hilton of Sierra College recovered bones from an early Cretaceous dinosaur. The fossils are remains of a small herbivore about the size of a deer belonging to a group called *hypsilophodonts* (which also includes the "duckbill" dinosaurs). The fossils were found in a marine formation, suggesting that the carcass of this terrestrial animal probably drifted out to sea. Other fossils of land-dwelling dinosaurs have recently been found in similar geologic settings in other parts of California. Remains of a large duckbill known as *Saurolophus* were discovered in western Fresno County, and hadrosaur remains were unearthed in southern California near San Diego. During the 1990s a dinosaur footprint was also discovered in the Jurassic Aztec sandstone in the easternmost Mojave Desert Province. Fossils of swimming reptiles have also been found in Cretaceous rocks in California, including ichthyosaur remains near Los Banos and Redding. A plesiosaur fossil and remains of a pteranodon were found in the Moreno Formation in western Fresno County.

Microscopic fossils of radiolaria have proved to be among the most useful Mesozoic fossils for California geologists. Radiolaria are microscopic, single-celled protozoans that build their skeletons of silica. In the ocean basins, sedimentary layers can consist entirely of radiolaria, which, when lithified, form a sedimentary rock known as radiolarian chert (Fig. 3-10). Radiolaria can remain recognizable even after metamorphism, so they have proved to be very valuable to paleontologists and geologists interested in the origins of California's accreted oceanic terranes.

Diatoms are also microscopic organisms composed of silica. Diatoms, a type of floating plant with a siliceous cell wall, live in fresh and marine water (see Fig. 16-23). Diatoms are found in California sedimentary rocks of late Cretaceous to modern age. Cenozoic deposits composed almost entirely of diatoms (diatomite) are mined from several localities in California (see Chapter 16). Diatomite is used for soil amendments and filters. Diatoms are also important to California's geologic history because they are the likely source of much of California's petroleum (see Chapter 16). Microscopic *foraminifera*, single-celled protozoans that construct their shells from calcium carbonate, are also important microfossils for rocks of Cretaceous and Cenozoic age.

Cenozoic Fossils

During Cenozoic time, much of California emerged from the sea, creating a variety of land-based environments that supported an amazing assortment of organisms. The Cenozoic is the era of mammals, and many of California's best-known fossil sites are localities bearing the remains of mammals (Fig. 3-11). Many sites also preserve fossil plants. Near Elsinore, in the Peninsular Ranges, fossil leaves and ferns are found in rocks of Eocene age. At the Petrified Forest near Calistoga, in the Coast Ranges of Sonoma County, redwood trees of Pliocene age were perfectly preserved when they were buried by a volcanic eruption from the nearby Sonoma volcanic field (Plate 6).

The oldest known land animal fossils in northern California are those at Blackhawk Ranch on the southern side of Mt. Diablo, in the Diablo Range east of San Francisco Bay. At this site, fossils of rhinoceros, antelope, camels, horses, rodents, oreodonts, and many others are found in rocks of Miocene and Pliocene age. Paleontologists continue to excavate fossils from this site, and many of the finds

Fig. 3-10 Photomicrograph of Jurassic radiolaria in chert from the central Coast Ranges. The white bar scale is 0.1 mm in length. (Source: Oscarson, R., and Murchey, B. U.S. Geological Survey.)

are on display at local museums. Just southeast of the Blackhawk site, in the city of Fremont, is the world-renowned Irvingtonian mammal site, where a community of land mammals excavated from the gravels is the best early Pleistocene assembly found in North America. Fossils found at the quarries in Irvington include deer, camels, horse, saber-toothed tigers, a wolf, and the oldest mammoth elephant discovered in the western hemisphere. Many residents in the Irvington suburbs today have no idea that the area was home to such a collection of animals in the not-too-distant past.

The California fossil locality best known to nongeologists is the Rancho La Brea (Spanish for *tar*) tar pools, which are now part of Hancock Park in Los Angeles. The natural pools of tar trapped animals and plants of all kinds (Fig. 3-13A),

Fig. 3-11 Miocene mastodon tracks, Copper Canyon. (Source: Mc-Mackin, M. San José State University.)

preserving them for excavation by 20th-century paleontologists. More than half a million specimens of mammals, birds, reptiles, insects, and plants—representing more than 200 species—have been excavated from Rancho La Brea. Extinct species of bison, mammoth, mastodon, and camel are among the mammal species. Also present are fossils of the extinct saber-toothed tiger, *Smilodon*, California's official state fossil. Many of the specimens from La Brea are on display at the George Page Museum of Natural History in Los Angeles.

It is clear that fossils provide information far beyond establishing the ages of rocks. Fossils provide valuable information about past environments. Sensitive marine organisms like diatoms or foraminifera can provide information about water temperature, salinity, depth, and amount of light (Fig. 3-14). For example, solitary corals are found in 80,000-year-old sediments in marine terraces on San Nicolas Island off southern California. Today these species are found only far to the north, suggesting that the water temperatures off southern California once were cooler than today's, even though global temperatures were higher at that time.

Fossils with limited geographic distribution can provide information about the ancient latitudes of oceanic terranes. Fossils also give information about ecosystems and communities and their adaptations to changing habitats. By studying extinctions, we have learned about major crises in the history of life on Earth. By studying past changes in groups of organisms, we can better appreciate how organisms evolve and how their environments change. From such studies, we can make projections about future changes in Earth's environment and evaluate possible effects of humans on other organisms. Finally, fossils provide us with a broad perspective of our place on Earth and an appreciation of the expanse of geologic time.

THE EXILED ELEPHANTS OF THE CHANNEL ISLANDS

The Channel Islands are an east–west chain that lies 20 to 50 kilometers off the coast of southern California near Santa Barbara (see Chapter 16). They are separated from the California mainland by water up to 200 meters deep. Today some of the Channel Islands are a popular recreational destination, and others are wildlife sanctuaries. During late Pleistocene time, they were apparently the home of an unusual group of land mammals.

In the late 1800s the remains of extinct Pleistocene elephants were first discovered on Santa Rosa Island. Since that time, geologists and paleontologists have found a number of additional specimens on Santa Rosa Island, plus additional fossil elephants on San Miguel and Santa Cruz Islands. The most recent discoveries were made on Santa Rosa Island in 1994. From what we know about elephants, it is clear that they must have arrived in the Channel Islands during a time when sea level was lower, a time when the islands were either separated by narrower bodies of water or connected by land bridges. Worldwide sea level was lower during several periods of Pleistocene glaciation (see Chapter 12), so it is likely that the original elephants arrived in the Islands during one of these periods. The most recent period when sea levels were low was about 18,000 to 20,000 years ago.

What is most amazing about the fossil elephants of the Channel Islands is that they are dwarf species of the more common fossil elephants, *Archidiskodon imperator* and *Parelephas columbi*. The dwarf species of the Channel Islands were only about 2 to 3 meters high at the shoulders, in comparison to mainland species, which stood about 4 meters high (Fig. 3-12). Paleontologists recognize the Channel Islands fossils as belonging to a separate species, *Elephas exilis*, or exiled elephant. It appears that once the population was isolated from the mainland by rising sea levels, the smaller elephants survived better because of limited food supplies. Over time, the process of natural selection favored the smaller animals. Fossils of dwarf elephants and other species have been discovered in other parts of the world on isolated islands.

Fig. 3-12 Fossil pygmy mammoth from Santa Rosa Island. (Source: Rockwell, T., San Diego State University.)

A

B

Fig. 3-13 The Rancho La Brea Tar Pits: **A,** Excavation of La Brea. **B,** Section of Rancho La Brea mural by Charles Knight, depicting Pleistocene life in southern California. (Source: **A,** U.S. Geological Survey; **B,** Reproduced with permission of the George C. Page Museum, Los Angeles.)

Fig. 3-14 Changes in foraminifera species with different water depths (bathyal = ocean depths of 180–1,800 meters). (Source: Ingle, J. 1978. U.S. Geological Survey, Open-File Report 78-446.)

SUMMARY

1. The principles of superposition, original horizontality, and cross-cutting relations allow geologists to determine the relative ages of a series of layered rocks. Relative dating also allows geologists to establish the ages of events such as fault movements or intrusion of igneous rocks by applying the same principles.
2. The modern landscape is younger than the rocks and sediments beneath it.
3. Fossils are the remains of ancient life. They can provide important clues about the sedimentary environments that existed when an organism was living.
4. Marine organisms with hard parts are preferentially represented in the fossil record.
5. Paleontologists have used major changes in the record of life on Earth to establish the geologic time scale. Evolution of life forms and extinctions of species mark the major boundaries between geologic eras and periods. The Pre-Cambrian, Paleozoic, Mesozoic, and Cenozoic eras are the major divisions of the time scale.
6. Radiometric dating, based on the half-lives of naturally occurring radioactive isotopes, is now used to determine the absolute ages of rocks, geologic events, and the major geologic time periods.

7. Paleozoic, Mesozoic, and Cenozoic age fossils have all been found in California. Because much of California is composed of Mesozoic and younger rocks, older Paleozoic fossils are found only in southeastern California and in parts of the Sierra Nevada and Klamath Mountains provinces. Mesozoic marine fossils are common, and rare bones from dinosaurs have also been found in marine sedimentary rocks. Cenozoic fossil localities are widespread and include areas where fossils of land-dwelling mammals are abundant.

REFERENCES

BEDROSSIAN, T.L. 1971. Fossils: The living past. *California Geology* 24(12): 227–239.

BEDROSSIAN, T.L. 1975. Vertebrate fossils and the history of animals with backbones. *California Geology* 28(11): 243–259.

California Geological Survey. 2002. *Fossils.* Note 51, 4 pp.

DECOURTEN, F. 1997. Dinosaurs in California. *Pacific Discovery, California Academy of Science* 50(3): 26–31.

GOULD, S.J. 1987. *Time's Arrow, Time's Cycle.* Cambridge, MA: Harvard University Press, 222 pp.

HILTON, R.P. 2003. *Dinosaurs and Other Mesozoic Reptiles of California.* Berkeley: University of California Press, 348 pp.

LEVIN, H.L. 2003. *The Earth Through Time,* 7th ed. Hoboken, NJ: Wiley, 563 pp.

Web Sites and Media Resources

Smithsonian Institute, Museum of Natural History: *http://www.mnh.si.edu*

University of California, Berkeley, Museum of Paleontology (site includes links to other museums of natural history): *http://www.ucmp.berkeley.edu/*

Buena Vista Museum of Natural History, Kern County Fossils: *http://sharktoothhill.com/kern.html*

San Diego Natural History Museum, Field Guide, Fossils of Our Region: *http://www.sdnhm.org/fieldguide/fossils/index.html*

Page Museum, Los Angeles, La Brea Tar Pits: *http://www.tarpits.org/*

CHAPTER

4

Geologic Information

GEOLOGIC MAPS AND CROSS SECTIONS

The purpose of most geologists' work is to understand how and where the rocks or sediments in an area were formed, what forces have affected them since they formed, and when these events occurred. Chapter 2 presented an overview of how rocks and sediments are described, classified, and interpreted, and Chapter 3 provided a brief summary of how geologic materials and events are dated. After materials are identified, dated, and interpreted, geologists assemble a three-dimensional view of the rocks or sediments in an area to help them understand the area's geologic history. Understanding both the type and rates of changes in geologic conditions through time is also critical to solving many geologic problems. The three-dimensional picture of geology changes gradually, adding a fourth dimension—time—to the picture. Because this ever-changing geometric arrangement of geologic materials is so important for interpretation, geologists rely heavily on visual techniques for portraying information.

Maps are the fundamental tool for displaying geologic information. A geologic map can cover a large area, such as the entire state of California, or it can portray in detail the geology of a small area, such as a housing development. Regardless of scale, geologic maps show four fundamental types of information: (1) the types of rock or sediment at the surface; (2) the ages of the materials; (3) the orientation of the rocks or sediments, including the inclination or *dip* of layered rocks; and (4) the types of boundaries separating the different materials. On a generalized map such as Endpaper 1 (inside the front cover), only major rock types and ages are portrayed. However, many geologic maps are very detailed, with rock and sediment types subdivided on the basis of physical characteristics, slight differences in composition, or environments of origin. Faults are typically portrayed by heavy black lines, differentiating them from other types of *contacts* between different rock or sediment units, which are shown by light black lines.

Formations

Map units are frequently shown by different colors and abbreviated symbols. The fundamental geologic map unit is a *formation*. A geologic formation is a distinctive, recognizable unit of rock or sediment. A formation is recognizable and mappable by its *lithology*, rock or sediment type. On most geologic maps, the map symbols denote first the age of the unit, followed by the nature of the material or the formation name. For example, Quaternary alluvium (almost always shown in yellow on traditional geologic maps) is generally denoted by the abbreviation *Qal*. On many California geologic maps, *Tm* denotes the Tertiary Monterey Formation, *KJf* the Cretaceous and Jurassic Franciscan Assemblage, *Tv* Tertiary volcanic rocks, and so on.

All maps portray an area as if the viewer were suspended above the Earth looking down at the surface (Fig. 4-1). Because a map shows an area as viewed from above, it shows only a two-dimensional view of what is at the surface. However, specialized maps can include some information about the vertical dimension. For example, on many geologic maps, information about the landscape is also included because the geologic information is mapped on a *topographic* base map. A topographic map depicts the shape of the land surface with contour lines connecting points of equal elevation. Special symbols on geologic maps can also give the reader information about the inclination of beds, folds, or faults beneath the surface.

A geologic *cross section* is a view of the geology beneath the surface to some depth, constructed as if one had sliced through the earth along a line (Fig. 4-1). To construct a cross section, a geologist must understand the three-dimensional geometry of the map units and be able to project the contacts, including faults, from the surface. Although at least partly inferred, a completed cross section provides a clear view of the relative ages of map units, because one can use the principles of

Fig. 4-1 Map view, cross section, and block diagram of a birthday cake. Note that a map shows only the distribution of the surface layer, the frosting. The cross section shows the vertical arrangement of frosting, cake, and filling. The block diagram provides a perspective view of the top and the layers.

Map View

Block Diagram

Line of Cross Section

Cross Section

Frosting

Filling

Angel Food Cake Layers

cross-cutting relationships and superposition (see Chapter 3). To a geologist, a cross section can provide an interpretive summary of geologic events in the area. Many published geologic maps show cross sections across one or more lines along the map area, with the symbols and colors on the cross section coordinated with those on the maps (Fig. 4-2).

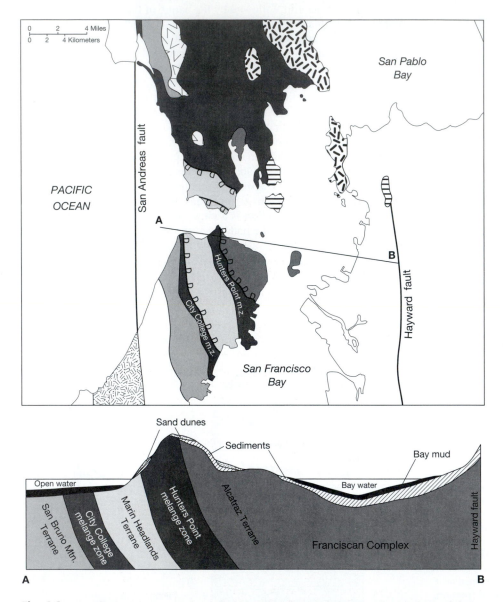

Fig. 4-2 Simplified geologic map (top). Cross section (bottom) showing the geology of the San Francisco area. The line A–B on the map shows the position and orientation of the cross section. (Source: Adapted from Elder, W., 2001, Field Trip 3, in *Geology and Natural History of the San Francisco Bay Area*. U.S. Geological Survey, Bulletin 2188.)

In many parts of California, particularly in urbanized areas at risk from geologic hazards, Earth scientists have prepared specialized maps useful for planners and other government officials. Specialized maps can show flood-prone areas, potential and active landslides, active fault zones, or areas prone to liquefaction during earthquakes. The California fault map (Endpaper 2) is an example of such a map. Other types of hazard maps show the relative risk of certain events in an area, such as the probability of volcanic debris flows around the flanks of Mt. Shasta.

Geospatial Data

Earth scientists increasingly use global positioning satellite (GPS) systems and geographic information systems (GIS) to collect and portray spatial information. As a result, it is possible to obtain large digital data sets and portray different types of data on maps and in geographic databases. Examples of these digital data sets include earthquake epicenters, elevational changes, geologic maps, and water-sampling data.

SOURCES OF GEOLOGIC MAPS AND DATE

The California geologic map is available from the California Geological Survey (CGS) in a larger format than the one included in this book (Endpaper 1). More detailed geologic maps of California, a series of 27 map sheets, at a scale of 1:250,000, are also available from the CGS. Each map sheet is named for a city within the area: for example, the San Bernardino sheet covers an area including the city of San Bernardino. These regional maps are excellent references, and most include cross sections, fault maps, and source lists for additional information.

More detailed geologic maps are available for many areas of California. These include maps of individual topographic quadrangles or other restricted areas. Specialized geologic maps often emphasize materials of certain ages or types. Many of these maps are published by the U.S. Geological Survey or the CGS. Other detailed or specialized maps are included in graduate theses. County planning offices are an excellent source for obtaining hazard maps, as are the sources for geologic maps.

Maps showing earthquake epicenters and active faults are of particular interest to Californians. The California fault map, available from CGS, shows all known and suspected active faults. This is the map used to define the Alquist-Priolo (AP) zones of special study for seismic hazards. The AP zone maps for California are available on CD-ROM from the CGS. The U.S. Geological Survey also has a series of maps showing earthquake epicenters plotted on raised-relief topographic maps. Specialized hazard maps showing landslide areas, floodplains, liquefaction-prone areas, or volcanic hazard zones are published by the U.S. Geological Survey for areas where these hazards have been evaluated.

The California Water Map, published by the Water Education Foundation in Sacramento, shows the state's rivers, lakes, dams, aqueducts, and other water features. County soil surveys, published by the U.S. Department of Agriculture's Soil Conservation Service, are available for most counties. Topographic maps, which show the elevations of the ground surface with contour maps, are available from the U.S. Geological Survey and at many sporting and recreation stores.

GENERAL SOURCES FOR MAPS AND PUBLICATIONS

U.S. Geological Survey: *http://pubs.usgs.gov*

Earth Science Information Service	*or*	Publications Office
Western Region		Denver Federal Center
345 Middlefield Road		Lakewood, CO 80225
Menlo Park, CA 94025		

California Geological Survey: *http://www.consrv.ca.gov/cgs*

Publication Sales
1059 Vine St., Suite 103
Sacramento, CA 95814-0321

Web Sites, Books, and Periodicals About California Geology

This book is intended to provide a general overview of California's geology for those with little or no background in Earth science. Because the emphasis is on geologic processes, the chapters only highlight particular aspects of the geology in each part of California; they are in no way comprehensive. Many excellent sources of information are available to readers looking for more detailed treatment of particular areas or subjects. The following list should be considered a starting point for those wishing to go beyond the material presented in this book.

Online Resources

Many Earth science institutions in California have information that can be accessed by computer. The U.S. Geological Survey, University of California's Paleontology Museum, Southern California Earthquake Center, and Monterey Bay Aquarium are examples of institutions with easily navigated home pages. Many of the sites listed here and at the end of each chapter also contain links to additional sites.

U.S. Geological Survey: *http://www.usgs.gov*

National Park Service: *http://www.nps.gov*

U.S. Geological Survey and National Park Service: What is a geologic map?
http://www2.nature.nps.gov/grd/usgsnps/gmap/gmap1.html

National Weather Service: *http://nws.noaa.gov*

EROS Data Center for satellite images and maps: *http://edcwww.cr.usgs.gov/*

California Geological Survey: *http://www.consrv.ca.gov/cgs*

California Resources Agency: *http://www.ceres.ca.gov*

California Office of Emergency Services: *http://www.oes.ca.gov*

Association of Bay Area Governments: *http://www.abag.ca.gov*

Southern California Association of Governments: *http://www.scag.ca.gov*

Northern California Earthquake Data Center: *http://quake.geo.berkeley.edu*

Southern California Earthquake Center: *http://scec.org*

Museum of Paleontology, UC Berkeley: *http://www.ucmp.berkeley.edu*

Monterey Bay Aquarium Research Institute: *http://www.mbari.org*

Scripps Institute of Oceanography: *http://www.sio.ucsd.edu*

BOOKS AND PERIODICALS ABOUT CALIFORNIA GEOLOGY (GENERAL REFERENCES)

ALT, D.D., AND HYNDMAN, D.W. 1975. *Roadside Geology of Northern California.* Missoula, MT: Mountain Press, 244 pp.

California Geology, published bimonthly by the CGS (formerly the California Division of Mines and Geology) until December 2002.
   This periodical contains articles of interest to beginning geology students, and most are written at an appropriate level for beginners. In later years, the publication included a "Teacher Feature" section for precollege teachers. Each December issue contains both a topical and regional index. Issues can be ordered from the Web site given previously.

Older publications from the California Division of Mines and Geology:
JAHNS, R.H., ed. 1954. *Geology of Southern California.* Bulletin 170.
BAILEY, E.H., ed. 1966. *Geology of Northern California.* Bulletin 190.
   Both of these publications predate the acceptance of plate tectonics as the ruling theory of geology, and for that reason, they are somewhat out of date. However, both contain excellent summary discussions of different parts of California and include discussion of important localities.

ERNST, W.G. 1981. *The Geotectonic Development of California.* Englewood Cliffs, NJ: Prentice Hall.
   This volume is the first summary of California's plate tectonics history. Although many of the detailed studies presented are now somewhat out of date, the book is still a useful reference.

Geology of North America Series, Decade of North American Geology (DNAG), 1988 and later. Boulder, CO: Geological Society of America, including:

BURCHFIEL, B.C., LIPMAN, P.W., and ZOBACK, M.L., eds. 1992. *The Cordilleran Orogen: Conterminous U.S.* Volume G-3, 724 pp.
   This volume is an exhaustive summary of the geologic history of the western U.S. as understood in the early 1990s. This is a comprehensive and current treatment of the geology of the western U.S., but anyone other than its intended audience of professional geologists might find it difficult reading.

HILL, M.L., ed. 1987. *Centennial Field Trip Guide:* : Cordilleran Section: Boulder, CO: Geological Society of America, Centennial Field Guide, vol. 1, 490 p.
   These field trip guides to important geologic localities are intended for geologists. However, many are appropriate for beginning geologists, and all include directions to the sites and reference lists.

HUBER, N.K. 1987, *The Geologic Story of Yosemite National Parks.* U.S. Geological Survey, Bulletin 1595. (Also available from the Yosemite Association.)

McPHEE, J. 1993. *Assembling California.* New York: Farrar, Straus, & Giroux.
   This is an engaging account of the plate-tectonics history of California. It also provides an insight into the working life of field geologists.

NORRIS, R.M., and WEBB, R.W. 1990. *California Geology,* 2nd ed. New York: Wiley, 541 pp.
   A comprehensive look at California geology, with detailed information about the rocks and history of its provinces.

OAKESHOTT, G.B. 1978. *California's Changing Landscape*. New York: McGraw-Hill, 379 pp.

SHARP, R.P. 1993. *Field Guide to Southern California*. Dubuque, IA: William C. Brown, 181 pp.

SHARP, R.P., and GLAZNER, A.F. 1993. *Geology Underfoot in Southern California*. Missoula, MT: Mountain Press, 224 pp.

SHELTON, J.S. 1966. *Geology Illustrated*. San Francisco: W.H. Freeman, 434 pp.

WALLACE, R.E. 1990. *The San Andreas Fault*. U.S. Geological Survey, Professional Paper 1515, 283 pp.

Geologic Highlights of California

Chapters 5 through 17 will introduce readers to the major geologic features found in different regions of California. The intent is not to give a comprehensive description of all geologic materials and events in each region, but rather to highlight the most important and interesting geologic features. Some features, like the California coastline or the San Andreas fault system, and some processes, like earthquakes, cross over region boundaries and are covered by individual chapters. The provinces are discussed in an order that presents less complicated areas and processes first, so the later chapters can logically build from them.

California's Geomorphic Provinces

Even a brief examination of California's geology, faults, and landscape features allows the reader to see that California has some distinct regional differences (see Endpapers 1–4 inside this book's covers). When these patterns are combined with California's patterns of regional climate, the state can be divided into distinct natural regions. Even before satellite imagery allowed geologists to view California's regions from space, these areas had been identified on the basis of their distinctive geology, climate, and landscape. Each of these regions, or geomorphic provinces (see the map on the next page), displays a distinctive landscape character. The amount of relief, the types of landforms, the orientation of valleys and mountains, and the type of vegetation in each province are elements of its geomorphology. Because geology is one of the controlling factors that shapes the landscape of each province, it is appropriate to discuss California's geology in the context of its geomorphic provinces.

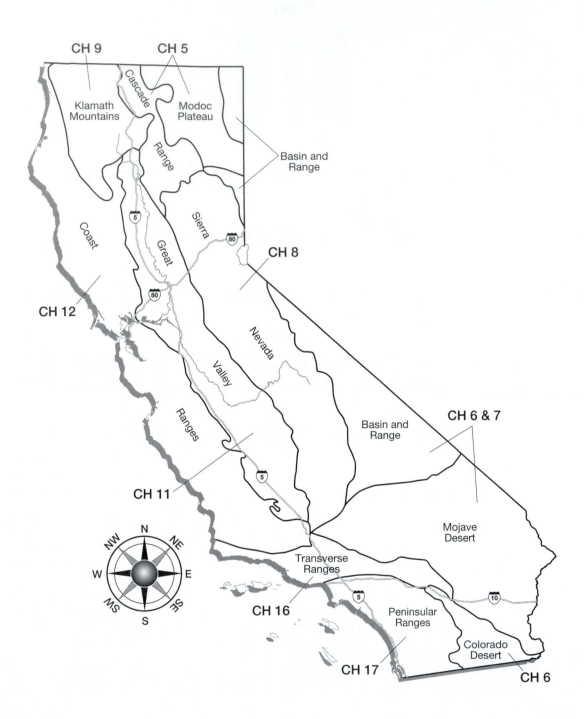

CH 9

CH 5

Klamath
Mountains

Cascade

Modoc
Plateau

Basin and
Range

Range

Coast

Sierra

Great

CH 8

CH 12

Nevada

Valley

Basin and
Range

CH 6 & 7

Ranges

Mojave
Desert

CH 11

N

NW   NE

W   E

SW   SE

S

Transverse
Ranges

CH 16

Peninsular
Ranges

Colorado
Desert

CH 17

CH 6

Plate 1

Plate 2

Plate 3

Plate 4

Rocks, Sediments, Fossils, and Stratigraphy

Plate 1 Late Cenozoic alluvial sediments and the 3.5-million-year-old ash of Mesquite Springs, western Death Valley. A person standing at the base of the ash layer near the right edge of the cliff provides scale. (Andrei Sarna-Wojcicki, U.S. Geological Survey.)

Plate 2 The oldest rocks exposed in the cliff are tilted sandstone and mudstone of the Late Cenozoic Purisima Formation. A younger channel and deposit of alluvial gravel can be see in the upper cliff; the contact between the two units is an unconformity. The two people are standing on an elevated marine terrace. San Gregorio Beach, San Mateo County. (Photo by author.)

Plate 3 Close-up view of Mesozoic plutonic rocks at Half Moon Meadow near Ten Lakes, Sierra Nevada. Bodies of darker, more mafic rock are enclosed within gray granitic rock formed from a more silica-rich magma. The white, quartz-rich veins cut through the darker rock types, indicating that they crystallized last. (Andrei Sarna-Wojcicki, U.S. Geological Survey.)

Plate 4 An outcrop of serpentinite along the beach south of the Golden Gate Bridge in San Francisco, showing both sheared and more massive bodies of rock and light veins of dolomite. (Andrei Sarna-Wojcicki, U.S. Geological Survey.)

Plate 5

Plate 6

Plate 7

Plate 8

Plate 5 A sample (about 20 cm across) of mercury ore from New Idria mine, San Benito County. This type of cinnabar is known as "strawberry ore" for its resemblance in color and shape to the fruit. (Edgar Bailey [deceased], U.S. Geological Survey.)

Plate 6 Petrified redwood tree, Petrified Forest, Sonoma County. (Photo by author.)

Plate 7 The shell of a freshwater clam embedded in fine-grained lake deposits, Cronese Dry Lake, Mojave Desert. (Howard Wilshire, U.S. Geological Survey.)

Plate 8 A layer of volcanic ash within the sand and layers of the Rio Dell Formation at Centerville Beach, south of Eureka in Humboldt County, provides an age for the sediments. From its chemical fingerprint, the ash is identified as one erupted from the central Cascade Range about 1.2 million years ago. (Photo by author.)

Plate 9

Plate 10

Plate 11

Plate 12

Volcanic Features

Plate 9 Lassen Peak viewed from the southwest, Lassen Volcanic National Park. The light colored area in the middle distance was covered by volcanic debris flows in 1915. (Andrei Sarna-Wojcicki, U.S. Geological Survey.)

Plate 10 A climber ascends the high-silica pyroclastic flows at Pinnacles National Monument. Differential weathering along the cracks will eventually form pinnacles like those seen to the left of the climber. (Amy Jordan, Friends of Pinnacles.)

Plate 11 Columnar joints in volcanic rocks at Devil's Postpile. The columns result from weathering along the shrinkage cracks created as the lava cooled. (Freeberg, U.S. Geological Survey.)

Plate 12 Ubehebe Crater, Death Valley National Park. Dark-gray ash along the rim of the crater overlies tilted, white and red sedimentary rocks. (Andrei Sarna-Wojcicki, U.S. Geological Survey.)

Plate 13

Plate 14

Plate 15

Plate 16

Desert Provinces

Plate 13 Tufa towers at Mono Lake. (Andrei Sarna-Wojcicki, U.S. Geological Survey.)

Plate 14 View of the steep, faulted eastern front of the Sierra Nevada from New York Canyon in the southern Inyo Mountains. The Owens River can be seen in the middle distance, and the Alabama Hills at the base of the mountain front. (Photo by author.)

Plate 15 Pleistocene lake beds of Lake Tecopa near Shoshone, east of Death Valley. (Andrei Sarna-Wojcicki, U.S. Geological Survey.)

Plate 16 A view of Death Valley looking northwest from the Black Mountains, Death Valley National Park. The flat playa surface and white salt deposits can be seen in the center of the photo, and alluvial fans can be seen at the base of the mountains. (Andrei Sarna-Wojcicki, U.S. Geological Survey.)

Plate 17

Plate 18

Plate 19

Plate 20

Coast Ranges and Faults

Plate 17 In this color infrared Landsat image of the San Francisco Bay area, the faults of the San Andreas system are clearly revealed as linear features on either side of the Bay. Irrigated fields of the Great Valley show as a distinct patchwork at the right edge of the photo, and Mt. Diablo is the teardrop-shaped mountain just south of the eastern part of the Bay.

Plate 18 The Golden Gate Bridge, looking northeast from Bakers Beach State Park. The Marin Headlands terrane of the Franciscan Complex forms the rugged mountains at the northern (left-hand) end of the Golden Gate Bridge.

Plate 19 Just south of the Cienega Valley winery (formerly the Almaden winery) in San Benito County, a drainage ditch crosses the creeping section of the San Andreas fault. In 1961, the ditch showed notable right-lateral offset. (U.S. Geological Survey.)

Plate 20 In 1996, 35 years later, the amount of right-lateral offset has markedly increased. (Robert Sorenson.)

Plate 21

Plate 22

Plate 23

Plate 24

Plate 21 The Alves Quarry near Pittsburg, Conta Costa County. At this locality, the Miocene Neroly Formation has been mined for sand and gravel, an increasingly scarce resource in urbanizing California. View north toward Suisun Bay. (Andrei Sarna-Wojcicki, U.S. Geological Survey.)

Sierra Nevada

Plate 22 Yosemite's Half Dome viewed from Olmsted Point, Yosemite National Park. (Andrei Sarna-Wojcicki, U.S. Geological Survey.)

Plate 23 A view of Hetch Hetchy reservoir showing the rounded domes typical of the glaciated granitic rocks of the High Sierra. (Andrei Sarna-Wojcicki, U.S. Geological Survey.)

Plate 24 The Sevehah Cliffs, a roof pendant of folded Paleozoic marble near Convict Lake, eastern Sierra Nevada. (Calvin Stevens, San José State University.)

Plate 25

Plate 26

Plate 27

Plate 25 Glaciated surface of granitic rock, Yosemite National Park. (Dan Sarna.)
Plate 26 A caver explores an unimproved solution cave formed in metamorphic limestone of the Calaveras Formation, central Mother Lode. The formations decorating the cave formed by the precipitation of the mineral calcite from bicarbonate-laden dripping water. (John Tinsley, U.S. Geological Survey.)
Plate 27 A vernal pool in the western Sierra Nevada foothills near Merced. (Alisa Durgarian Ellsworth.)

Plate 28

Plate 29

Plate 30

Beaches, Transverse Ranges

Plate 28 Typical features of an emergent coast, including sea stacks, steep cliffs, and coastal landslides, can be seen in this view along the Humboldt County Coast. Trinidad Head is the large point in the middle distance. The formation of sea stacks is enhanced by the nature of the Franciscan bedrock here, because a zone of melange containing resistant blocks parallels the coastline. (Photo by author.)

Plate 29 Natural arches formed by wave erosion at Natural Bridges State Park near Santa Cruz. In 1976, wave erosion collapsed the right-hand arch. The flat surface on top of the sandstone marks a marine terrace approximately 105,000 years old. (Andrei Sarna-Wojcicki, U.S. Geological Survey.)

Plate 30 Hogbacks formed by resistant, dipping sandstone beds of the Cajon Formation near Cajon Pass, San Bernardino County. (Photo by author.)

CHAPTER

5

Young Volcanoes: The Cascades, the Modoc Plateau, and the Long Valley Caldera

Cascade Range and Modoc Plateau Highlights

Cascade Range

Major geographic feature:	Cascade Mountain Range
Highest points:	Mt. Shasta: 4,319 m (14,162 ft.)
	Lassen Peak: 3,188 m (10,457 ft.)
Predominant rock type:	Late Cenozoic volcanic, mainly intermediate
Plate tectonic events:	Active subduction along Cascadia subduction zone, volcanic eruptions
Recent eruptions:	Lassen 1914–1921
Other geologic events:	Pleistocene glaciation
Resources:	Pumice, aggregate, decorative rock
National Park to visit:	Lassen Volcanic National Park
State Parks of interest:	Shasta, Castle Crags, Burney Falls

Modoc Plateau

Major geographic feature:	Modoc Plateau: avg. elevation 1,370 m (4,500 ft.)
Other landscape features:	Medicine Lake Highlands
	Warner Range
Highest point:	Eagle Peak, Warner Range: 3,043 m (9,983 ft.)
Predominant rock types:	Late Cenozoic volcanic, mainly basalt; also silicic
Tectonic setting:	Basin Range extension (see Chapter 7)
Other features of interest:	Lava Tubes
National Park to Visit:	Lava Beds National Park

Fig. 5-1 Map showing the location of the Cascades and Modoc Plateau Provinces.

GEOGRAPHIC HIGHLIGHTS

The Cascade Range is an 800-kilometer-long (500-mile) chain of young volcanic mountains that reaches from British Columbia to the northernmost part of California (Figs. 5-1 and 5-2). California's Cascades Province is the southern tip of this chain, and the geology of the Cascades Province is more similar to that of the Cascades of the Pacific Northwest than to other parts of California. Of the 12 major volcanic cones in the range, California's Mt. Shasta (elevation 4,319 m) is the second highest (behind Washington's Mt. Rainier) and the largest in volume (see Fig. 5-2). The towns of Weed, Mount Shasta, and McCloud sit at the base of the volcano. The other large cone in California is Lassen Peak (elevation 3,188 m), which last erupted during the period between 1914 and 1921. Mt. Shasta and Lassen Peak are spectacular landmarks that can be seen from as far as 150 kilometers away along Interstate 5 in the Sacramento Valley (Fig. 5-3 and Color Plate 9). The summits of both mountains are snow-covered bare rock, while the slopes are forested. Sparsely vegetated volcanic tablelands of fairly uniform topography flank the volcanic cones.

The volcanoes that make up the Cascade Ranges are the result of active subduction of the Gorda and Juan de Fuca Plates beneath North America. The rocks formed by eruptions from this chain during the past 3 million years are known as the High Cascades Series. Older Late Cenozoic volcanic rocks predate the present volcanic cones. Known as the Western Cascades Series, these rocks are evidence that active volcanoes were present in the California Cascades 20 to 30 million years ago. Alluvium, lake deposits, and glacial deposits are associated with the young volcanic rocks of the Cascades Province.

The Modoc Plateau lies east of California's Cascade Range in the northeastern corner of California. On the east, it is bordered by the Basin and Range Province, which is discussed in Chapter 7. As the name "plateau" indicates, the Modoc Plateau

Fig. 5-2 Map and cross section showing the plate tectonic setting of active volcanoes of the Cascade Range. (Source: Dengler, L., and others. 1992. Sources of North Coast seismicity. *California Geology* 45:40.)

is high, relatively flat country (Fig. 5-4), although the province is cut by valleys. This part of California is arid because the high ranges to the west block Pacific storms (see Chapter 6). Like the Cascades, the Modoc Plateau is made up of young volcanic rocks. However, the volcanic activity in the Modoc Plateau has been less explosive than that in the Cascades and has produced mainly lava flows like those seen in Lava Beds National Monument. On the border between the Modoc Plateau and the Cascade Range is Medicine Lake Volcano.

Fig. 5-3 Mt. Shasta.

Fig. 5-4 Volcanic plateaus of Lava Beds National Monument, showing typical dark basalt flows and sparse vegetation.

The contrast between the volcanic histories of the Cascades and Modoc Plateau Provinces provides an excellent opportunity to investigate the types and causes of volcanoes. A third area of recent major volcanic activity in California is the Long Valley area in the Basin and Range province near Mammoth (Fig. 5-5). Although this area lies well beyond the Cascade Province, its geologic history is similar enough to that of Mt. Shasta and Lassen Peak to warrant its inclusion here.

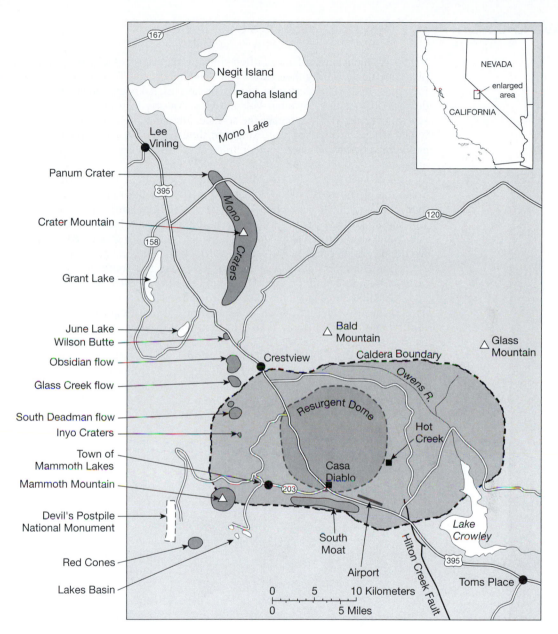

Fig. 5-5 Recently active volcanic centers in the Long Valley area east of the Sierra Nevada. (Source: U.S. Geological Survey, Long Valley Observatory.)

An evaluation of future volcanic hazards in California emerges from a review of the past volcanic events at these three major centers. The Coast Range Province and other parts of California's Basin and Ranges have also experienced geologically recent volcanic activity (see Chapter 7 and Chapter 12).

LAVA CHEMISTRY AND THE BEHAVIOR OF VOLCANOES

Visitors to the island of Hawaii often consider themselves lucky if an eruption takes place at Kilauea during their stay. Planes and helicopters fly over the lava flows and fountains, and bystanders take videos of lava as it knocks down trees and oozes down roads. At the Hawaiian Volcano Observatory, instruments predict eruptions several hours before lava breaks the surface, allowing geologists to travel to the likely site where lava will appear. Although thousands of homes have been overtaken and burned by lava flows, the predictions allow orderly evacuations from threatened areas.

Contrast the behavior of Kilauea with the events of 1980 at Mt. St. Helens in Washington and those of 1991 at Mt. Pinatubo in the Philippines. The May 18, 1980 eruption of Mt. St. Helens killed 60 people, including a U.S. Geological Survey geologist who was observing the mountain from what he thought was a safe site. If the area surrounding Mt. St. Helens had not been evacuated, the loss of life would have been far greater. A huge debris avalanche, traveling at speeds of up to 300 kilometers per hour, filled the North Fork of the Toutle River, 22 kilometers away, with debris and ice up to 45 meters thick (Fig. 5-6). More than 200 houses and 300 kilometers of roads were destroyed, 4 billion board feet of timber damaged or destroyed (Fig. 5-7), and 12 million salmon fingerlings killed. Almost 2 million cubic meters (more than 2 million cubic yards) of ash had to be removed from Washington highways and runways. Air and highway traffic was disrupted for weeks. The estimated cost of the eruption was $1.1 billion, a sizable amount for what was considered by geologists to be a small eruption. At Mt. Pinatubo, the June 1991 eruption produced about 5 to 7 cubic kilometers of debris, compared with about 1 cubic kilometer erupted from Mt. St. Helens in 1980. The Pinatubo eruptions killed 350 people, heavily damaged two U.S. air bases, and left thousands homeless.

Fig. 5-6 Geologic map showing the volcanic deposits and the blast zone of the 1980 eruption of Mt. St. Helens. (Source: Tilling, R.I. *Eruptions of Mount St. Helens: Past, Present, and Future.* U.S. Geological Survey, Public Information Series.)

Outline of crater
Pyroclastic flow deposit
Mudflow deposits
Lateral blast zone
Debris avalanche deposit

0   5 Miles

Fig. 5-7 Downed timber within the blast zone of the 1980 eruption of Mt. St. Helens. (Source: Keller, J.)

TABLE 5-1   EFFECTS OF MAGMA COMPOSITION ON VOLCANIC PROCESSES

Lava Type	Temperature of Lava	Fluidity of Lava	Gas Content	Events	Cone Type	California Examples	Rock Name(s)
Silicic High silica (−70% Si, 0) Low Fe, Mg	800°C	Viscous	High: explosive	Ash/steam Pyroclastic flows Gas clouds Volcanic Debris flows Glass flows	Dome	Mono Craters	Rhyolite Obsidian Pumice
Intermediate Intermediate silica (60%–65%)	1000°C	Intermediate			Strato-volcano (Caldera)	Mt. Shasta Mt. Lassen (Crater Lake, OR)	Andesite
Mafic Low silica (50%) High Fe, Mg	1200°C	Fluid	Lower: less explosive	Fluid lava flows Cinder cones Spatter cones	Shield volcano Fissure eruptions	Medicine Lake	Basalt

What makes volcanoes like Mt. Pinatubo and Mt. St. Helens so deadly? Differences in the violence of volcanic eruptions can be explained by the chemistry of the magma feeding the volcano (Table 5-1). By sampling and analyzing molten lava and volcanic rocks, geologists can determine the chemical composition of magmas. Although their chemistry is highly variable, lavas can be grouped into three broad categories. In all three, silica (see Chapter 2) is the predominant component.

• *Mafic* (sometimes called basaltic) lava contains relatively less silica (silicon and oxygen) and has high iron and magnesium content. A mafic lava contains about 50 percent silica and 15 to 20 percent iron and magnesium. About 90 percent of all of the lava erupted worldwide is mafic or basaltic lava. Most volcanic rock formed from mafic lava is called **basalt**, which is dark-colored because of the high content of dark, mafic silicate minerals (Fig. 5-8).

• *Intermediate or andesitic* lava has a silica and iron-magnesium content that is intermediate between mafic and silicic lavas. Typical volcanic rock formed from intermediate lava is termed **andesite**, and it is typically gray or brown (see Fig. 5-8).

• *Silicic or felsic* lava has high silica content and is low in iron and magnesium. Silicic lava contains about 70 percent silica and only a small percentage of iron and magnesium. Volcanic rock formed from silicic lava is called

Fig. 5-8 Volcanic rocks ranging in composition from mafic (*left*) to silicic (*right*). (Source: Sarna-Wojcicki, A. U.S. Geological Survey.)

Fig. 5-9 Obsidian from Sonoma County (*left*) and from the Mono Craters area, Inyo County (*right*). Because of its superior cutting ability, obsidian had great value to early Native Americans in California (*top* and *center*).

rhyolite, which is typically white or pink because of the absence of mafic minerals (see Fig. 5-8). Volcanic glass formed when silicic lava cools too quickly for minerals to crystallize is called *obsidian* (Fig. 5-9).

The chemical differences between the major types of lava make them behave very differently. In a silicic lava, the silica molecules tend to clump together, or polymerize, even though the lava is still a fluid. The clumping makes the lava *viscous* or sticky. In contrast, the iron and magnesium in a mafic lava act to keep the silica molecules separated, thus preventing the clumping action. As a result, mafic lava is very fluid and flows freely. In addition, silicic and intermediate lavas also contain more gas than mafic lavas. The gas is largely steam and carbon dioxide, but sulfur, nitrogen, and chlorine gases might also be present in small amounts.

An intermediate or high-silica lava moves sluggishly and might even become stuck in a volcanic vent because of its high viscosity. In contrast, mafic lavas are very fluid, sometimes behaving almost like water as they stream down valleys (Fig. 5-10). Extremely fluid basalt flows can move as fast as 16 kilometers per hour, and beneath the active volcanoes of Hawaii, basaltic magma can rise to the volcanic vents as fast as 1 kilometer per day.

The viscosity and the high gas content of intermediate and silicic lavas make eruptions explosive and potentially devastating. Because of the "sticky" movement within the volcanic vent, eruptions are also unpredictable. In extreme cases, the buildup of pressure is so great beneath the plug of lava that the eruption explodes with enough force to destroy an entire mountain. In contrast, eruptions of mafic lava low in gas produce relatively well-behaved lava flows and fountains. Because of its greater fluidity, mafic lava does not generally get stuck in the volcanic vents, eruptions are rarely very explosive, and they occur with predicable regularity. We can now see that whether a volcano is a tourist attraction or a deadly menace can be explained by small differences in the chemical composition of silicate lavas. Mt. St. Helens and Mt. Pinatubo are both fed by intermediate to silicic magma, and both have had a long history of violent eruption.

Fig. 5-10 Fluid basalt flow, Hawaii. (Source: Griggs, J. U.S. Geological Survey.)

ORIGIN OF LAVA

Around the world today, chains of active volcanoes are concentrated near plate boundaries. As discussed in Chapter 1, volcanoes occur at both divergent boundaries and along subduction zones. Some volcanic centers, like the Hawaiian chain or the Yellowstone volcanic area, are found within plates; these reflect localized melting over hot spots in the mantle. The hundreds of volcanoes surrounding the Pacific Plate, often referred to as the "Ring of Fire," are mostly of andesitic or intermediate composition. In fact, andesite is named for one of these volcanic chains, the Andes Mountains. These volcanoes are associated with active subduction zones, some between an oceanic and a continental plate and others between two oceanic plates.

Geologists agree that water is a necessary ingredient for the production of andesite. Sediments that have accumulated in a deep-ocean trench or on the floor of the ocean basin are naturally saturated with the sea water that surrounds them. Water is also present within the crystalline structure of silicate minerals such as clays and mica. As a plate is pushed beneath the crust during subduction, it heats up and this water is released. The water acts like a catalyst, lowering the melting temperature of minerals to enable partial melting in the portion of the mantle that overlies the subducting plate.

Mafic magma rises at divergent boundaries and at zones where the crust is being extended within plates. At these localities, the magma is generated by partial melting of the mantle beneath the Earth's crust. Magma rises relatively rapidly along pathways created as the crust is stretched. The magma is lower in silica, and because it is fluid and thus less dense than the surrounding rocks, it rises freely to the surface. Because it does not interact with crustal rocks on its journey, the magma retains its low-silica composition.

Volcanoes erupting only silicic lavas are found only within continental plates. These lavas are thought to originate from the partial melting of continental crustal rocks rich in silica as magma rises from beneath the crust. Because of its very high viscosity, high-silica magma tends to rise very slowly and cool beneath the surface to form plutonic rocks. As a result, silicic lavas are less common than either andesitic or basaltic lavas. However, an enormous eruption of silicic lava occurred in the Long Valley area in Inyo County 760,000 years ago.

Some volcanoes, including Medicine Lake in the Cascades Province and the Coso volcanic field east of the Sierra Nevada (discussed in Chapter 7), produce both mafic and silicic eruptions. In these areas, rising basaltic magma can heat adjacent crustal rocks and partially melt them to form silicic magma. At the same time, basalt can form directly from sources in the mantle. The result is a volcanic area that contains basalt and rhyolite of the same age, erupted from different source magmas. It is important to point out that an almost infinite variety of lavas is possible, considering the diverse types of rock that can be partially melted and the possibilities for mixing of different magmas.

Geologists can use the lava and young volcanic rocks found at a locality as a clue to the tectonic setting of the area:

- The Cascade Province, which is dominated by intermediate volcanism, shows the characteristics of a subduction zone setting. North of Cape Mendocino,

Mt. Shasta and Lassen Peak are being fed by magma that is generated as the Gorda Plate is subducted beneath the North American Plate.

- The Modoc Plateau, where volcanic rocks are largely mafic, is an area where crustal extension is occurring. As the crust stretches, magma rises from beneath the North American Plate to form the volcanic centers of the Basin and Range.

- Volcanic activity in the Long Valley area is also thought to be related to extension in the Basin and Range Province. In this area, however, silicic lavas are produced as basaltic magma rises into a more easily melted, thick section of Sierra Nevada granitic rocks, melting them and mixing to form the silicic magma beneath Long Valley.

VOLCANIC EVENTS AND FEATURES

Mafic

Lava flows are the most common product of basaltic volcanic eruptions. Lava can erupt from a central vent or from a *fissure* or crack along the flanks of the volcano (Fig. 5-11). Lava fountains or curtains of fire are common occurrences at Hawaiian volcanoes. Because of its fluidity, basaltic lava moves relatively freely down valleys, filling the low places in the landscape. Streams of lava can flow continuously in a channel for hours or days.

Although eruptions from low-silica volcanoes are less explosive than those from other volcanoes, some more explosive events can occur. The most common product of more explosive eruptions from basaltic cones is **scoria** or **cinders**, which are porous fragments, pebble-sized or larger, thrown from the volcano. The fragments typically have a frothy appearance because of the abundance of **vesicles**, or bubbles left by escaping gas.

Fig. 5-11 Lava fountain from fissure eruption, Iceland.

Fig. 5-12   Medicine Lake Volcano, northeastern California. Medicine Lake is a typical shield volcano formed by repeated eruptions of basaltic lava flows. (Source: Donnelly-Nolan, J. U.S. Geological Survey.)

Over a period of centuries or more, repeated eruptions of mafic lava from a central vent can create a *shield volcano* like Medicine Lake (Fig. 5-12). Shield volcanoes are broad and gently sloping, and they are often less dramatic landscape features than other volcanoes. Repeated eruptions along fissures can produce broad plateaus of lava, like those in the vicinity of Lava Beds National Park, rather than a volcanic cone. *Cinder cones* are common features of all active volcanic terrains. They are usually fairly small cones, composed of fragments thrown from a central vent. After the eruption, a funnel-shaped crater remains at the peak of the cone (Fig. 5-13). Small *spatter cones*, formed when blobs of lava are thrown from a vent and then fall in heaps, are also typical features of basaltic volcanic terrains.

Intermediate

During a large eruption from an andesitic volcano, huge volumes of gas and instantly chilled volcanic glass are released, along with chunks of rock from the volcano itself. An explosive eruption produces a wide variety of *pyroclastic* materials, or fragmented debris produced by fire (the same word root appears in *pyromaniac*). Great clouds of steam and other gases are released as the eruption releases the pressure within the vent, in much the way that opening a shaken bottle of carbonated soda sends jets of carbon dioxide and soda from the bottle. Some of the gas clouds can be deadly: one or two breaths of pure carbon dioxide is sufficient to asphyxiate a person. During the eruption of Mt. Vesuvius in 79 AD, many citizens of Pompeii dropped in their tracks as they inhaled the deadly gas streaming from the volcano (Fig. 5-14).

Fig. 5-13 Quarry excavated in a cinder cone near Manton, Lassen County. (Source: Sarna-Wojcicki, A. U.S. Geological Survey.)

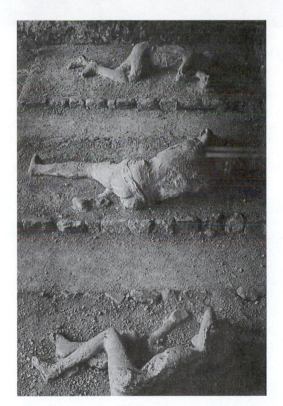

Fig. 5-14 Petrified remains of victims of the 79 AD eruption of Mt. Vesuvius, Pompeii, Italy. (Source: Bradley, W.C.)

As the lava blows out from the vent, it is instantly chilled to form volcanic glass, and the mixture of debris and gas is sent high into the atmosphere. The pyroclastic material carried into the atmosphere is called *tephra*. Much of the tephra is very small particles known as *volcanic ash*, but some tephra fragments, known as *lapilli*, are centimeters in size, and others, known as *volcanic bombs*, are a half-meter or

Fig. 5-15 Large pumice blocks erupted from Mt. St. Helens in 1980. (Source: Sarna-Wojcicki, A. U.S. Geological Survey.)

more in diameter (Fig. 5-15). When gas trapped in the vent is violently separated from the lava, **pumice** is produced. Pumice is volcanic glass riddled with holes or vesicles. Because the vesicles make pumice fragments very light, they can be carried long distances from the source of the eruption.

Hot ash and other debris flow down the flanks of the erupting volcano, often releasing deadly gas as the lava travels. The flows can glow from incandescent lava fragments, and these deadly avalanches known as glowing clouds or *nuee ardente* can reach speeds as great as 300 kilometers per hour. In 1902, the glowing clouds produced by the eruption of Mt. Pelee in the Caribbean killed 30,000 people. ***Pyroclastic flows***

Fig. 5-16 Welded tuff, Sonoma volcanic field. (Source: Sarna-Wojcicki, A. U.S. Geological Survey.)

can be hot enough to partly weld the fragments within them (Fig. 5-16). When volcanic ash or pyroclastic flows are lithified to form rocks, they are referred to as *tuff*.

The steep sides of an andesitic volcano are prime sites for debris avalanches, a type of landslide that can reach speeds of hundreds of kilometers per hour. A debris avalanche can occur in fairly dry materials, or it might involve some water. Huge debris avalanches can send debris tens of kilometers down the volcano's flanks.

Among the most deadly hazards during a large eruption of an andesitic volcano are *volcanic debris flows*. Consider a snow-capped volcanic peak, especially one with glaciers on its flanks, like Mt. Shasta. The flanks of an intermediate volcano are littered with volcanic ash, and early eruptions can produce a fresh covering of fine ash. A major eruption instantly melts the snow and ice, mixing in huge amounts of ash and debris, and sends the saturated mixture down the steep sides of the mountain at incredible speed. This is the recipe for disaster that has been repeated in recent years. At Mt. St. Helens in 1980, debris avalanches became saturated debris flows that traveled about 35 kilometers from the mountains into the Cowlitz and Toutle Rivers, raising river levels as much as 5 meters and filling the channels with debris (Fig. 5-17). The village of Armero, 30 kilometers from Nevada Ruiz in the Colombian Andes, was destroyed by debris flows in September 1985; 23,000 people were buried by the flows. At Mt. Pinatubo, debris flows continue to be a problem for villages in the valleys on the sides of the volcano; to date, hundreds of people have been killed by the flows.

Over a period of several thousand years, eruptions of intermediate lava can form a huge volcanic mountain known as a *stratovolcano* or *composite cone*. These are the "textbook" volcanoes, with summit craters and steep symmetrical sides. The shape of a stratovolcano reflects the fact that it is composed of viscous pyroclastic material as well as lava flows.

A stratovolcano might be partly or completely destroyed by a catastrophic eruption. About 7,000 years ago, an Oregon Cascades volcano, named Mt. Mazama by geologists, was destroyed by a series of huge explosions. After the magma was removed from the vent, the entire top of the mountain collapsed to form the depression known as Crater

Fig. 5-17 Effects of volcanic debris flows, Toutle River, Washington, 1980. (Source: Miller, C.D. U.S. Geological Survey.)

Fig. 5-18 Glass Mountain obsidian flow, Medicine Lake Highlands. (Source: Sarna-Wojcicki, A. U.S. Geological Survey.)

Lake. Crater Lake is an example of a **_caldera_**, a circular depression formed by the collapse of an area above a volcanic vent. Volcanic centers might undergo a cycle of volcano building and destruction, with new cones occupying the site of a destroyed mountain. For example, the north side of Mt. St. Helens was blown away in 1980, and since then, a new dome, called a **_resurgent dome_**, is being built in the summit crater.

Silicic/Felsic

Like the andesitic volcanoes, volcanic centers erupting high-silica lava can also produce explosive eruptions, creating many of the hazardous events typical of andesitic eruptions. Rhyolite pumice, tephra, and ash flows are common products of explosive eruptions from high-silica sources.

Eruptions of very high-silica lava also commonly produce **_obsidian domes_** and glass flows. Because high-silica lava is thick and pasty, it oozes slowly from vents and does not travel very far before it solidifies. Domes and flows formed from high-silica lava are generally steep-sided (Fig. 5-18). The glassy flows and domes near Medicine Lake volcano and in the Mono-Inyo volcanic chain, both discussed later in this chapter, are striking examples of high-silica volcanic features.

ERUPTIVE HISTORY OF CALIFORNIA'S ACTIVE VOLCANOES

Mt. Shasta

Mt. Shasta is a huge stratovolcano built by eruptions from four main volcanic vents (Fig. 5-19). Geologists estimate that it has taken more than 100,000 years to construct the present volcano. During that time, eruptions have produced andesitic lava flows, pyroclastic flows, and debris flows, but few deposits of ash. Much of the volcano's

TABLE 5-2 VOLCANIC ERUPTIONS IN CALIFORNIA

Date	Source Area	Volcanic Events
1914–1922	Mt. Lassen	Ash and steam clouds, debris flows
200–400 years BP	Paoha Island, Mono Lake	Most recent ash eruptions
200–300 years BP	Medicine Lake	Most recent lava flows, Glass Mountain area
Few centuries BP	Cima Dome, Mojave Desert	Youngest basalt flows and domes
1325–1365 AD (650 years BP)	Panum Crater, Mono Craters Chain	Eruption of tephra ring; Obsidian dome in crater
9300–9700 years BP	Shasta area	Construction of Shastina Cone
40,000 years BP	Coso volcanic field	Youngest basalt flow
300,000–380,000 years BP	Shasta area	Giant debris flow, Shasta Valley
600,000, (new date) years BP	Lassen area	Rockland ash eruption
760,000 (new date) years BP	Long Valley	Caldera-forming eruption, Bishop ash and tuff, 500 km^3 ash erupted
3.4 million years BP	Lassen area	Nomlaki tuff
3.4 million years BP	Sonoma (Coast Ranges)	Putah tuff (dated age), but older than Nomlaki, which is found above it in the Sacramento Valley

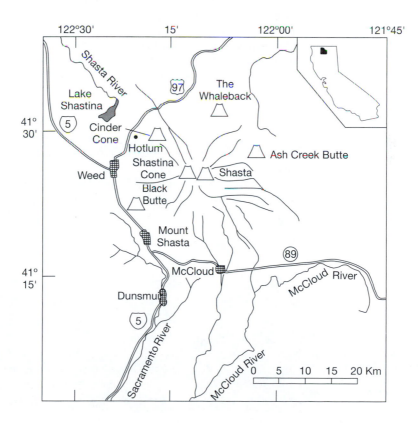

Fig. 5-19 Map showing eruptive centers at Mt. Shasta. (Source: Adapted from Miller, C.D. 1980. U.S. Geological Survey, Bulletin 1503, p. 2.)

construction has taken place during the past 10,000 years, when at least 13 eruptions have occurred. During the past 4,500 years, the average interval between eruptions has been only 600 years. Today two small centers of hot springs remain near the summit.

Two of Mt. Shasta's eruptive centers have been very active during the past 10,000 years. Hotlum Cone at the summit has been constructed during the past 8,000 years. During that time, it has erupted at least eight times, producing pyroclastic flows, andesitic lava flows, and debris flows that have traveled down the mountain's flanks. A more explosive eruption about 9,600 to 9,700 years ago also produced the Red Banks tephra, a layer of ash and pumice fragments found today north and east of the summit. About 200 years ago, several debris flows originated from the Hotlum area. These might have been produced by an eruption that was apparently observed in 1786 by explorer LaPerouse from a ship in the Pacific Ocean.

Shastina Cone, a subsidiary cone on the western flank of Mt. Shasta (Fig. 5-20), was built by eruptions between 9,300 and 9,700 years ago. These eruptions produced andesitic lava flows and pyroclastic flows; the town of Weed is built on one pyroclastic flow that erupted from Shastina about 9,500 years ago. Shastina's summit is composed of several *dacite* domes (a material with a composition between andesite and rhyolite) built at the end of its eruptive period. A subsidiary center of recent volcanic activity is Black Butte, a group of dacite domes about 13 kilometers west of Mt. Shasta. At both Shastina and Black Butte, the collapse of volcanic domes resulted in pyroclastic flows about 9,500 years ago.

Prior to the cone-building activity at Shastina and Hotlum Cones, volcanic centers at Misery Hill, just south of the summit, and Sargeants Ridge, on the south flank of the mountain, were responsible for most of the activity. Geologists can identify older eruptive centers by tracing volcanic deposits back to their source, where they are thickest. By dating the deposits, geologists can determine the time period when that center was active. At Mt. Shasta, glaciers that were present until about 15,000 years ago eroded the two older cones. In contrast, Shastina and Hotlum Cones show no evidence of glaciation.

Fig. 5-20 View of Black Butte, a young volcanic cone north of Chico. This is not the Black Butte just east of Interstate 5 south of Redding. (Source: Harwood, D. U.S. Geological Survey.)

A 300,000-YEAR-OLD CATASTROPHE

The western two-thirds of the valley of the Shasta River, on the northwest side of Mt. Shasta, is a maze of mounds, hills, and ridges (Fig. 5-21) that have puzzled geologists for more than 80 years. The blocks of volcanic rock that form the hills are up to hundreds of meters long, and they are strewn chaotically about the valley. Although the volcanic rock is similar to the andesite found in the modern Mt. Shasta cones, a variety of volcanic rock types are found in the hills and mounds. One of the original interpretations of the deposits of Shasta Valley was that they are deposits left by glaciers or by streams washing out from the glaciers that once occupied Mt. Shasta.

After a careful study of Shasta Valley, U.S. Geological Survey geologist Dwight Crandell came up with a startlingly different conclusion: The deposits are the remains of a giant debris avalanche from the sides of an ancestor of Mt. Shasta. Based on the location of volcanic blocks, Crandell determined that the debris avalanche traveled 49 kilometers from its source and covered 675 square kilometers of the Shasta Valley. As estimated from records of water wells drilled in the valley, the debris avalanche filled the valley to a depth of 50 meters. The estimated volume of 45 cubic kilometers makes it the largest volcanic debris avalanche measured by geologists.

The age of the debris avalanche can be bracketed fairly closely. Some of the blocks within the deposit contain the 600,000-year-old Rockland ash, indicating that the debris avalanche postdated that eruption. Also contained within the deposits are blocks of andesite about 360,000 to 380,000 years old. Overlying the debris avalanche material at some sites is a lava flow dated at about 300,000 years old. Using the principles of relative dating outlined in Chapter 3, these dates bracket the age of the debris avalanche at between 300,000 and 600,000 years. Considering that modern Mt. Shasta's four cones are all younger, the debris avalanche must have come from the sides of an ancestral cone.

A

B

Fig. 5-21   **A,** Photo and **B,** map of the remains of a 350,000-year-old debris avalanche from Mt. Shasta. (Source: **A,** Hotz, P. U.S. Geological Survey; **B,** Crandell, D.R. 1989. U.S. Geological Survey, Bulletin 1861, p. 2.)

Medicine Lake Highlands and Lava Beds National Monument

Within the uplifted plateau known as Medicine Lake Highlands lies Medicine Lake volcano, a recently active shield volcano on the border between the Cascades and Modoc Plateau Provinces about 50 kilometers northeast of Mt. Shasta (see Fig. 5-12). Volcanic activity in the Medicine Lake area has been varied, and it is an excellent place to view the results of basaltic to silicic volcanic eruptions. Many of the volcanic features formed within the past few thousand years, and the youngest flows took place no more than a few hundred years ago. During the summer of 1989, a flurry of tens of thousands of earthquakes indicated that a new eruption might occur, but none materialized and the seismic activity ceased.

The earliest volcanic activity at Medicine Lake, about 700,000 years ago, was the eruption of basalt that can be seen today at the base of the Highlands. The eruptions were similar to those that formed other parts of the Modoc Plateau, including Lava Beds National Monument on its northern flank (Fig. 5-22). Cinder cones, explosion pits, and spatter cones are found near the vents. Higher on the slopes of the volcano are younger and more silica-rich volcanic rocks. These include Glass Mountain and Little Glass Mountain, produced by rhyolite lava flows erupted from a number of vents. Some of these eruptions occurred only 200 to 300 years ago. Associated with the flows are domes of obsidian, pumice, and rhyolite. The obsidian of Glass Mountain and Little Glass Mountain was the source for tools and trade materials used by the Modoc Indians. On the north side of the volcano is the Callahan andesite flow, dated at about only 1,110 years old. The youngest basalt flow, at Mammoth Crater, is probably less than 100 years old. Medicine Lake itself sits within a crater at the summit of the Highlands. The entire shield volcano complex, which appears gentle and nonthreatening, actually has a greater volume than towering Mt. Shasta.

To the north of Medicine Lake lies the bare, flat, but rough-surfaced country of Lava Beds National Monument. This is a land of dark basalt flows, sparsely vegetated due to the desert climate. Only ranches and the town of Tulelake break up the open spaces of this part of California. At the northeastern end of Lava Beds is Tule Lake itself, a large lake that straddles the California–Oregon border.

Lava Beds National Monument is famous for its network of tunnels and caves (Fig. 5-23). The tunnels are *lava tubes*, some of which carried mafic lava 25 to 35 kilometers from vents on the north side of Medicine Lake. As fluid lava streams along a channel, a solid crust or roof might form on the surface and sides of these channels, because the outer surfaces cool more rapidly. Lava can continue to flow within the channels, just as water can flow beneath an ice-covered stream. The lava can drain out to leave an open lava-tube cave, or the tunnel might be partly or completely filled by later lava flows. At Lava Beds, parts of the roofs of many of the tubes have collapsed, providing access to the tunnels (Fig. 5-24).

The last major war between the U.S. government and Native American tribes was fought in the Modoc Plateau in 1872–1873. Led by Captain Jack, a band of about 150 Modoc Indians managed to elude as many as 650 Army troops for 7 months by taking advantage of the rough volcanic terrain of the Modoc Plateau. The blocky lava, lava tubes, collapse pits, and low areas behind lava flows provided shelter and escape routes for the outnumbered Modocs. They were captured only after being betrayed by fellow Modocs.

Fig. 5-22 Map showing recent volcanic flows and domes at Medicine Lake volcano. (Source: Donnelly-Nolan, J. 1987. *Geological Society of America Centennial Field Trip Guide No. 66*, p. 290.)

Fig. 5-23 Caves and lava tubes at Lava Beds National Monument (C = cave). (Source: Waters, A., and others. 1990. U.S. Geological Survey, Bulletin 1673.)

Fig. 5-24 Entrance to lava tube, Lava Beds National Monument. (Source: Skapinsky, S. San José State University.)

Lassen Volcanic National Park

It was 5 PM on Memorial Day, May 30, 1914, and Ike McKenzie of Chester happened to be gazing toward Lassen Peak. To his astonishment a huge cloud of steam rose from the center of the mountain, signaling the beginning of the first volcanic eruption in the United States since the 48 states had been united (Fig. 5-25). Few people lived in the area around Mt. Lassen in 1914, but many local residents made careful notes of the eruptions that took place between 1914 and 1921. During the first year, 170 eruptions of steam and ash were recorded. Many locals made the climb to look into the crater of the erupting volcano, where most of the fragments ejected were lapilli and small rocks. This foolish behavior accounted for the only injury during the 7 years of eruption. Lance Graham and two friends were standing on the crater when the volcano blew out a shower of ash and rock. One rock hit Graham, knocking him unconscious and breaking his collarbone.

In the spring of 1915, Lassen picked up its activity. "Fire," which was actually glowing lava, could be seen coming from the crater on the night of May 19; the next day, viewers could see the dark mass of a new lava flow extending 2,000 feet down the west side of the peak. On the eastern slope, a massive debris flow swept down Lost Creek, picking up enough speed to jump over a divide into Hat Creek. The flow washed away buildings and timber and carried 20-foot boulders along its path. Barking dogs had alerted residents to the oncoming hazard, allowing them to escape. On May 22, a glowing cloud leveled and scorched trees in what is now known as the Devastated Area (Fig. 5-26). Today in and near Lassen Volcanic National Park, hot springs and *fumaroles*, areas where volcanic gas and steam boil from vents, remind visitors of the recency of volcanic activity at Lassen Peak (Fig. 5-27).

Until the eruption of Mt. St. Helens in 1980, the Lassen eruptions of 1914 to 1921 were the only eruptions of the 20th century in the contiguous 48 states. By

Fig. 5-25 Ash eruption from Mt. Lassen, photographed by B.F. Loomis. (Source: National Park Service, Loomis Museum.)

Fig. 5-26 The Devastated Area, Lassen National Park, 1915. (Source: Model Studio, Anderson. National Park Service, Loomis Museum.)

Fig. 5-27 Active fumaroles, Bumpass Hell area, Lassen National Park. (Source: Sedlock, R. San José State University.)

studying the deposits of past eruptions from Lassen, as well as comparing them with eruptions from other stratovolcanoes, geologists know that the eruptions of 1914 to 1921 were minor events. The geology of the Lassen area reveals a history of repeated catastrophic eruptions during the past few million years.

Lassen Peak itself is a plug dome of dacite that rose on the eastern flank of a huge hole created by a very large blast about 200,000 years ago. That eruption, centered west of today's Lassen Peak, almost totally destroyed a large stratovolcano, which geologists have named Mt. Tehama (Fig. 5-28). Brokeoff Mountain, appropriately named, marks one edge of the hole. An earlier explosion took place in the same general area 600,000 years ago. This event sent ash, named the Rockland ash

Projected former height of Brokeoff Mountain (Mt. Tehama)

1915 Lava

LASSEN PEAK

Fig. 5-28 Cross section of Lassen Peak and the remains of Mt. Tehama, which was destroyed by a major eruption about 600,000 years ago. (Source: Harris, S. 1988. *Fire Mountains of the West.* Missoula, MT: Mountain Press.)

by geologists, downwind for at least 800 kilometers, covering much of northern California and western Nevada. Near the source, the eruption produced pyroclastic breccia containing pumice fragments as well as fine ash. Here the pyroclastic flows and ash were up to 12 meters thick. As far away as Humboldt County, 23 centimeters of ash fell, and ash fell in San Francisco and in the Santa Clara Valley. The eruption was at least as large as, and perhaps even greater than, the eruption of Mt. Mazama, which produced Crater Lake about 7,000 years ago. By comparison, the 1980 eruption of Mt. St. Helens produced about 1/50 to 1/70 the volume of ash and pumice (Fig. 5-30).

The Mt. Lassen area has been the source of at least one other major explosion in the recent geologic past. Approximately 3.4 million years ago, an even earlier stratovolcano experienced a catastrophic eruption that sent ash flows down the Sacramento Valley for at least 130 kilometers. The rhyolite flows, which contain large pieces of pumice carried by the ancient Sacramento River, reached as far south as what is now the town of Willows. Today they are found as a volcanic tuff, named the Nomlaki tuff, which is up to 90 meters thick at some localities east of Redding. Along the east side of the Sacramento Valley, the Nomlaki tuff is interlayered with volcanic flows and alluvial deposits of the Tuscan Formation (Fig. 5-29). On the western side

Fig. 5-29 Sediments of the Tuscan Formation, interlayered with the Nomlaki tuff, which was erupted from the Lassen area about 3.4 million years ago. (Source: Harwood, D. U.S. Geological Survey.)

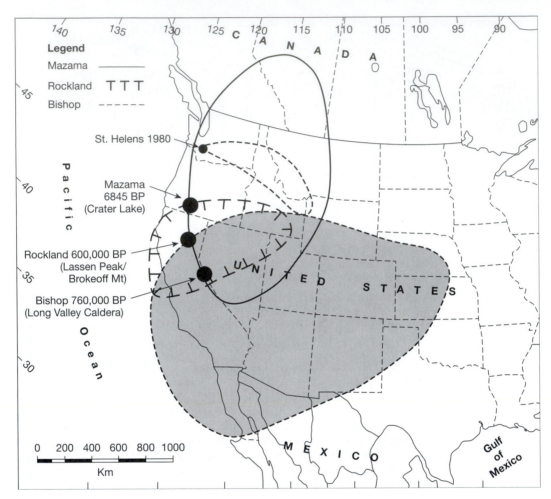

Fig. 5-30 Comparative distribution of some volcanic ashes erupted in the western United States (BP = years before present). (Source: Sarna-Wojcicki, A. U.S. Geological Survey.)

of the Sacramento Valley, the Nomlaki tuff is found within alluvial deposits of the Tehama Formation. Much farther south, pumice and ash from the eruption are found in marine sedimentary deposits of the San Joaquin Formation in the Kettleman Hills near Coalinga (see Chapter 12).

Long Valley Caldera and the Bishop Tuff

The Long Valley area is located on the western edge of the Basin and Range Province along the base of the Sierra Nevada Range (Fig. 5-1, Fig. 5-31). Volcanic activity began there about 3.6 million years ago, when mafic and andesitic lavas were erupted over a fairly wide area. These flows were followed by the eruption of more silicic domes and flows near the northern edge of the present caldera. About 2 million years ago, a series of high-silica eruptions began near Mono Glass Mountain, resulting in the formation of a series of domes and flows surrounded by

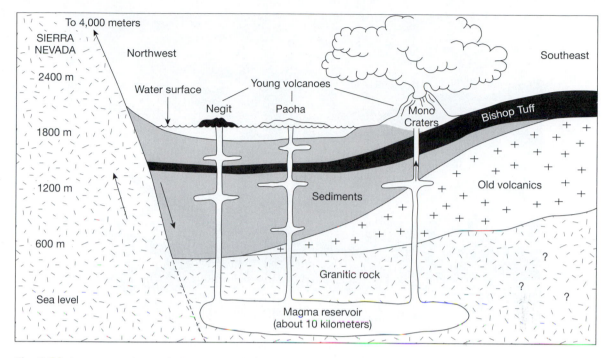

Fig. 5-31 Cross section beneath the Mono Lake area. (Source: Adapted from Shearin, R. 1989. *Mono Lake Guide-book.* Lee Vining, CA: Kutsavi Books.)

ash and ash flows. The change in the lava chemistry occurred as a large, fairly shallow magma chamber developed beneath Long Valley.

About 760,000 years ago, the roof of the Long Valley magma chamber ruptured, causing a catastrophic eruption of 600 cubic kilometers of rhyolite magma. Most of the lava formed hot, glowing ash flows that covered 1,500 square kilometers of the surrounding area to a depth of up to 200 meters. Ash was also blown high into the atmosphere and carried long distances by prevailing winds. Today airfall ash from this explosion is found in sediments as far east as Kansas and off the coast of California (see Fig. 5-30). The deposits from the eruption are known as the Bishop tuff, named for the town of Bishop in Inyo County (Figs. 5-31, 5-32). The sudden emptying of the magma chamber caused the ground above it to collapse, forming an oval depression 2 to 3 kilometers deep known as the Long Valley Caldera, which measures about 17 by 32 kilometers. The resort area of Mammoth Lakes lies within its rim, and U.S. Highway 395 crosses through the caldera.

After the roof of the magma chamber collapsed, eruptions continued within the caldera. Pyroclastic flows and obsidian flows erupted onto the caldera floor. A resurgent dome then formed in the center of the caldera floor, accompanied by obsidian flows and rhyolite lava flows. Following a relatively quiet period, eruptions began again about 500,000 years ago and continued at roughly 200,000-year intervals. No eruptions have occurred in the past 100,000 years, but magma appears to remain beneath the central dome at fairly shallow depths. The presence of this magma 5 to 15 kilometers beneath the surface is suggested by seismic surveys and unusual earthquake activity. Between 1980 and 1986, thousands of earthquakes occurred in the area, alarming residents and

Fig. 5-32 Airfall ash (lower cliffs) and ash flow (upper cliffs) units of the Bishop tuff, north of Bishop. (Source: Sarna-Wojcicki, A. U.S. Geological Survey.)

Fig. 5-33 Trees killed by carbon dioxide gas near the north shore of Horseshoe Lake in the Long Valley Caldera near Mammoth. (Source: U.S. Geological Survey, Long Valley Ovservatory.)

geologists alike. The activity appears to have been the result of magma movement within the chamber. The activity has decreased since 1986, but the area continues to show signs of unrest. The resurgent dome in the center of the caldera is rising at a rate of 2 to 3 centimeters per year and has risen a total of 60 centimeters since the current activity began. In 1995, carbon dioxide gas leaked from the southwestern part of the caldera, killing large stands of trees (Fig. 5-33). Numerous hot springs in the Long Valley Caldera also suggest that magma is present beneath the surface.

Mono-Inyo Volcanic Chain

Another center of recent volcanic activity on the east side of the Sierra Nevada near the Long Valley Caldera is the Mono-Inyo volcanic chain. The Mono-Inyo chain is a system of vents 45 kilometers long that overlaps with the Long Valley Caldera at the southern end and extends north to Mono Lake (see Chapter 6). The chain of volcanic vents is relatively linear, and the volcanic activity might be controlled by a fault in the underlying rocks. Volcanic activity at the Mono-Inyo chain is more recent than most of the Long Valley eruptions.

The Mono Craters chain of at least 30 rhyolite and obsidian domes, flows, and craters began to form only 40,000 years ago. The highly viscous lava from these vents flowed very slowly and maintained very steep front edges; these obsidian flows are striking features of the Mono-Inyo chain (Fig. 5-34). The most recent eruption at Mono Craters took place about 650 years ago, between 1325 and 1365 AD. The Inyo Craters chain, also obsidian and rhyolite domes, flows, and craters, extends south of the Mono Craters chain. The youngest eruptions from the Inyo Craters chain occurred as recently as 520 years ago in about 1472 AD. Within Mono Lake, several cinder cones and rhyolite flows form Negit Island and part of Paoha Island. The youngest explosions from these centers might have taken place as recently as 150 years ago.

Fig. 5-34 Domes and craters of the Mono Craters chain; view southwest, Sierra Nevada in the distance. (Source: Miller, C.D. U.S. Geological Survey.)

VOLCANIC ASHES RECORD A MOMENT IN GEOLOGIC TIME

As we have seen, the chemistry of lavas and the composition of the volcanic rocks formed from them are highly variable. Although all volcanic ashes erupted from the Cascades volcanic chain show the general characteristics of subduction-zone volcanoes—they typically have intermediate amounts of silica and mafic materials—not all Cascades lavas are exactly the same. Each eruptive source has a distinctive chemistry, and ashes erupted from each source show slight variations in composition. For example, volcanic ashes erupted from Mt. St. Helens have different compositions than those erupted from Long Valley (Fig. 5-35), and even though all volcanic ash erupted from Mt. St. Helens is generally similar, each individual eruption is slightly different. When an eruption occurs, the composition of the lava remaining in the magma chamber changes slightly. Over time, the magma chemistry can also change because new molten material is produced.

Because it is instantaneously cooled lava, volcanic glass gives geologists a particularly accurate look at the chemistry of the lava at the time of an eruption (Fig. 5-36). Crystals of silicate minerals formed during the slow cooling of the magma prior to a volcanic eruption might also be present in the ash. Because they could have formed long before the eruption, these crystals are a less instantaneous record of the lava chemistry. The volcanic glass in each ash has a unique chemical "fingerprint" and can be identified even when the source volcano has long since been destroyed by eruption or erosion.

Geologists analyze the chemistry of volcanic ash using a variety of techniques. Using a microscope, geologists can observe the shapes of the glass fragments and their vesicles and the types of minerals within the glass. An electron microprobe can be used to measure the abundance of some of the chemical elements in the glass. For very detailed comparisons, neutron activation analysis, which involves use of a nuclear reactor, can be used to measure rare elements present in concentrations of only a few parts per billion.

What can be learned from "fingerprinted" volcanic ashes? Geologists can trace the source of each chemically distinct ash by comparing its thickness at many places and pinpointing the area where it is thickest. By mapping out its general distribution and thickness, geologists can also reconstruct the size of an ancient eruption. The size estimates are only a minimum, because areas distant from the source would receive only a fine covering of ash that probably would not be preserved in the geologic record. Frequently, the volcanic rocks near the source can be dated using radioactive isotopes (see Chapter 3). For each source area, then, geologists can estimate the number and size of eruptions over a time period of thousands of years. Volcanoes with a history of frequent, major eruptions, like Mt. St. Helens, can be identified. The characteristic frequency of major eruptions, or the *recurrence interval*, can be estimated from each active volcano. This information allows a much more accurate assessment of volcanic hazards than the picture obtained by the historic record of eruptions over the past two centuries.

Another important use of volcanic ash is for dating and correlating the sediments interlayered with the ash. Many sedimentary deposits lack fossils or materials that can be dated using radiometric methods. Because an ash layer represents a moment in geologic time, it can provide an excellent *marker bed* that can be used to correlate sediments from different areas. For example, the 600,000-year-old Rockland ash discussed earlier is found in marine sediments off Cape Mendocino, in alluvial deposits in the Santa Clara Valley, and in lake and alluvial fan deposits in eastern California and western Nevada.

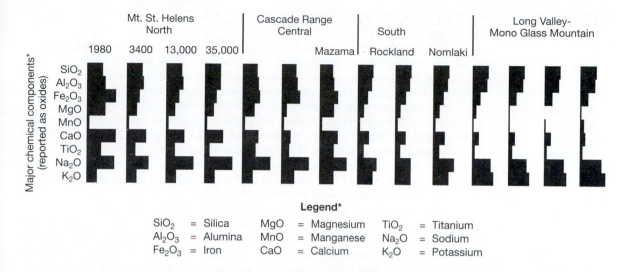

Fig. 5-35 Chemical comparison of volcanic ash erupted from various volcanoes. The length of the bars for each component shows its relative abundance. Note the general similarity among glasses from the same source area. (Source: Sarna-Wojcicki, A. U.S. Geological Survey.)

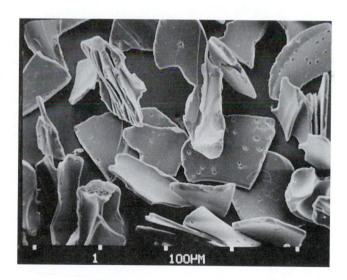

Fig. 5-36 Scanning electron micrograph of volcanic glass. (Source: Sarna-Wojcicki, A. U.S. Geological Survey.)

FUTURE ERUPTIONS? VOLCANIC HAZARDS IN CALIFORNIA

In 1980, the newspapers in California came alive with news of earthquake swarms near the resort town of Mammoth on the eastern side of the Sierra Nevada. The thousands of small to moderate earthquakes were occurring at fairly regular intervals, a pattern known as ***harmonic earthquakes***. Believing the earthquakes to be a signal that magma was rising beneath the Long Valley Caldera, seismologists at the U.S. Geological Survey issued an alert for the area. Leveling instruments detected a bulging of the Earth's surface on the floor of the caldera, giving strength to the idea that magma was on the move. However, the flurry of activity at Long Valley calmed down in 1986, causing geologists to conclude that magma movement had ceased for the time being. Since 1989, geologists and seismologists have closely watched renewed seismicity and ground movement (Fig. 5-37).

At Long Valley, Mt. Shasta, Lassen Peak, and Medicine Lake, networks of seismographs and ground-leveling stations are monitored as part of the U.S. Geological Survey's volcanic hazards program. The purpose of the monitoring programs is to detect precursors of impending volcanic eruptions, enabling government agencies to plan for evacuations if the signals indicate that eruption is imminent. However, because of the unpredictable nature of eruptions from the Cascade volcanoes and the Long Valley area, precise warnings of eruptions might not be possible.

After detailed studies of each of the known active volcanoes in California, geologists are able to piece together a record of the characteristic eruptive history for each volcanic center. By examining and dating the volcanic deposits formed around each volcano, geologists predict the type, size, and frequency of likely future events. Based on

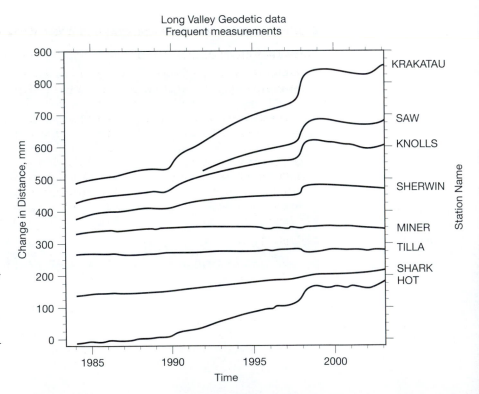

Fig. 5-37 Repeated laser surveys show changes in the ground surface at Long Valley; the changes might reflect upward movement of magma beneath the caldera. (Source: U.S. Geological Survey.)

Fig. 5-38 Volcanic hazards map for Mt. Shasta, showing the areas likely to be affected by eruptions of volcanic ash. The outline of the hazardous area was based on the distribution of the 9,600-year-old Red Banks tephra, found today at localities within the shaded circle. (Source: Crandell, D.R., and Nichols, D.R. 1989. U.S. Geological Survey, public information pamphlet.)

the geologic information, ***volcanic hazards maps*** can be prepared for each area. These maps outline the areas likely to be impacted by future volcanic events (Figs. 5-38 and 5-39). The maps can then be used by government agencies as they debate the advisability of future developments in the vicinity of known active volcanoes. In addition, evacuation routes and special high-risk activities and structures can be pinpointed (Fig. 5-40).

COMMERCIAL USES OF VOLCANIC ROCKS

Anyone who has purchased decorative red rock at the local building supply center to use in a planter box or in the yard has contributed to California's volcanic rock industry. Late Cenozoic volcanic eruptions have supplied California with abundant cinders, pumice, ash, and volcanic glass that are used in a variety of industries. These materials have been mined at quarries in the Cascade and Modoc Plateau Provinces, as well as in the northern Coast Ranges and southeastern California. In 1990, $5 million worth of pumice alone was mined at quarries in the Cascade and Modoc Plateau Provinces, as well as from cinder cones in southeastern California.

Fig. 5-39 Volcanic hazards map for Mt. Shasta, showing the areas likely to be affected by volcanic debris flows. Zones designated by letters show relative likelihood of being affected by debris flows: zone A is most likely and zone C is least likely to be affected. No debris flow hazard exists on high areas within or beyond the zones. The hazard decreases everywhere within the zones with greater height above stream channels and greater distance from Mt. Shasta (Cr. = creek; R. = river). (Source: Crandell, D.R., and Nichols, D.R. 1989. U.S. Geological Survey, public information pamphlet.)

Fig. 5-40 Reported encounters of aircraft with erupting volcanic ash. (Source: Wright, T.L., and Pierson, T.C. 1992. *Living with Volcanoes*. U.S. Geological Survey, Circular 1073, p. 27.)

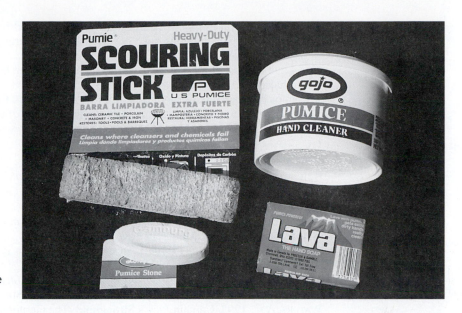

Fig. 5-41 Products made from pumice.

Volcanic cinders are reddened fragments of basalt thrown from a volcanic vent and often piled up into cinder cones. They are used as base material for roads and railroad beds, as decorative rocks, and to make lightweight concrete blocks. Pumice is used as an aggregate in lightweight concrete and in acoustical insulating materials. It has long been known to be an effective abrasive for cleaning everything from oven surfaces to skin (Fig. 5-41). Pumice can be added to soil as a conditioner and is used in some absorbent materials. Volcanic ash is used as a base material in agricultural pesticides and insecticides as well as in some ceramics.

Throughout California's early history, volcanic glass was a valuable commodity. Discoveries of obsidian tools found at archaeological sites hundreds of kilometers from their volcanic sources prove that Native Americans traded obsidian across much of the western United States. Even today, some surgeons prefer obsidian blades to steel scalpels because of obsidian's sharpness. Hydrated volcanic glass known as ***perlite*** is used in aggregates for plaster and fire-resistant wallboard. It is also used in insulating materials.

SUMMARY

1. The chemistry of lavas determines the types of volcanic eruptions, their predictability, and the types of materials that form during an eruption. Lavas are grouped into three general types: mafic, intermediate, and silicic.
2. Volcanic eruptions in the Cascade Province are caused by the subduction of the Gorda Plate beneath North America. Mt. Shasta and Lassen Peak are the two active volcanoes formed along the Cascadia subduction zone in California. Lavas erupted in the Cascade Province are predominantly intermediate.
3. Recent volcanic activity in the Modoc Plateau Province and in the Basin and Range Province is a result of crustal extension. The lavas erupted in these areas are both mafic and silicic. Medicine Lake Highlands is the major volcanic center in the Modoc Plateau Province, and basalt flows from this area have formed Lava Beds National Monument.
4. The most recent major eruptions from Mt. Shasta occurred about 9,000 years ago.

5. The most recent eruptions from Lassen Peak occurred between 1914 and 1921, and much larger eruptions have occurred in the general Lassen area during the past million years.
6. A huge eruption from the Long Valley area formed the Long Valley Caldera and produced the Bishop tuff about 760,000 years ago.
7. Geologists continue to monitor California's volcanoes for signs of future eruptions, which could have major impacts on the citizens of the state.

REFERENCES

BAILEY, R.A. 1989. *Geologic Map of the Long Valley Caldera, Mono-Inyo Craters Volcanic Chain, and Vicinity, Eastern California*. U.S. Geological Survey, Map I-1933, scale 1:62,500.

HARRIS, S.L. 1988. *Fire Mountains of the West: The Cascade and Mono Lake Volcanoes*. Missoula, MT: Mountain Press, 379 pp.

HILL, M.R. 1970. Mount Lassen is in eruption and there is no mistake about that. *California Geology,* November, pp. 211–227.

MILLER, C.D. 1980. *Potential Hazards from Future Eruptions in the Vicinity of Mount Shasta Volcano, Northern California*. U.S. Geological Survey Bulletin 1503.

WATERS, A.C. 1992. Captain Jack's Stronghold: The Geologic Events That Created a Natural Fortress. *California Geology,* pp. 135–144.

WATERS, A.C., DONNELLY-NOLAN, J.M., and ROGERS, B.W. 1990. *Selected Caves and Lava-Tube Systems in and Near Lava Beds National Monument, California*. U.S. Geological Survey Bulletin 1673, 102 pp.

WRIGHT, T.L., and PIERSON, T.C. 1992. *Living with Volcanoes*. U.S. Geological Survey, Circular 1073, 57 pp.

Web Sites and Media Resources

University of North Dakota: Volcano World
http://volcano.und.nodak.edu
U.S. Geological Survey, Long Valley Observatory
http://lvo.wr.usgs.gov
U.S. Geological Survey, Cascades Volcano Observatory
http://vulcan.wr.usgs.gov
Includes sites for Lassen, Shasta, Medicine Lake-Lava Beds
U.S. Geological Survey, Cascades Volcano Observatory: America's volcanic past – California:
http://vulcan.wr.usgs.gov/LivingWith/VolcanicPast/Places/volcanic_past_california. html
California Geological Survey: Online Geologic Field Trip Guides of California
http://www.consrv.ca.gov/cgs/geotour
National Park Service, Lava Beds National Monument
http://www.nps.gov/labe
National Park Service, Lassen Volcanic National Park
http://www.nps.gov/lavo
National Park Service, Devils Postpile National Monument
http://www.nps.gov/depo

California's Deserts Climate, Changing Environments, and Resources

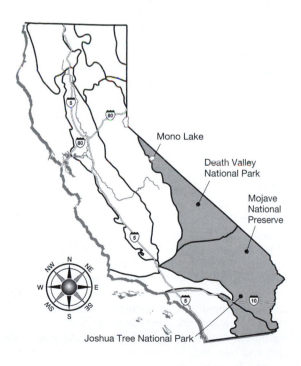

Fig. 6-1 Map showing the location of the Basin and Range, Colorado Desert, and Mojave Desert provinces.

California Desert Highlights

Named deserts:	Colorado, Great Basin, Mojave
Highest points:	Boundary Peak, White Mts.: 4,341 m (14,242 ft.)
	Telescope Peak, Panamint Mts.: 3,366 m (11,045 ft.)
Lowest point:	Badwater, Death Valley, −86 m(−282 ft.)
Hottest temperature:	134°F (57°C), Furnace
Average precipitation:	3 cm/year (1.63 in.), Death Valley
Most important natural process:	Running water
Unique desert landforms:	Alluvial fans, playas, dunes, tufa towers
Resources:	Evaporite minerals, rare earth elements (see Chapter 7)
Other geologic features:	Pleistocene lakes

Colorado Desert

Major features:	Salton Sea, Colorado River (see also Chapter 10)
	San Andreas fault (see Chapter 14)
National Parks:	Joshua Tree National Park
State Parks:	Anza-Borrego Desert State Park
Other:	Imperial (Algodones) Dunes

Great Basin Desert

Major features:	Death Valley, Inyo Mountains, Panamint Valley, Owens Valley, Owens Lake, Mono Lake
National Parks:	Death Valley National Park
State Parks:	Red Rock Canyon
Other:	Mono Lake County Preserve

Mojave Desert

Major features:	Providence Mountains, Cima volcanic field (see Chapter 7)
	Mojave River
National Parks:	Mojave National Preserve
Other parks:	Rainbow Basin National Landmark

THE DESERT ENVIRONMENT

Covering more than 25 million acres, southeastern California's deserts encompass as much land as the entire state of Ohio. Together, the deserts of the southern Basin and Range Province, the Mojave Desert, and the Colorado Desert account for about 1/5 of California's landscape (Figs. 6-1 and 6-2). More than 20 million people live within a few hours' drive of the California desert, but the population there remains relatively low because of the lack of available water. However, Californians have made extensive use of the desert, removing billions of dollars in mineral wealth since the late 1800s. The clear skies, open landscape, and low population have made the desert ideal for military training facilities since World War II. Perhaps the open spaces of the desert will find their most important role as the growing urban population looks to these areas for recreation.

Fig. 6-2 Map showing California's deserts, including parks, monuments, and other areas included in the 1994 California Desert Protection Act. The Great Basin Desert extends eastward into Nevada.

A desert is an area where average evaporation is high relative to precipitation, resulting in scarce vegetation and low population. California's deserts are the result of location and topography. The winter storms that deliver most of California's water originate in the Pacific Ocean near Alaska and approach the coast from the west. Moist air masses generally travel eastward across the state, cooling as they rise over the Coast Ranges, the Sierra Nevada, the San Bernardino Mountains, and the ranges to the east. Because the cooler air cannot hold as much moisture, condensation occurs on the western slopes of the mountains, relieving the air masses of much of their moisture. This pattern, which results from the topography of the ranges, is known as an ***orographic effect***, and it creates a ***rain shadow*** on the eastern side (Fig. 6-3). The rain shadow is most pronounced on the eastern side of the Sierra Nevada because the high range is a major barrier. However, a less pronounced rain shadow exists in the western Great Valley east of the Coast Ranges.

The high mountains of the Sierra Nevada and, further south, the San Gabriel and San Bernardino Mountains are so effective in blocking moist Pacific air that they have created the desert environment of southeastern California. In the Basin and Range Province, east-traveling air masses must cross three additional ranges to reach Death Valley, by which time they are usually stripped of almost all moisture (Endpaper 3). The average annual precipitation on Death Valley's floor is about 3 centimeters, but

Fig. 6-3 The *orographic effect* of a mountain range on precipitation. In most of California, moist air masses travel eastward from the Pacific Ocean, rising first over the Coast Ranges and then over the higher Sierra Nevada, where the strongest orographic effect, or rain shadow, is created.

West

Prevailing wind

East

Dry slope in "rain shadow"

Pacific Ocean   Coast Ranges   Great Valley   Sierra Nevada   Basin and Range

some years pass with no precipitation. The town of Bagdad in the Mojave Desert experienced 767 consecutive days from 1909 through 1912 without a single drop of precipitation. Summer temperatures are fierce throughout California's deserts, particularly in the low-lying areas. The hottest temperature ever recorded in the western hemisphere was 134°F (57°C) at Furnace Creek in Death Valley on July 10, 1913.

One of the distinguishing features of a desert is the scarcity of vegetation. The plant species that can survive are well adapted to the dry environment. Many spectacular desert plants—the Joshua Tree, the palo verde, the smoke tree, and various cacti—are found nowhere else in California (Fig. 6-4). Near springs, streams, and lakes, and on the higher slopes of some of the ranges, the ground surface is vegetated. However, throughout most of the California desert, the rock and soil at the surface are exposed to the forces of weathering and erosion.

Chemical weathering of rocks proceeds slowly in the desert because of the lack of water and vegetation. Because decayed plant material is generally absent, organic acids, which are highly effective chemical weathering agents, are also absent. Without

Fig. 6-4 Joshua trees in the Mojave Desert. (Source: Nakata, J. U.S. Geological Survey.)

Fig. 6-5 Desert varnish and desert pavement have formed on the surface of an alluvial fan near Las Vegas, Nevada. (Source: Photo by author.)

sufficient water to dissolve them, soluble minerals such as salt (halite), gypsum, and calcite persist at the ground surface and in desert soils. One major effect of desert weathering is the formation of ***desert varnish*** (also called rock varnish), a black iron- and manganese-oxide coating on the rocks. Over tens of thousands of years, the varnish slowly turns the rock surfaces dark (Fig. 6-5). Early desert dwellers chipped through the varnish coatings to carve intricate ***petroglyphs***, rock drawings that remain visible today in several places in the California desert (Fig. 6-6).

WIND AND SAND DUNES

Wind is an important geologic agent in the desert regions of California. Strong winds can be generated by local temperature variations between the mountains and basin floors of the desert, as well as by more regional storms. Enormous dust storms occur periodically in the Mojave desert when the Santa Ana winds blow from the desert toward the coast. During these storms, sand particles can be carried as high as 2 meters above the ground surface, sandblasting the windshields of travelers unlucky enough to be caught in the storms (Fig. 6-7). Finer dust is carried higher and can be transported across the Los Angeles Basin as far as the Channel Islands west of Santa Barbara.

Evidence of wind erosion is abundant in the California desert. One of the features produced by the removal of fine sediment by wind a process termed ***deflation*** is ***desert pavement***. Removal and downward settling of fine particles leaves behind the larger rock fragments. The larger rocks then become coated with desert varnish, and the ground surface becomes armored with the desert pavement. The surface is then protected from further wind erosion, unless the pavement is disturbed by erosion or human impact (see Fig. 6-5).

Fig. 6-6 Aerial view of Native American intaglios at Giant Desert Figures Historic Landmark, along the Colorado River near Blythe. (Source: Wilshire, H. U.S. Geological Survey.)

Fig. 6-7 A dust storm on the bed of Owens Lake. (Source: Great Basin Air Pollution Control District, Bishop, California.)

Fig. 6-8 Algodones Dunes (also known as Imperial Dunes), east of El Centro, Imperial County. Note the vehicle tracks in the foreground for scale. The rock in the distance approximately marks the Mexican border.

Fig. 6-9 The Kelso Dunes, Mojave National Preserve. (Source: Aurora & Quanta Productions Inc.)

Sand dunes come to most people's minds when they think of a desert, but dunes make up a very small part of even the sandiest desert. In Death Valley, sand dunes cover only 2 percent of the total area. The California deserts, however, contain some spectacular dune fields. The Algodones Dunes, one of the largest dune fields in the U.S., stretch for 70 kilometers along the eastern edge of the Salton Sea near the Mexican border, reaching heights of 100 meters in some places (Fig. 6-8). In the Mojave Desert south of Soda Lake, Kelso Dunes are as high as 200 meters. The Kelso Dunes (Fig. 6-9) are also famous for their "booming" or "singing" noises, produced when the sand is in motion. Large dune fields can also be found in Death Valley at Stovepipe Wells, in the Eureka Valley, and at Edwards Air Force Base east of the Rogers Lake Bed. Cat Mountain in San Bernardino County takes its name from a large isolated dune shaped like a cat. For a sand dune to accumulate, three conditions must be met:

1. Wind speed in an area must be adequate to lift and transport sand grains.
2. The area must have little or no vegetation: Plant roots are effective anchors that stabilize the sand.
3. Sand-sized sediment (1/16 to 2 mm) must be available for transport. Smooth, polished rock surfaces will not give rise to sand dunes, even in the fiercest winds.

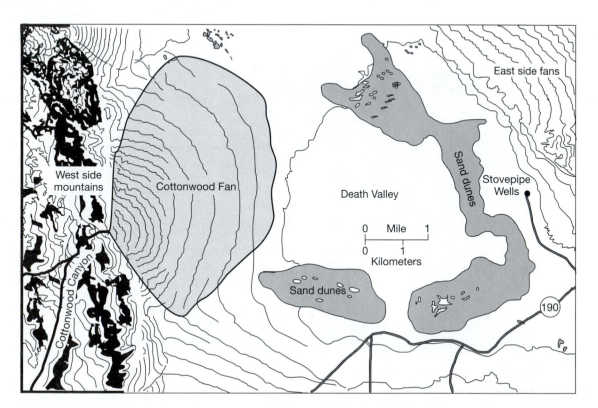

Fig. 6-10 The topographic setting that favors dune formation at Stovepipe Wells in Death Valley. Prevailing winds from the west pick up sand particles from the Cottonwood fan, carrying them eastward until they encounter the mountainous barrier along the eastern edge of the valley.

The valleys of southeastern California are ideal for dune formation because they are generally unvegetated and windy. The sandy alluvium in dry washes and alluvial fans are ideal sources of sand. Sand is picked up by the prevailing winds, which typically blow from west to east across the valleys. As the air masses reach the front of a mountain range, the wind speed decreases and the sand piles up against the mountain front. For example, sand is carried eastward across the Panamint Valley, through Emigrant Pass, and deposited at Stovepipe Wells at the mouth of Death Valley's Emigrant Canyon (Fig. 6-10). At Algodones Dunes, sand has been transported from the unvegetated sandy delta of the Colorado River southeast of the dunes.

RUNNING WATER

Even in the driest desert, running water is the primary force that shapes the landscape. In the California desert, infrequent rainstorms can be intense, particularly during summer when storms move north from the Gulf of California in Mexico. In addition, summer thunderstorms can produce locally intense rainfall in the mountains surrounding the desert valleys. Several centimeters of rain can fall within an hour during these storms. Because the mountain slopes are steep and unvegetated, little of the rainfall soaks into the ground. Water quickly collects in steep mountain channels, resulting in *flash floods* in desert canyons. In larger canyon systems, floodwaters can travel so quickly that streams begin to flow without warning.

When intense rains result from thunderstorms in the mountains, the skies in downstream areas can be blue and cloudless. Flash floods are a significant hazard in the desert. In Clark County, Nevada, near the California border, nine people were killed by a destructive flash flood in 1974. Desert hikers learn to take precautions, checking the weather continuously while exploring steep canyons (Fig. 6-11) and never leaving their vehicles in a dry wash. When floodwaters are carried from the steep, confined canyons or *arroyos* (water-carved channels) of the mountains onto the basin floors, the water spreads out. As the velocity decreases, the sediment carried by the stream, usually a mixture of sand and gravel, is deposited at the mouth of the canyon. The channel shifts back and forth at the canyon mouth, and after repeated floods, a fan-shaped deposit of alluvium accumulates. Throughout the California deserts and along the fronts of other ranges in semiarid environments, these *alluvial fans* are prominent features of the landscape (Fig. 6-12). The base of the fan is at the canyon mouth, and the fan slopes toward the basin floor (Fig. 6-13). In general, the size of an alluvial fan is proportional to the size of the drainage network that feeds it. Along the front of some ranges in the California desert, the alluvial fans coalesce along the range front to form a broad, sloping surface known as a *bajada*.

Alluvial fans can be built by both flash floods and debris flows or mudflows. The shifting of channels within the fan can occur suddenly as a channel becomes blocked during a flood. Because all active surfaces of alluvial fans are vulnerable to periodic floods, construction on fans is risky. Floods might occur only once in a century, or even less frequently, on average. However, the potential magnitude of those floods is very large, as indicated by the size of boulders left behind during past events (Fig. 6-14).

Because of the high temperature and relatively permeable soil along the mountain fronts, floodwaters from the desert mountains quickly evaporate or sink into the desert floor. Even during floods, most streams lack the water to flow beyond the confines of the basin floor. A system of unconnected streams flows into each basin. In many areas, the streams are isolated by the ranges that separate the basins, a pattern typical of most of the California desert in today's climate and known as *interior drainage*.

A

B

Fig. 6-11 A, Flash floods in steep-walled desert canyons can pose a serious threat to unsuspecting hikers. Note the flood-deposited boulders at the mouth of this canyon, as well as the light-colored high water line on the cliff to the hiker's left (Jumpup Canyon, southern Utah). **B,** New homes destroyed by a flash flood in the Mojave Desert, summer 1984. (Source: Marith Reheis, U.S. Geological Survey.)

Fig. 6-12 A portion of the U.S. Geological Survey Cucamonga topographic map, showing the outlines of two alluvial fans along the margins of the San Bernardino Mountains near Ontario. This map was published in 1903, long before the area was extensively developed.

PLAYAS AND ANCIENT LAKES

During major flash floods, the water that runs off from the canyons can cover the basin floor. Because most of the basins have no outlet, the water collects in a shallow lake on the basin floor. The flat basin floors are known as ***playas***, from the Spanish word for *beach* or *shore*, and the lakes that periodically form in the basins are known as ***playa lakes*** (Fig. 6-15). A series of unexpected storms can flood roads for weeks, leaving unprepared tourists stranded until the lakes recede. During times without runoff, the water in the lakes evaporates, leaving behind fine-grained lake sediment

Fig. 6-13 Alluvial fan at Badwater, Death Valley. (Source: U.S. National Park Service.)

Fig. 6-14 Large boulder in alluvial fan gravel along the San Andreas fault near Palm Springs.

Fig. 6-15 Shallow lake, Racetrack playa, Death Valley. (Source: Bradley, W.C.)

and minerals crystallized from the evaporating lake water. Because many of the *evaporite* minerals ("salts") are light-colored, the dry surfaces of the playas are white and highly reflective. When viewed from a nearby mountain summit, the basins of the California desert appear flat and white. These white flats shimmer in the glare of the hot summer sun as light reflects off the lake beds. From some angles, the flats reflect the blue of the clear desert sky, creating a mirage of clear blue water. Tired and thirsty pioneers making the journey across the desert on foot commonly saw sailing ships, green forests, and flocks of birds floating on these blue seas. However, as they came closer, the mirage faded into the reality of rocky outcrops along the dry valley floors.

A few of the basins in the California desert contain permanent lakes today. The Basin and Range Province of northeastern California includes Goose Lake, along the California–Oregon border, and Honey and Eagle Lakes in Lassen County. Mono Lake, in Inyo County, is fed by Sierra Nevada snowmelt (Fig. 6-16), as was Owens Lake until 1913, when the Owens River was diverted into the Los Angeles Aqueduct (see Chapter 10). At that time, Owens Lake was as much as 10 meters deep and covered 275 square kilometers, but today it is dry except during periods of high runoff.

High winds blowing across the dry bed of Owens Lake generate enormous dust storms, and the dust from the Owens Lake bed makes it the worst particulate pollution source in the United States (see Fig. 6-7). Because the dust from the lake bed contains high levels of toxic elements, including arsenic, the Environmental Protection Agency, the Los Angeles Water and Power Association, and other government agencies recently began a program to cover the lake deposits and stabilize the surface. Some areas will be covered with gravel, others will be vegetated, and some lower portions of the lake will be covered by water by raising the lake level.

Another example of desert streams is the Mojave River, the single stream that flows through the Mojave Desert. Most of the time, the floor of the Mojave River is dry. However, every few years, intense rainfall brings enough water to the river that it flows into Soda Lake, near the town of Baker in San Bernardino County. Further south, the Salton Sea also contains water year-round (see Chapter 10 for a discussion of the Salton Sea's origin).

MARK TWAIN AND MONO LAKE

Not all visitors to California's deserts are captivated by their beauty, as is evident in this excerpt from Mark Twain's *Roughing It*. However, Twain did observe features such as the Mono Craters volcanic chain, and he exaggerates the good use made of the alkaline lake water:

Mono Lake lies in a lifeless, treeless, hideous desert, eight thousand feet above the level of the sea, and is guarded by mountains two thousand feet higher, whose summits are always clothed in clouds. This solemn, silent, sailless sea—this lonely tenant of the loneliest spot on earth—is little graced with the picturesque. It is an unpretending expanse of grayish water, about a hundred miles in circumference, with two islands in its center, mere upheavals of rent and scorched and blistered lava, snowed over with gray banks and drifts of pumice stone and ashes, the winding sheet of the dead volcano, whose vast crater the lake has seized upon and occupied.

The lake is two hundred feet deep, and its sluggish waters are so strong with alkali that if you only dip the most hopelessly soiled garment into them once or twice, and wring it out, it will be found as clean as if it had been through the ablest of washerwomen's hands. While we camped there our laundry work was easy. We tied the week's washing astern of our boat, and sailed a quarter of a mile, and the job was complete, all to the wringing out. If we threw the water on our heads and gave them a rub or so, the white lather would pile up three inches high.

Source: Twain, M. 1872. *Roughing It*, Chapter XXXVIII. Hartford, CT: American Publishing Co.

Fig. 6-16 Mono Lake and tufa towers along the shore. The steep eastern front of the Sierra Nevada can be seen at the edge of the lake. (Source: Bradley, W.C.)

Throughout California's deserts and across the Basin and Ranges of Nevada, Utah, and Arizona, geologists have found conclusive evidence of landscapes dotted with lakes as recently 11,000 years ago. In the basins containing modern lakes, shorelines can be seen as high as hundreds of feet above the modern lake levels (Fig. 6-17). In many dry basins, some of which have no history of even rare flooding, shorelines are preserved like bathtub rings after the water has been drained. Some of the ancient lakes were so large that they spilled over to connect basins that are separated today. Throughout the desert basins, lake beds interbedded with layers of evaporite minerals and volcanic ashes are found, all associated with ancient shorelines in areas where shorelines have been erased by later erosion or deposition. When the ancient lake deposits and the shorelines are used to reconstruct the area in California covered by these lakes, the picture is an amazing contrast to today's deserts (Fig. 6-18). From Utah's Great Salt Lake, which pioneering geologist Grove Karl Gilbert recognized as a small remnant of ancient Lake Bonneville, to California's Mono Lake, whose larger predecessor is Lake Russell (named for geologist I.C. Russell), water covered almost all of the low parts of the landscape. This would have been a considerably more hospitable environment for the Native Americans and grazing animals that lived along the lake shores than today's barren deserts.

Fig. 6-17 Shorelines marking former positions of Mono Lake near Lee Vining.

Fig. 6-18 Map showing the basins filled by Pleistocene lakes in southeastern California and (inset) the connected system during times of overflowing lake levels. Today only Mono Lake remains. (Source: Adapted from Blackwelder, E. 1954. California Division of Mines, Bulletin 170.)

TUFA

Scattered along the shores of Mono Lake are clusters of rounded white towers that look like the pieces of a giant chess game or, as described in a Mono Lake guidebook, "giant towers of cemented cauliflower" (Fig. 6-19). The towers are a form of calcium carbonate known as *tufa*. Tufa forms in zones where water from springs that discharge into the lake mixes with the lake water. The two waters differ in their chemistry: The springs carry dissolved calcium from nearby rocks, and the alkaline lake water contains abundant carbonate. The mixed water is saturated with calcium and carbonate, leading to the crystallization of the mineral calcite ($CaCO_3$). Some tufa contains cells of algae, indicating that the algae might aid the process. Tufa towers form at or near the shore of the lake, whereas tufa-cemented sand forms beneath the lake (Fig. 6-20).

Tufa towers have been left standing along the shoreline long after the lake level has dropped. Along Highway 167, ancient tufa towers, grayed by weathering and covered with orange lichens, can be seen several miles north of the present lake. These towers mark the level of Pleistocene Lake Russell approximately 16,000 years ago. For geologists studying the fluctuations of Pleistocene lakes, the tufa provides an important marker because it can be dated using radiocarbon methods.

The Pinnacles near Trona, along the southwestern shore of Searles Lake, are another example of tufa in California. The towers stand more than 100 feet above the now-dry basin (Fig. 6-19). Tufa can also be found along the ancient shores of Pleistocene lakes at Honey Lake, Indian Wells Valley, Panamint Valley, the Salton Sea, and in Nevada.

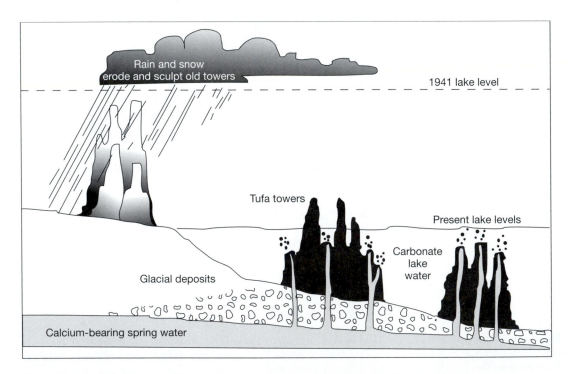

Fig. 6-19 The formation of tufa towers at Mono Lake. (Source: *California Geology*, 1992.)

Fig. 6-20 Tufa pinnacles along the edge of Pleistocene Searles Lake, near Trona. (Source: Bradley, W.C.)

Geologists have established the ages of the ancient lakes by carbon-14 dating of scarce wood fragments, shells, and tufa deposits, and by fingerprinting volcanic ash layers interbedded with lake beds (Fig. 6-21). Although the exact ages of the ancient lakes are not yet known, it appears that lakes were periodically extensive during Pleistocene time, until about 11,000 to 13,000 years ago throughout the southwestern United States The lakes began to shrink as the global climate warmed following the most recent glaciation. High lake levels occurred during or at the end of the last glacial period, an observation made by Gilbert more than a century ago. The ancient shorelines and lake beds record high lake levels during a time when temperatures were cooler and evaporation rates much lower than present. The lakes in eastern California and western Nevada were fed principally by precipitation and meltwater from the Sierra Nevada.

In basins where lake history has been studied in detail, it appears that lakes have grown and shrunk episodically, reflecting major changes in the climate during the past 1 to 3 million years. Mono Lake did not overflow during its most recent high level 13,000 years ago or during its maximum level during the most recent glacial episode between about 100,000 and 18,000 years ago (Fig. 6-22). However, at some time during the previous glacial period, Mono Lake spilled eastward into Adobe Valley, then into Owens Lake. Water from Owens Lake spilled into the China Lake Basin, Searles Lake, the Panamint Valley, and into Death Valley (see Fig. 6-18). The lake that occupied Death Valley has been named Lake Manly for an early explorer to the region. At its height, Lake Manly was up to 200 meters deep. Today, the ancient shorelines and lake beds of Lake Manly, including ancient beaches and gravel bars, can be seen at Shoreline Butte, along the Black Mountains, and along the highway between Death Valley and Beatty, Nevada.

In the Mojave Desert, a much more vigorous Mojave River once flowed beyond its present mouth at Soda Lake all the way to the Colorado River, passing through a chain of overflowing lakes on its way (see Fig. 6-18). Today these lakes—Bristol, Cadiz, Danby, and Chuckwalla—are dry playas. The original watercourse

Fig. 6-21 Volcanic ash and pumice erupted from Mono Craters, interbedded with lake sediments of Pleistocene Lake Russell, Mono Lake. (Source: Sarna-Wojcicki, A. U.S. Geological Survey.)

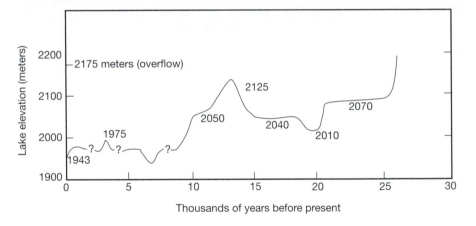

Fig. 6-22 Graph showing fluctuations in the level of Mono Lake during the past 30,000 years, based on the elevations of ancient shorelines, lake beds, and tufa. The elevation in 1975, at the 0 point on the graph, was 1,943 meters. (Source: LaJoie, K.R. 1984. *Friends of the Pleistocene Field Trip Guidebook.*)

has been broken up by recent faulting and lava flows as well as by the drying up of the river and lakes. Isolated lakes were present throughout other parts of California's deserts and in the southern San Joaquin Valley, as indicated by shorelines and lake beds preserved there. In the Imperial Valley, Pleistocene Lake Cahuilla, periodically filled by floods from the Colorado River, occupied the basin now filled by the Salton Sea.

Because the desert lakes are excellent recorders of past climatic conditions, they are being intensively studied today as researchers attempt to better understand the Earth's global climate system. Deep cores have been sampled from several of California's desert lakes to determine the history of lake fluctuations. Because they are originally horizontal, ancient shorelines can also be used like long-term level lines to determine the amount of faulting and tilting that have occurred in a basin since the shoreline was cut by the lake (Fig. 6-23).

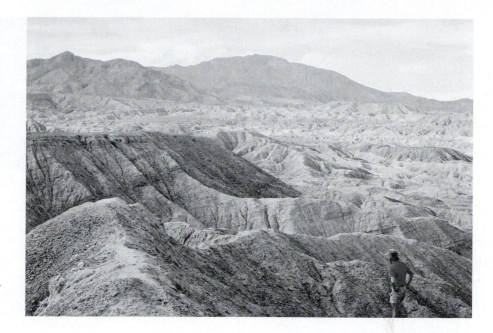

Fig. 6-23 Lake beds of Pleistocene Lake Cahuilla, near today's Salton Sea, Imperial County.

RICHES IN THE DESERT LAKE BEDS

By the 1880s, gold and silver miners had become disillusioned with the mountains of the California desert. Under the harshest of conditions, fortunes had been made, and many more lost, extracting silver and gold from the area. The hopeful turned their sights to the common white coverings on the dry lake beds. Two minerals discovered in the playas in 1873 were sodium borates (Fig. 6-24): ulexite (chemical formula: $NaCaB_5O_9 \cdot 8H_2O$), also known as "cottonball," and tincal or borax ($Na_2B_4O_7 \cdot 10H_2O$). A large deposit of a third borate mineral, colemanite ($Ca_2B_6O_{11} \cdot 5H_2O$), was discovered in 1880 in ancient lake bed deposits east of Death Valley's floor. These discoveries brought prosperity to the desert, and borate minerals were mined from Death Valley from 1882 until 1928.

At the time of the Death Valley discoveries, borate of soda was used by glassmakers and metalsmiths as alloys, by potters to give ceramics a lustrous finish, and by meat packers as a preservative. Used in enamel, borax created the smooth surfaces on newly fashionable bathtubs, toilets, and sinks. Soap companies had discovered that borax was an excellent water softener and buffer. In the late 1800s, borax was sold in apothecaries for as much as 50 cents per ounce. The world's supplies of borax came originally from Tibet, and by the 1800s, from Tuscany in Italy, from Turkey, Chile, Argentina, and near Clear Lake in the northern Coast Ranges.

The most significant problem in the mining operation was transporting the borax across 275 kilometers of dry, steep mountains and salt marshes between Death Valley and the nearest railroad station at Mojave. Many stretches of the journey were entirely without water, and there were only three springs over 225 kilometers of the distance. To make the mining of borax profitable, huge weights had to be hauled in each trip. However, the weight of the loads created problems both climbing and descending the steep ranges. Crossing the thin salt crust over the playa was treacherous: Even a hiker choosing the wrong route could fall through it. One route with a firm surface

Fig. 6-24 Borate mineral specimens collected from evaporite deposits in southeastern California.

was found, but it was covered with jagged salt crystals. Building a road through this terrain, chopping through the maze of salt crystals by hand in temperatures greater than 110°F (43.3°C), was a difficult job accomplished by Chinese laborers. Over this road passed enormous wagons designed specially to carry the borax. The wagons stood almost two stories high, and each could carry half a freight car's worth of borax. The wagons were hitched in trains of two wagons and a water tank, and each train was pulled by 20 mules. These were the famous 20-mule teams whose picture appeared for years on borax boxes in American households (Fig. 6-25).

DRIVING THE MULE TRAINS OF DEATH VALLEY

The grade down the mountain shoulder near Granite Spring was the most hazardous stretch. It was straight; it was steep; it was miles long. Here was no sand to drag at the wheels; the surface was hard packed and the great wheels rolled down it all too easily. The difficulty in maneuvering that grade was in braking the wheels. And many times even the mighty brake blocks were not heavy enough to cope with thirty tons of borax, wagon, and water pushing from behind. As the wagons gained momentum the skinner threw his weight on the front brake; the swamper did the same on the trailer. Groaning, screeching and sliding, the cumbrous monsters lumbered on faster and faster.

The race with death began when the brakes no longer held and the wagons were out of control. Then the full dexterity of the skinner was called into play. There was no provision in the harness hitch for hold-backs; the animals simply had to be kept ahead of the load to avoid being run down. Their quickstep became a trot, the trot a kind of canter, the canter a dead run, a lope and a gallop.

All precaution was thrown to the winds. Like a succession of pistol shots, the skinner's whip cracked over the heads of the frantic mules. Bowling behind, the wagons thundered their threat. They groaned under their weight, bounced and careened.

. . . Often they made it to the bottom of the grade. Sometimes not. And a wagon would come along two days, four days, a week later, to shovel away the drifts of spilled borax from the tangle of wreckage . . .

Source: Lee, W. S. 1963. *The Great California Deserts* (pp. 114–115). New York: Putnam & Sons.

A

B

Fig. 6-25 A, Box of Boraxo hand soap, showing the 20 mule team trademark. **B**, The 20 mule team wagons, Death Valley National Park.(Source: **A**, Photo by author; **B**, McMackin, M., San José State University.)

In 1913, a homesteader named John Suckow, while drilling a water well 60 kilometers east of Mojave, discovered colemanite in the cuttings brought up by the drill bit. After a series of exploratory wells were drilled, mining engineers concluded that a major borate deposit lay beneath a cover of younger stream gravels. The area is known as the Kramer Borate District, and the town of Boron grew up around the mining operations. Along with tincal, ulexite, and colemanite, a new mineral was discovered: kernite ($Na_2B_4O_7 \cdot 4H_2O$), also referred to as rasorite. In 1928, the Pacific Coast Borax Company shut down operations at Death Valley and began production from the Kramer deposits.

The Kramer borate deposit is found within lake beds deposited in a basin during Miocene time, long before development of the present topography and the development of the Pleistocene lakes (these older basins are discussed in Chapter 7). The Miocene lake beds, including the borate deposits, were deposited on a series of basalt flows. Geologists believe that the borate originated from the volcanic vents in the area and then migrated to the lakes in the circulating groundwater. The borate deposits formed when the lake water evaporated to the point that it became supersaturated with borate, causing the minerals to crystallize.

The process of evaporation and precipitation of minerals is responsible for producing evaporite deposits of several types. The minerals that crystallize are saline minerals (salts), including sodium chloride, or common table salt. In the California deserts, salts have been mined from lake beds and ancient lake deposits in several areas. These include borates produced at Death Valley, Kramer, and Owens Lake; potassium, bromine, and lithium salts at Searles Lake; and sodium and calcium chloride at Bristol Lake near Amboy. Most of the world's supply of boron comes from the Kramer deposits and Searles Lake. In 1990, the value of California's boron minerals was more than $400 million. More than $2 billion worth of minerals have been extracted from the beds of Searles Lake alone.

THE FUTURE OF THE CALIFORNIA DESERT

As the population of southern California increases, the open spaces of the nearby California desert are increasingly appealing as a place for development and recreation. Today, the once-remote deserts are criss-crossed by roads and trails that bring one out of every three vacationing southern Californians to the desert for recreation and solitude. Conservationists have long been concerned about the impact of humans on the land surface, ecology, water resources, and scenery of the desert. One impact of concern is the disturbance of fragile vegetation and desert pavement by off-road vehicles. Studies of areas used for training tank operators during World War II have shown that vehicle tracks persist on the desert floor for more than 50 years. Although areas have been set aside for operation of off-road vehicles, much damage can be done by those few who stray from designated trails (Fig. 6-26). Controversy also exists about the extent of mining that should be permitted on desert lands and about the reclamation of existing mined lands. Excessive withdrawal of groundwater from desert basins for agriculture has caused loss of plants, salt damage to soils and water, and increased erosion in some areas.

Recognizing the unique value of the California desert, Congress designated 25 million acres as the California Desert Conservation Area in 1976 and authorized development of a comprehensive land use plan for the area. About 72 percent of this land is owned and managed by the federal government, including 3 million acres in military reservations, 2.5 million acres in Death Valley and Joshua Tree National Parks, and 12.5 million acres managed by the Bureau of Land Management. In 1994, Congress passed the comprehensive California Desert Protection Act. Under the Act, 9.2 million acres of land are protected from development to some extent, including 1.2 million acres added to Death Valley National Park and 3.5 million acres of Bureau of Land Management lands designated as wilderness. The bill also converted Death Valley and Joshua Tree from national monuments to national parks and created the 1.4-million-acre Mojave National Preserve.

Fig. 6-26 Tracks left by off-road vehicles on the Soda Lake playa, Mojave Desert. (Source: Wilshire, H. U.S. Geological Survey.)

SUMMARY

1. California's deserts occupy the southeastern 25 percent of the state. Three distinct deserts are present: the Colorado Desert, Mojave Desert and Great Basin Desert (which extends across Nevada).

2. Desert conditions occur because mountain ranges block these areas from the prevailing, moisture-producing air masses that travel eastward from the Pacific Ocean.

3. Running water is the most important force that shapes the desert landscape, although wind is locally important in forming sand dunes. Even though average precipitation is low in the desert, intense storms and flash floods are common events. Because mountain canyons are steep and unvegetated, flash floods carry large amounts of sediment to the edges of desert valleys. Here large alluvial fans have been built up. Dry lake beds (playas) found in the lowest parts of the desert valleys are also periodically flooded.

4. During times of global cooling, when glaciers covered the high parts of the Sierra Nevada, California's deserts were cooler and evaporation rates were lower. As a result, lakes formed in the valleys of the California deserts where only playas exist today. The most recent period of extensive lakes ended about 11,000 years ago.

5. Since the late 1800s, evaporite minerals have been mined from many of the dry lake beds and from ancient lake sediments in the California deserts.

6. Today, large areas of the California desert are protected as national and state parks and wilderness areas. The desert provinces are an important recreational resource for Californians.

REFERENCES

BLACKWELDER, E. 1954. Pleistocene lakes and drainage in the Mojave Region, southern California. In California Division of Mines and Geology. *Geology of Southern California*. Part V, pp. 35–44.

GAINES, D. 1989. *Mono Lake Guidebook*. Lee Vining, CA: Kutsavi Books, 104 pp.

HILL, M. 1984. *California Landscape, Origin and Evolution. California Natural History Guide 48*. Berkeley: University of California Press, 262 pp.

JAEGER, E. 1965. *The California Deserts*. Stanford, CA: Stanford University Press, 208 pp.

LEE, W.S. 1963. *The Great California Deserts*. New York: Putnam, 306 pp.

MUMFORD, R. 1954. Deposits of saline minerals in southern California. In California Division of Mines and Geology. *Geology of Southern California*. Chapter VIII, pp. 15–22.

RAE, C. 1992. *East Mojave Desert: A Visitor's Guide*. Santa Barbara, CA: Olympus Press, 206 pp.

RIEGER, T. 1992. Calcareous tufa formations, Searles Lake and Mono Lake. *California Geology* 45:99–109.

SCHAD, J. 1988. *California Deserts: California Geographic Series, No. 3*. Helena, MT: Falcon Press, 104 pp.

VER PLANCK, W.E. 1954. Salines in southern California. In California Division of Mines and Geology. *Geology of Southern California*. Chapter VIII, pp. 5–14.

WALKER, A.S. 1992. *Deserts: Geology and Resources*. U.S. Geological Survey, General Interest Publication, 60 pp.

Web Sites and Media Resources

National Park Service, Death Valley National Park *http://www.nps.gov/deva*

National Park Service, Joshua Tree National Park *http://www.nps.gov/jotr*

National Park Service, Mojave National Recreation Area *http://www.nps.gov/moja*

U.S. Geological Survey, Geology in the National Parks *http://www2.nature.nps.gov/grd/usgsnps/project/home.html*

U.S. Geological Survey, Geology of Mojave National Preserve *http://www2.nature.nps.gov/grd/usgsnps/mojave/mojave.html*

Mono Lake Mono Lake Committee Web Page *http://www.monolake.org*

California Geological Survey: Online Geologic Field Trip Guides of California *http://www.consrv.ca.gov/cgs/geotour*

The Basin and Range and Mojave Desert: Old Rocks and Young Faults

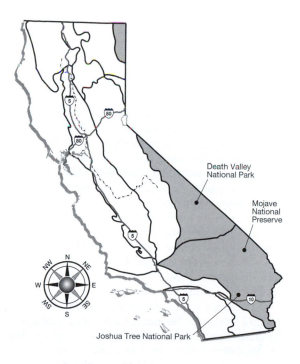

Fig. 7-1 Map showing the location of the Basin and Range Province in California.

Death Valley
National Park

Mojave
National
Preserve

Joshua Tree National Park

Basin-Range/Mojave Geologic Highlights (see Chapter 6 for additional highlights)

Major landscape feature:	Basin and Range topography
	White, Inyo, and Panamint ranges
	Owens, Panamint, Saline, Death valleys
High and low points:	Boundary Peak, White Mts.: 4,341 m (14,242 ft.)
	Badwater, Death Valley: –86 m (–282 ft.)
Predominant rock types:	Paleozoic carbonate and other sedimentary
	Late Cenozoic alluvium
Plate tectonic events:	Paleozoic passive margin
	Mesozoic subduction
	Extension since 22 million years ago
Faults:	Normal faults
	Detachment faults at depth
Famous earthquake:	1872 Owens Valley
Areas of recent volcanism:	Long Valley Caldera, Mono-Inyo Craters (see Chapter 5), Coso volcanic field, Cima volcanic field (Mojave), Death Valley
National Parks:	Death Valley National Park
	Mojave National Recreation Area
State Parks:	Red Rock Canyon

THE BASIN AND RANGE LANDSCAPE

Anyone who has traveled across Nevada, going either east or west, has encountered the major topographic characteristic of the Basin and Range Province. The climb to the summit of each mountain is followed by a rapid descent onto the flat desert surface. From each summit, you can view the highway stretching across a dry valley and up the next range, which might be more than 30 kilometers away. As the hours pass, you repeat this experience many times, traveling west until you reach the Sierra Nevada, or east until you reach Salt Lake City (Fig. 7-1). A relief map tells the same story: a series of ranges running north and south, each between 75 and 250 kilometers long, separated by parallel valleys. Most of the region is fairly high, with the valley floors standing more than 1,000 meters above sea level, and the highest peaks exceeding 4,000 meters.

The Basin and Range Province extends from eastern California to central Utah, from southernmost Oregon and Idaho on the north to southern Arizona and southwestern New Mexico and further south into the Sonoran state of Mexico (Fig. 7-2). The California portion of the province includes a small area in the northeastern corner of the state and a much larger area bordered by the Sierra Nevada on the west and the Mojave Desert on the east. Death Valley, perhaps the most famous of the basins in the entire Basin and Range Province, contains the lowest point in the United States, 86 meters below sea level. The highest point, east of Owens Valley in the White Mountains, stands 4,341 meters above sea level. Prominent California ranges include the Warner Range, the White and Inyo Mountains, the Panamint Range, and

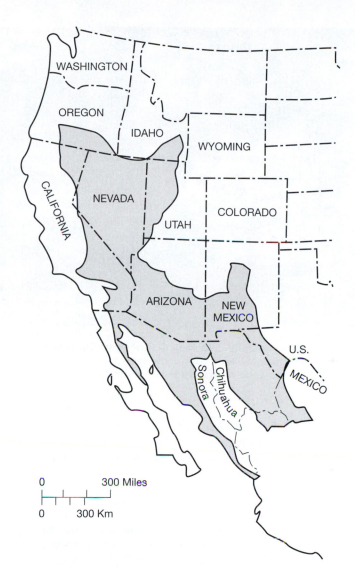

Fig. 7-2 The Basin and Range Province in western North America. (Source: Stewart, J. 1978, Geological Society of America, Memoir 152.)

the Funeral Mountains. The major valleys of California's Basin and Range Province are Owens Valley, along the eastern edge of the Sierra Nevada, and Saline, Panamint, and Death Valleys. Like all of California's deserts, the Basin and Range is an area of low rainfall and scarce vegetation (see Chapter 6). Only the higher parts of some ranges support a forest cover.

THE MOJAVE DESERT

South of California's Basin and Range province, the Mojave Desert is a large triangle bounded on the east by the Colorado River and the California–Nevada border, on

Garlock Fault

Fig. 7-3 Landsat satellite image showing the southern portion of Death Valley, the northwest-trending linear zone in the upper left. The Garlock fault, which separates the Basin and Range and Mojave Desert Provinces, crosses the image roughly through the center, from left to right, as shown by the arrows.

the north by the Garlock fault (see Chapter 14), and on the southwest by the San Gabriel and San Bernardino Mountains and the San Andreas fault (Fig. 7-1).

A comparison of the topography in the Mojave Desert with the Basin and Range shows some general similarity and some distinct differences. Basins and ranges are characteristic of both provinces, but the landscape is more subdued in the Mojave Desert. Most ranges of the Mojave Desert are shorter and lower than those in the Basin and Range. The basins are broader, particularly in the western Mojave Desert (Fig. 7-4), which is a fairly flat plain. The basins and ranges in most of the Mojave Desert are also less aligned than those of the Basin and Range to the north. Both the similarities and the differences in the landscapes of the two provinces can be explained by their recent geologic history.

EARTHQUAKES, RECENT FAULTING, AND BASIN-RANGE TOPOGRAPHY

On March 26, 1872, a major earthquake rumbled through the Owens Valley, leveling the entire town of Lone Pine and killing 23 of its 250 to 300 inhabitants. Most of the adobe and brick buildings in the region were completely destroyed, leading the local newspaper to point out that the deaths would not have occurred if the buildings had been constructed of wood. The earthquake, with an estimated magnitude of 7.8, was one of the largest in the history of the United States (see Chapter 13). The quake triggered huge rockfalls in Yosemite Valley, one of which was witnessed by famous naturalist John Muir. Shaking was felt as far away as Salt Lake City and

Fig. 7-4 View of Clark Mountain from the Kingston Range, looking to the south at a typical landscape of the eastern Mojave Desert. (Source: McMackin, M. San José State University.)

San Diego. The earthquake broke the ground along a 100-kilometer stretch of the Owens Valley fault system and separated the ground surface several meters in some places.

Today, some evidence of the 1872 quake is still well preserved. A prominent *fault scarp*, a break in the topography caused by surface faulting, can be seen west

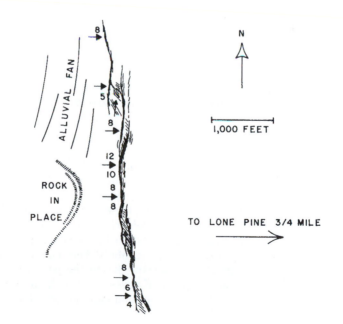

Fig. 7-5 A sketch map showing the surface ruptures produced during the 1872 Owens Valley earthquake, about 1.3 kilometers west of the town of Lone Pine. W.D. Johnson drew the original maps in 1907. At that time, the fault scarps measured as much as 12 feet high, as indicated by the numbers in the sketch (given in feet). (Source: *California Geology,* March 1972.)

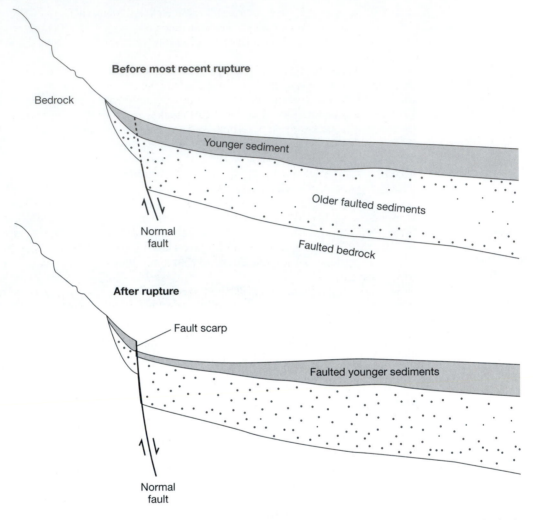

Before most recent rupture

Bedrock

Younger sediment

Older faulted sediments

Normal
fault

Faulted bedrock

After rupture

Fault scarp

Faulted younger sediments

Normal
fault

Fig. 7-6 Diagram showing the displacement of rock and sediment layers and the formation of a fault scarp along a normal fault.

of the town of Lone Pine along the Lone Pine fault (Fig. 7-5). A fault scarp forms when an earthquake ruptures the ground surface, displacing the landscape along a sharp break (Fig. 7-6). Following the earthquake, the scarp fades away as erosion and deposition smooth out the land surface. By careful study of the fault scarp and the young sediments broken by the fault near Lone Pine, geologists have estimated that the 1872 earthquake shifted the land surface about 1 to 2 meters vertically and as much as 6 meters horizontally. The fault scarp west of Lone Pine was built by three separate earthquakes during the past 10,000 years, each one moving the western side of the fault upward and to the northwest relative to the eastern side. Over geologic time, repeated earthquakes along the faults of the Owens Valley system have gradually elevated the Sierra Nevada 3,300 meters (10,000 feet) above the floor of Owens Valley. The occurrence of the 1872 earthquake, as well as smaller quakes west of Mono Lake in 1991, are dramatic proof that this process continues.

Throughout the Basin and Range province, ***range-front faults*** can be found along the base of the ranges. Some of these faults are clearly as active as the Owens Valley fault. Moderate earthquakes are relatively common along some faults of the Basin and Range Province. Two large earthquakes in 1954 in Dixie Valley, central Nevada, produced fault scarps as high as 5 meters that are well preserved today in the desert climate.

Sharp, well-preserved fault scarps along the base of many ranges in southeastern California indicate that earthquakes have occurred there during the past few thousand years. Along other range fronts, fault scarps are subtle or absent, indicating that a long period of erosion and deposition has passed since the last groundbreaking earthquake. Geologists can use the freshness of scarps to estimate the age of the most recent earthquake on a fault. This technique has been particularly successful in the Basin and Range and Mojave Desert where the type of faulting, the rock types, and the erosion rate are similar in many ranges. When detailed investigations are carried out along range-front fault scarps, the deposits of alluvial fans can provide important information about the timing of recent faulting. Many active range-front faults have broken through recent alluvial fans (see Fig. 14-49), whereas older fault scarps have been buried by young alluvial fans.

Along the faults of the Basin and Range Province, recent movement has elevated the ranges relative to the basins. Most of the range-front faults are ***normal faults***. On a normal fault, movement drops one side along the fault surface in the direction that one would expect by gravity, given the geometry of the fault, hence the term *normal* (see Fig. 7-6). It is repeated vertical movements along normal fault systems that produce the characteristic Basin and Range topography.

The overall shape of the mountain fronts and the sharpness of the basins and ranges also provide clues about an area's faulting history. The steep, straight mountain fronts of the Basin and Range province reflect the recent movement along normal faults. In contrast, the more subdued basins and ranges in most of the Mojave Desert are a clue that normal faulting is less recent there than in the Basin and Range. After normal faulting ceased in the Mojave Desert, erosion of the ranges and deposition in the basins softened the stark contrast between mountain and the valley. Over time, the straight mountain fronts have also become more ***sinuous*** as erosion wears back the ranges.

Many of the faults in southeastern California show lateral displacement (see Chapter 1), as well as normal movement. In the Basin and Range Province, the Owens Valley, Panamint Valley, Death Valley, and Furnace Creek faults all show lateral motion (Fig. 7-7). Recent lateral faulting has also occurred in three zones in the Mojave Desert: the San Andreas system along its southwestern edge, the Garlock fault along the northern boundary (see Chapter 14), and the eastern California shear zone in the central Mojave desert. As discussed in Chapter 1, lateral motion on faults east of the San Andreas system accommodates some of the relative motion between the Pacific and North American Plates that is not taken up by the San Andreas itself. The 1992 Landers earthquake and the 2000 Hector Mine earthquake have provided important new information about southern California faulting and they are discussed further in Chapter 14.

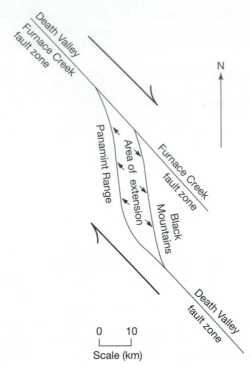

Fig. 7-7 Map showing recent faults in the Death Valley area. (Source: Burchfiel, P., and Stewart, J. 1966. *Geological Society of America Bulletin* 77[4].)

DETACHMENT FAULTS

Recent studies have demonstrated that many of southeastern California's normal faults are steep at the surface, but have a more shallow dip at depth. In some areas, multiple normal faults converge or are cut off along a single, gently dipping fault surface. Rocks on the upper surface move over those on the lower plate along these master faults, which geologists refer to as ***detachment faults*** (Fig. 7-8). The detachment faults can extend below the Earth's surface to great depths, perhaps more than 10 kilometers.

At depths where temperature and pressure are great enough, the rocks along a detachment fault can be greatly altered. The shearing motion of the fault blocks subjects the rocks in fault zone to great stresses, and a characteristic metamorphic rock known as ***mylonite*** can form (Fig. 7-9). Mylonite has a smeared texture, indicating that the rocks deformed in a plastic manner, rather than being crushed or broken. The rocks must therefore have been buried to great depths, where temperatures and pressures are great enough to make them deformable. In areas where mylonite is now seen at the Earth's surface along detachment faults, geologists can conclude that 8 to 15 kilometers of overlying rock must have been removed (Fig. 7-10).

As the crust in the Basin and Range and Mojave Desert regions stretched, the rocks at the surface responded by normal faulting (Fig. 7-11). Extension thinned the Earth's crust, and some areas became so extended that surface rocks no longer covered the extended area. In these regions, not surprisingly referred to as ***highly***

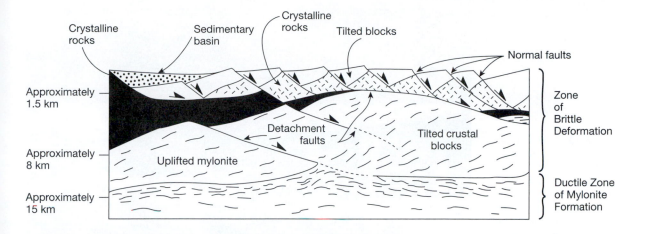

Fig. 7-8 Diagram showing the formation of detachment faults. (Source: Adapted from Pridmore, C., and Frost, E. 1992. Detachment faults. *California Geology* 45. Used with permission of the California Department of Conservation, Division of Mines and Geology.)

Fig. 7-9 Mylonite formed by ductile deformation along a detachment fault in the Basin and Range, Redtail Canyon. (Source: Mc-Mackin, M. San José State University.)

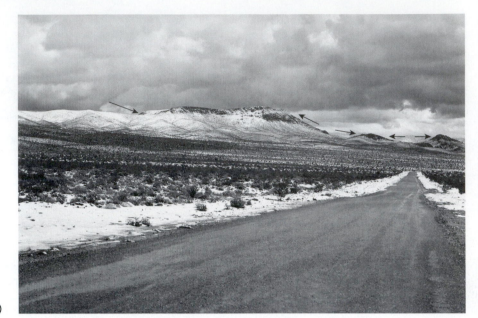

Fig. 7-10 View of a detachment fault, northern Mesquite Mountains (view to south). Arrows show detachment faults. Dark craggy areas at the top of the hills are Cambrian rocks of the upper plate. Proterozoic lower plate metamorphic rocks are snow-covered. (Source: McMackin, M. San José State University.)

Fig. 7-11 The development of detachment faults in response to crustal extension. (Source: Adapted from Pridmore, C., and Frost, E. 1992. Detachment faults. *California Geology* 45.)

extended terranes, the mylonite and other metamorphic rocks from the lower plate beneath the detachment fault can be seen at the Earth's surface.

Detachment faulting caused by Cenozoic extension might explain unusual surfaces along the eastern side of Death Valley. These features were named *turtlebacks* for their curved shapes (Fig. 7-12; see also Fig. 7-29), and geologists have long been puzzled about their origin. One model proposes that they are the surfaces of detachment faults and that the marble and sheared rocks along their surfaces are mylonite produced by the fault movement. The turtleback surfaces indicate that the region was extended in a northwest/southeast direction (see Fig. 7-7). The stretching caused the valley to open up, with younger sedimentary units moving downward. As extension progressed, the detachment faults, or the turtlebacks, became exposed.

Geologists have hypothesized that the Paleozoic rocks of the Panamint Range west of central Death Valley were originally located above or next to the pre-Cambrian rocks of the Black Mountains. Extension and detachment faulting carried them 80 kilometers to the northwest, exposing the older rocks in the Black Mountains. Another complex fault zone in the southern Black Mountains known as the Amargosa Chaos is now thought to have resulted from detachment faulting. Gently dipping normal faults that are likely detachment faults have also been recognized at several localities between the Panamint Range and the California–Nevada border, as well as along the southern boundary of the Basin and Range Province north of the Garlock fault.

Detachment Faults in the Mojave Desert

In many areas of the Mojave Desert, the upper crust moved along detachment faults sufficiently to expose the rocks beneath the detachments. In the eastern Mojave

Fig. 7-12 A turtleback along the eastern side of Death Valley. Another turtleback is shown in Fig. 7-29.

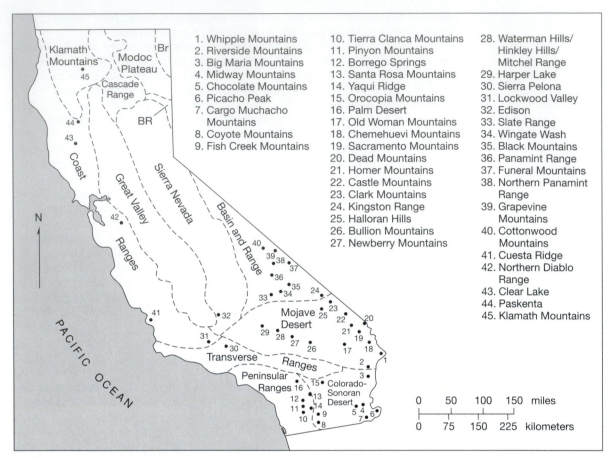

1. Whipple Mountains
2. Riverside Mountains
3. Big Maria Mountains
4. Midway Mountains
5. Chocolate Mountains
6. Picacho Peak
7. Cargo Muchacho Mountains
8. Coyote Mountains
9. Fish Creek Mountains
10. Tierra Clanca Mountains
11. Pinyon Mountains
12. Borrego Springs
13. Santa Rosa Mountains
14. Yaqui Ridge
15. Orocopia Mountains
16. Palm Desert
17. Old Woman Mountains
18. Chemehuevi Mountains
19. Sacramento Mountains
20. Dead Mountains
21. Homer Mountains
22. Castle Mountains
23. Clark Mountains
24. Kingston Range
25. Halloran Hills
26. Bullion Mountains
27. Newberry Mountains
28. Waterman Hills/ Hinkley Hills/ Mitchel Range
29. Harper Lake
30. Sierra Pelona
31. Lockwood Valley
32. Edison
33. Slate Range
34. Wingate Wash
35. Black Mountains
36. Panamint Range
37. Funeral Mountains
38. Northern Panamint Range
39. Grapevine Mountains
40. Cottonwood Mountains
41. Cuesta Ridge
42. Northern Diablo Range
43. Clear Lake
44. Paskenta
45. Klamath Mountains

Fig. 7-13 Map showing identified detachment faults in California. (Source: Pridmore, C., and Frost, E. 1992. Detachment faults. *California Geology* 45.)

Desert, detachment faults are exposed in the Old Woman Mountains, Chemehuevi Mountains, Whipple Mountains, and other ranges. In the central Mojave Desert, detachment faults can be seen in the Bullion Mountains, the Newberry Mountains, and the Waterman Hills (Fig. 7-13). In many of these areas, the detachment faults have uncovered metamorphic rocks that were 8 to 15 kilometers below the surface before the detachments formed.

EXTENSION AND VOLCANIC ACTIVITY IN SOUTHEASTERN CALIFORNIA

Beginning about 16 to 19 million years ago in Miocene time, lava erupted from several sources scattered throughout California's Basin and Range Province. Like detachment faulting, the volcanic activity is a result of extension of the crust (see

Chapter 5). Recent volcanic eruptions have left young cinder cones, craters, and lava flows. These include Ubehebe Craters in Death Valley, a series of cones and flows in the Owens Valley near Big Pine (Color Plate 12; Fig. 7-14), as well as the Bishop tuff and recent volcanic features in the Modoc Plateau discussed in Chapter 5.

The Coso volcanic field is a center of recent volcanic activity along the southern edge of Owens Lake, at the western edge of the Basin and Range Province. Lava has erupted intermittently in this area during the past 6 million years. During the past million years, both high-silica lavas and mafic lavas have been erupted (Fig. 7-15), a pattern known by geologists as ***bimodal volcanism***. Geologists believe that the mafic lavas are erupted directly from sources in the mantle. The high-silica lavas form when rising magma causes crustal rocks to melt. Eruptions from both mantle and crustal sources in the same area result in both basalt and rhyolite being formed in the same area. In the Coso volcanic field, 38 rhyolite flows and domes formed in the Coso volcanic field during the past million years. During the same time period, basaltic lava erupted from at least 14 volcanic vents. The youngest dated feature in the Coso field is a basalt flow erupted about 40,000 years ago.

An interesting feature of the eruptive history of the Coso field is the very regular pattern of eruptions during the past 500,000 years. By dating the rocks formed during individual eruptions and measuring the volume of erupted materials, geologists discovered a pattern. For both basalt and rhyolite, the volume of erupted material could be predicted by calculating the time that had passed since the previous eruption: More lava was erupted if more time had passed since the previous eruption.

From the pattern of volcanic eruptions in the Coso volcanic field, geologists concluded that the eruptions were controlled by the opening of ***feeder dikes*** that carried magma to the volcanic vents. Feeder dikes form when magma is injected into linear conduits. Geologists believe that extension of the crust created these pathways for the Coso eruptions.

Fig. 7-14 Rhyolite domes of the Coso volcanic field, viewed from Sugarloaf Mountain. (Source: Hill, M. U.S. Geological Survey.)

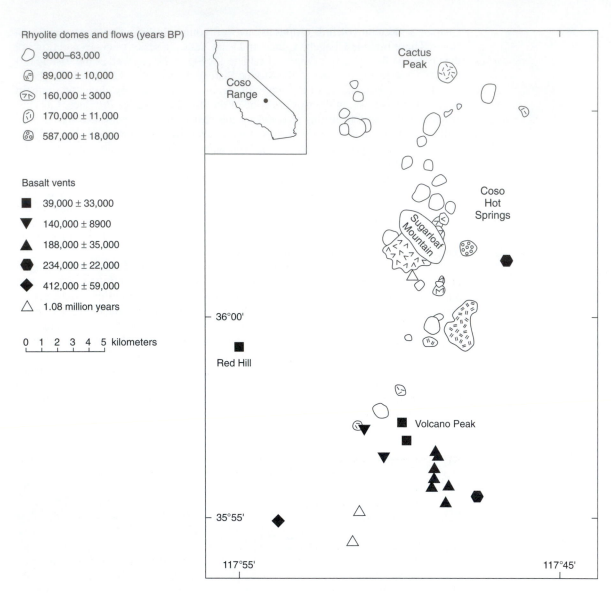

Fig. 7-15 Map showing volcanic eruptions in the Coso volcanic field during the past million years. (Source: Bacon, C. 1982. *Geology* 10: p. 66.)

RECENT VOLCANIC ACTIVITY IN THE MOJAVE DESERT

A string of young volcanic flows, cones, and craters stands out along a northwest-oriented zone between Barstow and Bristol Lake (Fig. 7-16). Volcanic eruptions have occurred along a depression in the Mojave landscape, and both the depression and the volcanic activity are thought to be caused by continuing extension. Basalt has been the dominant rock produced in the zone, as would be expected in

Volcanic areas

1. Cima

2. Amboy-Pisgah

3. Turtle Mountains

4. Pinto Basin-Salton Creek

5. Obsidian Buttes

6. Miscellaneous

Fig. 7-16 Map showing the distribution of young volcanic features in the Mojave Desert. (Source: California Division of Mines and Geology. 1954. Bulletin 170.)

an extensional environment (see Chapter 5). Many of the volcanic features appear unweathered, evidence of both recent activity and excellent preservation in the desert environment.

The Cima volcanic field, located in the Mojave National Preserve about 20 kilometers north of Kelso, contains 30 cinder cones and associated flows. Cones and flows in the northern part of the field are as old as about 10 million years, whereas the southern part of the field is much younger. There some of the eruptions are thought to have occurred only a few centuries ago, based on their excellent preservation (Fig. 7-17) and on radiocarbon dating of charcoal found beneath one of the flows. At the Aiken Mine, the red cinders have been mined for gardens, paths, and other decorative uses. Cinders are also mined from Pisgah Crater, a 250-foot-high cinder cone just south of Interstate 40 about 35 miles east of Barstow. Recent lava flows that covered part of Lavic Lake are associated with the crater. Three craters known as Sunshine Peak Craters, and their associated lava flows, are a few miles south of Pisgah.

Dish Hill contains two cinder cones erupted about 2 million years ago. Both at Dish Hill and in the Cima volcanic field, the volcanic eruptions have brought pieces of the lower crust and the mantle to the Earth's surface in the form of **xenoliths** found within the volcanic rock. These fragments or **inclusions** of already-crystallized rock are important because they provide samples of rocks from as deep as 70 to 80 kilometers below the surface, far beyond the reach of

Fig. 7-17 Aerial view of recent basalt flows and cones, Cima volcanic field, Mojave Natural Preserve. (Source: Nakata, J. U.S. Geological Survey.)

even the deepest drilling. Fragments of gabbro (mafic plutonic rock) suggest that the lower part of the crust in the Mojave Desert is composed of gabbro. Presumably, bits of rarely seen ultramafic plutonic rocks are representative of the underlying mantle.

Amboy Crater is a young cinder cone that rises about 75 meters above the surrounding terrain. Lava from Amboy Crater flowed onto the surface of Bristol Lake, probably within the past several thousand years. Volcanic features such as collapsed lava tubes, pits, and lava blisters can still be seen on the surface of the flow.

SEDIMENTARY BASINS AND EXTENSION

Throughout southeastern California, extension resulted in the formation of down-dropped basins between the ranges. Sediments eroded from newly raised mountains accumulated in those basins (see Chapter 2). By identifying and dating the oldest sedimentary rocks that were deposited in the basins, geologists can determine when crustal extension began. Throughout southeastern California, sedimentary rocks of this type began to accumulate during the Miocene period, between about 22 and 16 million years ago. Not surprisingly, the timing coincided with early volcanic eruptions caused by crustal extension in southeastern California.

As faulting formed topographic basins, the basins created a trap for alluvium eroded from nearby mountains and for sediments deposited in lakes within the basins. Later faulting has uplifted Miocene sedimentary units to expose them in several areas of southeastern California. The Miocene sedimentary units include alluvial fan gravels carried from the Miocene ranges into the ancient basins. The gravels

Formations

Funeral Basalt

Furnace Creek Formation, undivided

Upper sedimentary member

Upper pyroclastic member

Middle sedimentary member

Lower pyroclastic member

Upper submember

Lower submember
— Lower sedimentary member

Artists Drive Formation

Depositional Environment

Lake deposits

Volcanic

Lake deposits

Volcanic

Alluvial fan deposits

A

B

Fig. 7-18 A, A generalized stratigraphic column showing the members of the Miocene Artists Drive Formation and their inferred environments of formation. **B**, The Artist Drive Formation at Zabriske Point, Death Valley. (Source: A, adapted from *South Coast Geological Society Field Trip Guidebook*; B, Skapinsky, S. San José State University.)

contain recognizable pieces of older rocks recycled into Miocene sediments (see Chapter 2). One example of Miocene alluvial fan deposits is the Artists Drive Formation, 14 to 6 million years old, in Death Valley. As the Miocene basin developed, younger lake sediments and intermixed volcanic layers accumulated above the fan gravels formed earlier (Fig. 7-18).

BARSTOW BEDS

The Pleistocene lakes discussed in Chapter 6 demonstrate that the past environment in the California desert was vastly different than today's. Another contrast with the present is evident in the sedimentary layers preserved about 18 kilometers north of Barstow in an area known as the Rainbow Basin. A section of sedimentary rocks up to 650 meters thick records the history of deposition in and around a lake during Miocene time. Miocene extension created the down-dropped basin in which the sediments accumulated. The rocks, which have been named the Barstow Formation, are a mixture of mudstone, sandstone, and conglomerate, with some interbedded ash layers. Based on the ages of volcanic rocks above and below the sediments, the age of the Barstow Formation is placed between 16 and 13 million years ago.

The most interesting feature of the Barstow Formation is that it contains a diverse assortment of fossils. Since 1919, when the fossils were first described, paleontologists have identified the remains of dogs, bears, cats, mammoths, horses, camels, antelope, bison, sheep, turtles, shellfish, turtles, palm trees, and even flamingos.

RED ROCK CANYON STATE PARK

Travelers on State Highway 14 pass through Red Rock Canyon about 40 kilometers north of the town of Mojave. Even though many do not stop at Red Rock Canyon State Park, they still admire the beautiful red, orange, and pink rocks as they pass by (Fig. 7-19). Those who stop for a closer look find interlayered volcanic and sedimentary rocks. The sedimentary rocks—orange conglomerate and sandstone and white mudstone—were deposited in ancient rivers and lakes. Periodic volcanic eruptions added layers of white ash, pink pyroclastic flows, and dark andesite and basalt flows. Within the sedimentary beds are fossils of the same diverse animals that inhabited the Rainbow Basin.

The rocks at Red Rock Canyon belong to the Ricardo Group, deposited between 19 and 8 million years ago in a basin similar to the basin where the Barstow Formation accumulated. Like the Barstow Formation, the Ricardo Group records the extension of the crust in the Mojave Desert during Miocene time.

EXTENSION OF THE CRUST IN SOUTHEASTERN CALIFORNIA

Geologists and geophysicists working in the Basin and Range in the western United States agree that extension of the earth's crust began about 40 million years ago and continues today. In the California portion of the Basin and Range, extension began about 16 million years ago. Some areas, including the Death Valley region, might have been stretched by 100 percent, and geophysicists estimate that the entire Basin and Range province has been lengthened by about 250 kilometers

Fig. 7-19 Red Rock Canyon State Park, Highway 14, Kern County. (Source: Nakata, J. U.S. Geological Survey.)

during the past 40 million years. The 10-hour drive from San Francisco to Salt Lake City along U.S. Highway 50 would be shortened by at least 2.5 hours if the stretching could be undone! The present rate of extension is about 1 millimeter per year.

In the Basin and Range Province, the north–south alignment of the faults, the basins, and the ranges provides a clue to the forces that have operated during the past 20 million years. Only east–west extension would create this particular orientation of faults and the topography. However, the exact cause of this extension is not known at present. Geologists have proposed a number of mechanisms to explain Basin and Range extension. One of these is a change in the relative motion between the Pacific and North American Plates. Another hypothesis is that extension represents a giant-scale collapse of the crust in response to the tremendous thickening brought about by Mesozoic compression and crustal shortening. Of course, it is possible that several causes have operated together to produce the Basin and Range Province's most striking feature.

In the Mojave Desert, the topography is much less systematically oriented than in the Basin and Range. The landscape is also more subdued, with broader basins and less continuous mountain ranges. A major reason for the differences in the landscapes of the two provinces is that crustal extension in the Mojave Desert has been much less active since Miocene time. Recent right-lateral faulting is prominent in the central Mojave Desert, where the faults of the eastern California shear zone run parallel to the San Andreas fault (see Chapter 14). Range-front faults are also present in the northeastern Mojave Desert, where they are parallel to the Garlock fault. In the remainder of the Mojave Desert, the evidence of active faulting is much less widespread.

BEDROCK GEOLOGY: OLD ROCKS IN YOUNG MOUNTAINS

One result of normal faulting is that rocks are dropped down in the basins relative to the ranges (Fig. 7-20). Because the mountains are elevated, rocks exposed there are subject to erosion. Desert streams carry the materials eroded from the mountains down the steep, bare slopes of the ranges. These sediments are then deposited along the range fronts or on the flat basin floors. Over time, the older rocks in the basins are buried by these young sediments, whereas older rocks in the ranges are exposed by uplift and erosion. Examination of a geologic map of California's

Fig. 7-20 In this diagram, older rock layers are buried by younger materials in the down-dropping basins, as the ranges are uplifted and eroded.

Basin and Range and Mojave Desert Provinces confirms this pattern: Recent alluvium covers the basin floors, whereas a variety of Mesozoic and older rocks can be seen at the surface of the ranges (Endpaper 1). In some areas, drilling beneath the recent deposits in the basins has located the older rocks beneath them. This has allowed an estimate of the amount of vertical offset or separation of rock formations by the faults. However, the practical way to understand the geologic history of southeastern California prior to Basin and Range faulting is to study the rocks exposed in the ranges.

Few places in California provide less obstructed geologic views than the harsh, barren desert ranges like the Panamint Range, the Inyo Mountains, or the Funeral Range. No dense vegetation or human construction covers the outcrops here, and a

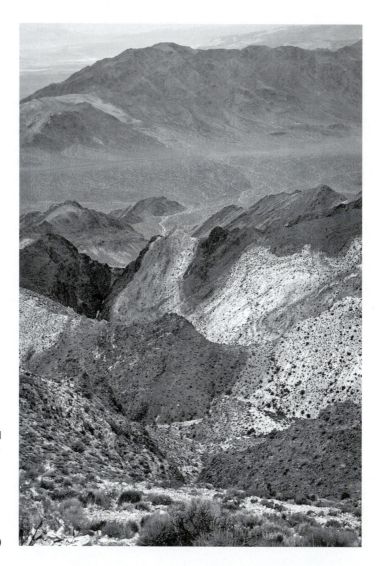

Fig. 7-21 A complex variety of folded and faulted sedimentary rocks are exposed in the Northern Panamint Mountains. View to the east toward Racetrack Valley, Death Valley National Park. (Source: McMackin, M. San José State University.)

view of any of the ranges reveals a perplexing array of rock types and structures (Fig. 7-21). After many decades of intense study, geologists have reconstructed an amazing picture of the environment that once existed in what is now North America's fiercest desert.

Exposed in most of the ranges in California's Basin and Range Province, as well as in the Marble Mountains, the Providence Mountains, and near Victorville in the Mojave Desert, are thick sequences of **carbonate** sedimentary rocks. Both limestone and dolomite (see Chapter 2) are abundant. Many formations contain fossils, including mats of fossil algae. Gastropods (Fig. 7-22) and coral in other formations indicate nearshore marine conditions, and other formations contain fossils of organisms that lived in deeper water.

All of the carbonate rocks and their fossils must have formed in a marine environment, in warm seas along a gently sloping continental margin. There were no nearby mountain ranges, and no streams carried large amounts of feldspar, mafic minerals, and rock fragments from the continent. Geologists know this because these sediments are rare in the sequence. Interlayered with the organically derived carbonate formations are layers of pure quartz sandstone, which might have formed in a beach or dune environment. To reconstruct the environment where the carbonate formations might have formed, we can locate marine environments where carbonates are forming today (see Chapter 3). As you stand on an outcrop of Paleozoic limestone, the picture that emerges is as different from the modern environment in the Basin and Range Province and the Mojave Desert as you could imagine The Paleozoic seas were similar in some ways to environments found today along continental margins such as the Bahama Banks or Australia's Great Barrier Reef.

Fig. 7-22 Fossils of giant shallow-water snails in the Badger Flat dolomite, southern Inyo Mountains.

MODERN CARBONATE PLATFORM

Imagine yourself snorkeling through a coral reef off the southern coast of Florida or along Australia's Great Barrier Reef. The water is warm and crystal clear, allowing you to view the amazingly diverse creatures that inhabit a coral reef (Fig. 7-23). No rivers empty into the area, allowing the coral reef to grow in the sediment-free waters. The sand on the nearby beaches and along the bottom is composed of coral and shell fragments. The water depth increases very gradually away from the coast, because the slope of the continental shelf is gentle.

Of course, a Paleozoic carbonate platform would be populated by a different assemblage of organisms than today's reefs. The carbonates of Cambrian time were built by sponge-like creatures called Archaeocyathids. Over time they caused layers of carbonate to precipitate from the surrounding sea, building masses known as **stromatolites** (Fig. 7-24). Later Paleozoic reefs or banks were built by corals and many other marine organisms (Fig. 7-25). Many of the Paleozoic animals, such as trilobites, horn coral, and jawed fish, are now extinct. Other forms like seals and whales had not yet evolved. Some organisms, including many mollusks, are remarkably similar to their ancestors of 350 million years ago.

Fig. 7-23 A modern coral reef. (Source: Tom Stack and Associates.)

Fig. 7-24 Preserved algal mounds in the Noonday dolomite, Do Not Pass Hills, southeastern Kingston Range, Death Valley National Park. (Source: McMackin, M. San José State University.)

Fig. 7-25 A reconstruction of a Middle Devonian reef. (Source: Smithsonian Institution.)

After removing the effects of faulting and folding that have tilted and separated the rocks since they were deposited, geologists estimate that the carbonate formations represent more than 7,000 meters of deposition (Fig. 7-26). Some individual formations can be found across most of the entire Basin and Range Province—for example, the Eureka Quartzite, a quartz sandstone named for the old mining town of Eureka in eastern Nevada. Fossils found within the formations indicate that the rocks formed from latest Proterozoic time through the entire Paleozoic era, a period of more than 350 million years between about 600 and 250 million years ago (see Fig. 3-3).

Fig. 7-26 A generalized stratigraphic cross section across the Basin and Range between Mono Lake and Las Vegas, Nevada, showing the general extent of the late Proterozoic and Paleozoic carbonate platform. (Source: Adapted from Burchfiel, P., and Davis, G. 1981. *Geotectonic Development of California.* Upper Saddle River, NJ: Prentice Hall.)

Because many formations can be traced for such great distances to the east, the ancient marine environments where they accumulated must have been fairly uniform over a large area. Geologists reason that the slope of the continental shelf was broad and very gently sloping, so that water depth increased only slightly westward. Any slight rise or fall in sea level would cause the shoreline to shift many miles east or west (Fig. 7-27).

A sample section of rocks in the southern Inyo Mountains records such changes in amazing detail. The type of sedimentary rock and the fossils provide the clues to each formation's environment of deposition. The Al Rose Formation contains intermixed mudstone and carbonate. As indicated by the presence of sedimentary features formed in the tidal zone, part of this unit must have formed in fairly shallow water. Above the Al Rose Formation is the Badger Flat Formation, dolomite containing huge snail fossils, formed in shallow marine water (Fig. 7-28; see also Fig. 7-22). By the time the next unit formed, the marine basin had filled in enough with sediment to expose this part of the continental margin above sea level. The sedimentary unit representing these conditions is the Eureka Quartzite, originally deposited on a beach or a sand dune. Finally, the sea rose again to create a shallow marine environment that produced the Ely Springs Dolomite, and then a deeper environment that produced the Vaughn Gulch Limestone. Thus, the rising and falling of sea level hundreds of millions of years ago can be reconstructed for a single location. By piecing together this kind of geologic information across the entire Basin and Range, geologists have come to understand some of the workings of an ocean that disappeared from southeastern California more than 200 million years ago.

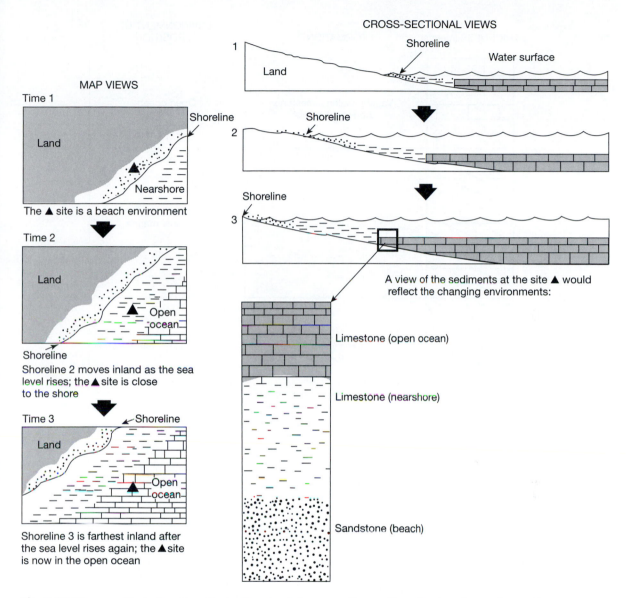

Fig. 7-27 Diagrams illustrating the effects of sea-level change on the position of a shoreline and on sedimentary environments. (Source: Adapted from Montgomery, C. *Physical Geology*. Dubuque, IA: Wm. C. Brown.)

THE PASSIVE CONTINENTAL MARGIN

West of ancient North America, the edge of the continental margin began to *subside* or sink about 600 million years ago, in late Proterozoic time. Geologists believe that the sinking resulted from rifting within the ancient North American continent. As the edge of the continent subsided, sediments gradually accumulated

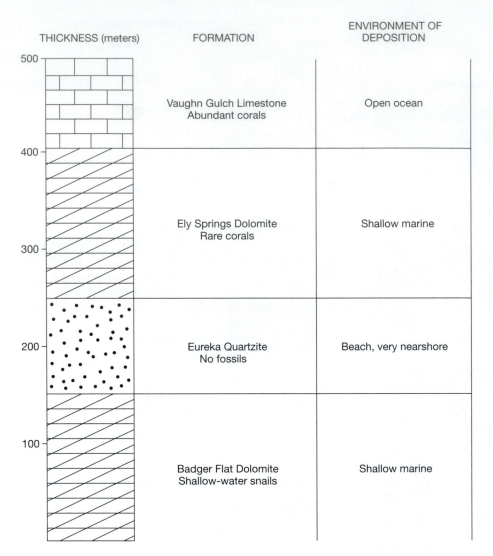

Fig. 7-28 Schematic column illustrating sedimentary formations, fossils, and ancient environments represented by older Paleozoic rocks in the southern Inyo Mountains. (Source: Stevens, C. San José State University.)

on what was then the continental shelf. The beginning of this 350-million-year-long period of subsidence and sedimentation is represented today by formations composed of quartz sandstone, siltstone, and shale. The early sediments washed onto the continental shelf from the ancient land mass, carried by rivers that flowed from the coastal plain. In the Death Valley area, the Johnnie Formation, Wood Canyon Formation, and the Zabriskie Quartzite are some of the recorders of these events (Fig. 7-29).

By earliest Paleozoic time, little sediment from the ancient continent to the east was reaching the area that is now easternmost California. As a result, sediments

Fig. 7-29 Proterozoic rocks in the Black Mountains, east of southern Death Valley, looking east. The view also shows the turtlebacks along the range front. (Source: Bradley, W.C.)

accumulating on the continental margin were composed mostly of carbonate produced by organisms, with only minor amounts of sandstone and siltstone. Today these sediments are preserved as the thousands of meters of Paleozoic carbonates seen in the ranges of the Basin and Range Province and the Mojave Desert, as well as in the San Bernardino Mountains. These rocks record the history of the **stable carbonate platform** that existed along the western edge of North America for 350 million years. During this time, no dramatic tectonic events occurred in the eastern part of California's Basin and Range or Mojave Provinces.

Active plate boundaries were apparently distant from this **passive margin** until the very end of the Paleozoic era. Oceanic plates were in motion somewhere offshore to the west of this margin during Paleozoic time, creating trenches, volcanic arcs, and subduction zones. As we will see in Chapter 8 and Chapter 9, these plates interacted with the North American continent in areas that are now part of the Klamath Mountains and northern Sierra Nevada.

It has taken scores of geologists and paleontologists a century to piece together the Paleozoic history of the Basin and Range and Mojave Desert. As the geology of each mountain range is better understood and compared to that of neighboring ranges, a coherent picture has evolved. For about 350 million years, easternmost California was a tectonically quiet, stable continental shelf. At no time before or since that period was California geology so uneventful, and in no other part of California was deposition of sediments less affected by active tectonics. However, when a visitor admires the broken and swirled rocks of the desert ranges, this picture is anything but obvious. The Paleozoic rocks have been greatly folded, faulted, and eroded during Mesozoic and Cenozoic time, and the imprint of this disruption on the rock outcrops is very strong.

BEFORE THE PASSIVE MARGIN: OLDER ROCKS IN SOUTHEASTERN CALIFORNIA

Less is known about the geologic environment of southeastern California before the development of the passive continental margin, because rocks older than latest Proterozoic age are found only in a few places. The rocks of the Pahrump Group in Death Valley, dated between about 1.2 billion and 700 million years old, appear to document the continental rifting that immediately preceded formation of the passive continental margin. Geologists believe that the rifting occurred as an ancient supercontinent broke apart during this time period (see Chapter 18). In Death Valley, the Kingston Peak Formation (Fig. 7-30), which is about 750 million years old, contains the types of coarse sediments produced along faults or in rift valleys. Abrupt changes in the types and thickness of the sediments indicate that faults were active at that time. Geologists reason that these rocks record some of the continental rifting that gave rise to the passive continental margin.

Rocks older than about 1.2 billion years old are also exposed in some of the ranges of southeastern California. These rocks are significant because they represent California's oldest known rocks, and they are confined to a very small portion of the state. Ancient rocks are found in the San Bernardino and San Gabriel Mountains, several ranges in the Mojave Desert, and in the Panamint Range near Death Valley. They include ancient metamorphic rocks that were metamorphosed about 1.7 billion years ago. Plutonic rocks in the same areas are dated at about 1.4 billion years old.

Fig. 7-30 Proterozoic rocks of the Kingston Peak Formation, eastern Kingston Range. (Source: McMackin, M. San José State University.)

MOUNTAIN PASS MINE

At the Mountain Pass Mine in San Bernardino County, some of California's oldest rocks provide elements necessary for some of our most advanced technologies. Located between Barstow and Las Vegas along Interstate 15, at the southern end of the Clark Range, the Mountain Pass area contains some very unusual plutonic rocks. One of these rocks is *carbonatite*, an igneous plutonic rock composed predominantly of carbonate minerals. As outlined in Chapter 2, most igneous rocks are composed of silicate minerals, and carbonate minerals are typically formed in sedimentary environments like those discussed earlier in this chapter. Because carbonatite magma is highly unusual, carbonatites are very rare. Geologists believe that carbonatite magmas formed very deep in the Earth's mantle. The carbonatite at Mountain Pass crystallized about 1.4 billion years ago, when magma intruded even older rocks.

About 9 percent of the carbonatite is the mineral *bastnasite*, a rare carbonate mineral. Bastnasite—chemical formula $(Ce,La)(CO_3)F$—contains varying amounts of the 15 chemical elements known as the *rare earth* or *lanthanide elements*. The primary rare earth elements in the bastnasite at Mountain Pass are cerium (Ce) and lanthanum (La). Many of the rare earth elements have important commercial uses (see Table 7-1), and the carbonatite at Mountain Pass mine is the largest concentration of rare earth elements in the Western Hemisphere. About 2,000 tons of rock are mined daily at Mountain Pass, and mining engineers estimate that enough bastnasite remains for another 50 to 60 years of production at the current rate.

TABLE 7-1 COMMERCIAL USES OF RARE EARTH ELEMENTS MINED AT MOUNTAIN PASS

Elements	Uses
Cerium	Polishing compounds for optical glass, camera lenses, and TV face plates, decolorizer in glass, catalytic converter ceramics
Lanthanum	Petroleum refining, optical manufacture of fibers, camera lenses, and X-ray intensification screens
Neodymium	Magnets, ceramic capacitors, enhancement of brightness and contrast in color TVs
Praseodymium	Magnets, yellow glaze for tile
Europium, yttrium	Red phosphor in color TV, fluorescent lighting

ACTIVE TECTONICS IN THE MESOZOIC

The tectonic quiet of the Paleozoic ceased in southeastern California at the beginning of Mesozoic time, about 245 million years ago. Throughout California, the passive margin gave way to an active boundary between the North American continent and plates of the Mesozoic ocean to the west. The evidence for this great change can be seen in the rocks of the Basin and Ranges and the Mojave Desert. As we will see in subsequent chapters, the development of an active plate boundary along the western margin of North America has probably been the dominant force shaping California's geology.

Granitic plutonic rocks and volcanic rocks similar to those found along modern subduction zones occur throughout the ranges of southeastern California. Most of the granitic plutonic rocks in the Basin and Range and Mojave Desert have been dated between 200 and 100 million years old. Most of the Mesozoic plutonic rocks are parts of the Sierran Batholith, which is discussed in Chapter 8. In eastern California, granitic magma intruded into the older sedimentary rocks and metamorphosed the rocks along the contacts with the plutons. In many areas, valuable mineral deposits of silver, gold, talc, and other commodities were formed in these contact zones and later extracted by miners.

Evidence for widespread compression during Mesozoic time is also preserved in southeastern California. The older sedimentary formations are greatly disturbed by folds and thrust faults. These structures can be dated as Mesozoic, because younger rocks are not affected by the folds and faults. In addition, the folds and thrust faults are themselves displaced by the normal faults discussed earlier in this chapter, as are the older rocks. Following the principles of cross-cutting relations outlined in Chapter 3, the normal faults must be younger than the structures created by Mesozoic compression. The Mesozoic igneous activity and the faults and folds produced by compression are typical events that occur along a subduction zone boundary.

SUMMARY

1. The mountain ranges of southeastern California are the only places in the state where rocks older than about 700 million years old (late Proterozoic age) occur. Little is known about California's plate tectonic setting at the time the oldest rocks formed, more than 1.2 billion years ago. By about 700 million years ago, during late Proterozoic time, the rocks of that age suggest that a period of continental rifting was occurring.

2. From about 600 million years ago throughout the entire Paleozoic era, southeastern California was a shallow marine environment on the western edge of the ancient North American continent. Marine sediments, mostly carbonates, accumulated along this tectonically stable marine platform.

3. At the end of Paleozoic time, about 245 million years ago, the tectonic setting of southeastern California changed dramatically from a passive margin to an active subduction zone. Mesozoic subduction generated granitic plutons in the Basin and Range and Mojave Desert Provinces, as well as in the Sierra Nevada (see Chapter 8 for further discussion).

4. Beginning about 22 million years ago, during mid-Cenozoic time, extension of the lithosphere began in the Basin and Range and Mojave Desert Provinces. The extension created detachment faults and generated volcanic eruptions. Continental sediments began to accumulate in the low-lying areas created by the faulting.

5. Normal faults are present along the major detachment surfaces of southeastern California. They produce the characteristic basin and range topography of both provinces. Many faults in the California Basin and Range and in the eastern Mojave Province are active, creating rugged mountains and flat basin floors.

6. A major earthquake occurred in 1872 on the eastern Sierra Nevada fault system near the town of Lone Pine. The earthquake had an estimated magnitude of 7.8.

REFERENCES

COLLIER, M. 1990. *An Introduction to the Geology of Death Valley*. Death Valley, CA: Death Valley Natural History Association, 60 pp.

DOKKA, R., AND OTHERS. 1991. Aspects of the Mesozoic and Cenozoic evolution of the Mojave Desert. In *Geological Excursions in Southern California and Mexico*. Geological Society of America Annual Meeting Guidebook, pp. 1–43.

FIERO, B. 1986. *Geology of the Great Basin*. Reno: University of Nevada Press.

HILL, M. 1972. The great Owens Valley Earthquake of 1872. *California Geology* 25(3):51–61.

HUNT, C.B. 1975. *Death Valley*. Berkeley: University of California Press, 234 pp.

OAKESHOTT, G.B., GREESFELDER, R.W., AND KAHLE, J.E. 1972. One hundred years later. *California Geology* 25(3):51–61.

PRIDMORE, C., AND FROST, E.G. 1992. Detachment faults. *California Geology* 45(1):3–17.

REYNOLDS, R. 1991. *Crossing the Borders: Quaternary Studies in Eastern California and Southwestern Nevada*. Redlands, CA: Movaje Desert Quaternary Research Center, Special Publication.

TROXEL, B.W., AND WRIGHT, L.A. 1976. *Geologic Features of Death Valley, California*. California Division of Mines and Geology, Special Report 106, 72 pp.

Web Sites and Media Resources

Also see the list of resources in Chapter 6

University of California, Museum of Paleontonology

Paleozoic fossil localities (clickable map for California sites) *http://www.ucmp.berkeley.edu/paleozoic/paleozoicloc.html*

8

The Sierra Nevada
Granite, Gold, and Glaciers

Fig. 8-1 Map showing the location of the Sierra Nevada province.

Sierra Nevada Province Highlights

Major geographic feature: Sierra Nevada Mountain Range
Other major landscape elements: Western Sierra Foothills

High points:
    Mt. Whitney: 4,418 m (14,495 ft.)
    10 other peaks over 14,000 ft.

Predominant rock types:
    Mesozoic granitic plutonic
    Mesozoic and Paleozoic metamorphic

Plate tectonic events:
    Mesozoic subduction and accretion

Other geologic events:
    Pleistocene glaciation (High Sierra)

Faults:
    Eastern Sierra Fault system

Resources:
    Water (snowpack and rivers)
    Gold (western foothills)

National Parks:
    Yosemite, Kings Canyon, Sequoia

State Parks:
    Emerald Bay (Tahoe),
    Empire Mine,
    Malakoff Diggins
    State Mining and Mineral Museum

Other:
    Mercer Cavern, Moaning Cavern

Stretching for 700 kilometers along much of California's eastern border, the Sierra Nevada has been the geographic key to California's history. The mountain range itself was a barrier to those seeking California, but the gold of the western foothills brought immigrants by the thousands. The mountains are responsible for the rain and snow dumped by Pacific storms (see Chapter 6), and without the Sierra Nevada's water, California could not support even a fraction of its 34 million people (see Chapter 10). The Sierra Nevada is also home to some of the world's most spectacular scenery, including Yosemite, Sequoia, and Kings Canyon National Parks, discussed later in this chapter.

The Sierra Nevada Province includes the Sierra Nevada range and a broad belt of foothills on its western flank. From its southern end east of Bakersfield to its northern end south of Lassen Peak, the Sierra Nevada trends north-northwest, having a slightly more north–south trend than the Coast Ranges (Fig. 8-1, Endpaper 3). The range is between 65 and 130 kilometers wide, with a broad region of foothills along the western slope.

In 1776, Father Pedro Font, a Spanish missionary and a member of the Anza expedition, which had journeyed from Mexico to settle along San Francisco Bay, made a sketch from a hill near the mouth of the Sacramento River. About 200 kilometers to the northeast he had seen *"una gran sierra nevada"*—a great snowy range—running from south-southeast to north-northwest. Font was not the first European to view the Sierra Nevada, which had been observed by Father Crespi 4 years earlier, but he is credited with naming California's most famous mountain range. Trapper Jedediah Smith became the first non–Native American to cross the range in 1827. Earlier that year, Smith and 17 others had crossed the Mojave Desert and arrived at the San Gabriel mission, where they were promptly invited to leave by the local authorities. After traveling north along the San Joaquin Valley, Smith and two companions struggled eastward through deep snow across what is today Ebbetts Pass.

PEAKS AND PASSES

For those unfamiliar with mountainous terrain, it is useful to review the geographic elements of a mountain range. A ***mountain range*** is a chain or group of mountains consisting of a number of individual ***peaks*** or mountains. In the Sierra Nevada, individual mountain peaks stand above the surrounding terrain—for example, Mt. Whitney or Mt. Coness (Fig. 8-2). The ***summit***, or highest point, of Mt. Whitney stands 4,418 meters (14,495 feet) above sea level. Some peaks in the Sierra Nevada stand out less obviously because the surrounding areas are of similar elevation.

An imaginary line connecting the summits and highest parts of the ridges along a mountain range is the ***crest*** of the range (Fig. 8-2). The crest of the Sierra Nevada, which includes its most spectacular summits, is highest in the southern part of the range near Mt. Whitney, where 11 peaks reach elevations greater than 4,267 meters (14,000 feet). At the lowest northern end of the range near the Feather River, the crest is between 1,800 and 2,450 meters. Along a 30-kilometer-wide band from north of Yosemite National Park to south of Mt. Whitney lies the part of the Sierra Nevada known as the High Sierra. Because the crest of a mountain range marks its highest points, streams flow from the crest to lower elevations on both sides of the crest. The crest of the Sierra Nevada is therefore also the ***divide*** between the watersheds of the streams that flow east and west from the crest (see Chapter 10 and Fig. 10-4 for further information about watersheds). The Sierra Nevada's crest was also chosen as the boundary between many of California's mountain counties, so it is easily traced on even a simple road map.

Fortunately for those traveling across the Sierra Nevada, the crest of the range has points of lower elevation, even in the higher southern part of the range. These low notches in the crest are the ***passes*** across the range. The major passes of the Sierra Nevada crest are at the heads of the larger river valleys, most of which have been shaped by glaciers. Some of them, including Mono Pass near Yosemite, were well traveled by Native Americans seeking obsidian at Mono Craters, trading with groups on the opposite side of the range, or escaping the summer heat. During the 1800s the passes became the key element of trails used by the pioneers headed for California. Because of the greater height of the southern crest, the trails were limited to the northern Sierra Nevada, and even today, no roads cross the highest part of the range. The most heavily traveled route today is Interstate 80, which crosses the Sierra Nevada at Donner Pass (elevation 2,175 meters), one of the lower passes through the range.

Fig. 8-2 View of the southern Sierra Nevada, looking westward. Mt. Whitney, California's highest peak, appears in the center of the image beneath the small cloud and behind the range of mountains in the foreground.

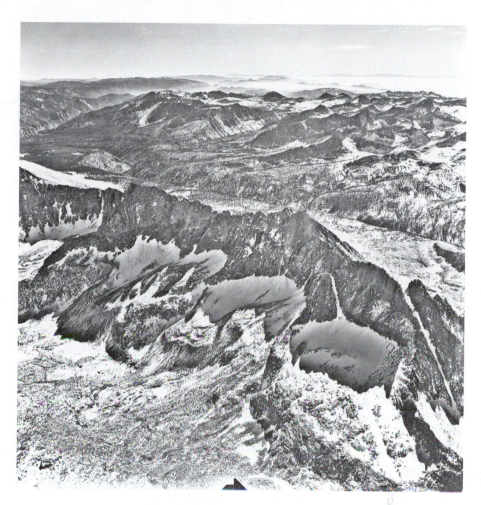

Fig. 8-3 Oblique aerial photograph of the Kaweah peaks area, southern Sierra Nevada, looking westward. The steep mountains in the foreground show prominent peaks and a sharp ridge crest. Note also the bowl-shaped glacial cirques in the foreground. (Source: U.S. Geological Survey.)

FORMATION OF THE SIERRA NEVADA RANGE

Any driver who crosses the Sierra Nevada going east on Interstate 80 learns firsthand that the mountain range is not symmetrical. From the base of the mountains near Roseville, the drive up the western slope to Donner Pass is about 130 kilometers, but the drive from Donner Pass down to Reno is only about 55 kilometers. Along the entire range north of the Kings River, the western slope is more gradual than the steep eastern front (Fig. 8-4). Active normal faults that mark the western edge of the Basin and Range Province account for the steepness of the eastern Sierra Nevada front (see Chapter 7). Uplift of the range along the eastern Sierra fault system has been accompanied by westward tilting, so that the entire Sierra Nevada has been tipped westward, producing the gently sloping western slope. However, faults are present along the western edge of the range as well, and some might be active. For example, a 1975 earthquake at Oroville along the western fault system was apparently triggered by the weight of the water filling the reservoir behind the Oroville dam.

Fig. 8-4 Uplift and westward tilting cause the asymmetry of the Sierra Nevada. (Source: Huber, N.K., and Wahrhaftig, C. 1987. *The Geologic Story of Yosemite National Park*. U.S. Geological Survey, Bulletin 1595.)

The age of the Sierra Nevada has long been a topic of study. As we will see in the next section of this chapter, the plutonic rocks that make up most of the Sierra are of late Mesozoic age. Following the principles outlined in Chapter 3, we know, therefore, that the mountains must be younger than the rocks beneath them. Based on earthquake activity and other measurements, we also know that the range is being uplifted relative to the neighboring provinces.

As discussed later, geologic evidence supports the idea that the present mountains and drainage patterns are relatively young, having formed within the past 9 to 10 million years. On the other hand, other studies of the uplift of the Sierra Nevada have used geophysical techniques and the temperature histories of minerals in the plutonic rocks to estimate the topography of the Sierra during late Mesozoic time. These studies suggest that major mountain ranges and steep river canyons have been present since about 70 million years ago, and that today's Sierra Nevada mountains could be significantly lower than their late Mesozoic ancestors. Although the age of the mountain range is not firmly resolved, most geologists agree that the current topography is relatively recent. What is unknown is whether mountainous terrain has been continuously present since late Mesozoic time, or whether there was an early Cenozoic period of relatively low topography.

Table Mountains and Inverted Topography

Prominent, flat-topped mountains can be found at several locations in the western Sierra foothills. These are examples of **table mountains**, landforms named for their resemblance to tables (Fig. 8-5). One of the best known in the Sierra Nevada, actually named Table Mountain, can be seen west of Sonora along Highway 108. Viewed from the air or on a geologic map (Fig. 8-6), Table Mountain looks strikingly similar to a sinuous river, with one major difference. Rather than winding along a valley bottom, it forms a flat-topped mountain.

Today Table Mountain is capped by 90 meters of volcanic rock, and underneath are sediments of the river channel that was filled with lava 9 million years ago. At the

Fig. 8-5 A sketch of Table Mountain near Sonora, taken from an 1886 guidebook to Yosemite Valley. Table Mountain is the flat-topped mountain in the center.

time of the volcanic flow, Table Mountain was a low-lying river valley, but once it filled with volcanic rock, it became resistant to erosion. Since the lava flow, the areas surrounding Table Mountain have been eroded at a faster rate, so that they are now lower than the lava-filled valley. As a result, the topography has become reversed or *inverted*, with former valley bottoms now the highest part of the landscape (Fig. 8-7). Other table mountains formed in the same way are found along the western flank of the Sierra Nevada; for example, San Joaquin Table Mountain and Oroville Buttes. Table mountains are excellent examples of the process of *differential erosion*, which is important in shaping all landscapes. Over time, areas underlain by weak rocks are eroded more than areas underlain by resistant rocks.

Late Cenozoic Uplift of the Sierra Nevada

The volcanic flows preserved at Table Mountain and other localities along the Sierra Nevada provide evidence that the present Sierra Nevada mountains are very young by geological standards. The eruptive sources of some of the volcanic flows were east of the present Sierran crest—one east of Sonora Pass and one northeast of Mono Lake. This was determined by geologic mapping and by determining the chemistry and ages of the flows, using methods similar to those used to fingerprint volcanic ashes (see Chapter 5). The volcanic flows traveled down stream channels into the ancient Stanislaus, Tuolumne, and San Joaquin Rivers (Endpaper 3). Clearly, no large mountain range blocked their path: At the time just prior to the eruptions, rivers flowed westward into the Central Valley across what is now the crest of the Sierra Nevada. Sediments from those ancient rivers were then buried by huge lava flows, ash flows, and volcanic mud flows about 9 to 10 million years ago (Fig. 8-7).

Fig. 8-6 A portion of the Sonora geologic map, showing the sinuous pattern of the Table Mountain volcanic flow. Areas covered by the flow are labeled *Tm* and *Tl*. On the location map (inset), locations of recent volcanic flows are shown in the V-pattern area. (Source: Adapted from Heyl, G., and others. 1955. *Geologic Map of the Sonora Quadrangle.* California Special Report 41.)

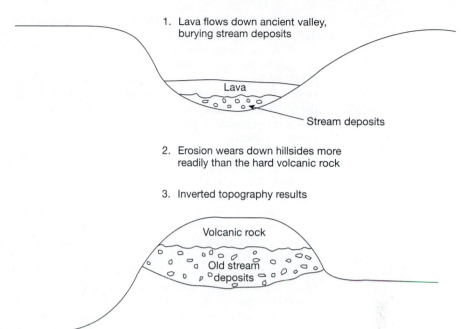

1. Lava flows down ancient valley, burying stream deposits

Lava

Stream deposits

2. Erosion wears down hillsides more readily than the hard volcanic rock

3. Inverted topography results

Volcanic rock

Old stream deposits

Fig. 8-7 The development of inverted topography at Table Mountain. This is a cross-sectional view with the stream flowing into (or out of) the page.

Geologists can reconstruct the general topography of the Sierra Nevada using the 9- to 10-million-year-old volcanic flows and the alluvial sediments they buried. Because only fragments of the original flows and sediments have been preserved, the remnants must be correlated to reconstruct the ancient valleys. By estimating the original slopes of the 10-million-year-old rivers, it is possible to determine the amount of uplift and tilting that have occurred since that time. The estimated uplift of the Sierra Nevada since the flows were deposited varies from about 1,830 meters (6,000 ft.) at Tioga Pass to 2,150 meters (7,050 ft.) at Deadman Pass at the head of the San Joaquin River.

THE SIERRAN BATHOLITH: ROOTS OF AN ANCIENT VOLCANIC ARC

Although today's Sierra Nevada mountains are most likely geologically young, the rocks of the Sierra Nevada are much older. Visitors to the High Sierra can see vast expanses of bare rock surfaces, recently scraped clean by glaciers and exposed at high elevations too cold for most vegetation (see Fig. 8-3 and Color Plates 22 and 23). These rocks provide evidence that, during Mesozoic time, an older mountain range must have stood in about the same position as today's mountains. The older mountains and the rocks beneath them were a direct result of the tectonic upheaval that began when the passive margin gave way to subduction along the western margin of North America. Although the Mesozoic mountains have long been eroded away, the rocks of the Sierra Nevada remain as testaments to their existence.

Today we have a firsthand look at the workings of the ancient magma chambers that were the roots of the Mesozoic mountains. Within the space of a 10-minute stroll, we can observe rocks that originally cooled and solidified 10 or more kilometers below the Earth's surface, at temperatures of about 750° to 800°C (1800° to 2000° F). This amazing view has been provided by the tremendous uplift of the Sierra Nevada and the erosion of thousands of meters of sediment and rock from above the crystallized magma. In fact, without the workings of the rock cycle (see Chapter 2) and the dynamics of plate tectonics, plutonic rocks would never be seen at the Earth's surface.

Even a very quick look at the California geologic map (Endpaper 1) reveals that Mesozoic plutonic rocks make up most of the Sierra Nevada. They make up the Sierran ***batholith***, a word combined from the Greek works *bathos* (deep) and *lithos* (rock). The plutonic rocks that make up the Sierran batholith (Fig. 8-8) crystallized from many different batches of magma to form hundreds of recognizable ***plutons*** that geologists have distinguished by their mineral components and ages. The magma emplacement and crystallization were hardly instantaneous, even by geologic standards. The oldest plutonic rocks are about 210 million years old, and the youngest plutons crystallized about 80 million years ago. A single body of magma might crystallize over thousands of years, allowing large crystals to grow slowly to several centimeters in length (Fig. 8-9 and Color Plate 3).

Almost all of the rocks in the batholith are ***granitic*** in their composition; that is, relatively high in silica. Because of the high silica content of the magmas from which they crystallized, granitic (or ***granitoid***) rocks contain abundant crystals of light-colored quartz and feldspar, minerals rich in silica. Set within the light minerals are randomly distributed dark silicate minerals, most commonly hornblende and biotite mica, resulting in a salt-and-pepper look—in more silicic rocks, mostly "salt," but in intermediate and mafic rocks, a more even mixture of light and dark (Figs. 8-10 and 8-11).

Because they crystallized from magmas of somewhat different compositions, granitic rocks contain variable proportions of silicate minerals. Depending on the relative amounts of quartz and the two feldspar minerals, the most common granitic rocks of the Sierra Nevada are classified as granite, granodiorite, or tonalite; granodiorite is the most abundant (Fig. 8-12). Plutonic rocks such as diorite, with more dark minerals and little or no quartz, are also present in the Sierra Nevada, but are much less abundant than granitic rocks.

Many decades of study in the Sierran batholith have led geologists to abandon the idea that it is a simple and homogenous body of granitic rock. Today, geologists continue to recognize distinct plutons of variable composition, structure, and age. The differences seen within the rocks of the batholith reflect the complexities of magma generation. Important factors that influence the composition and texture of rocks within the batholith include the origins of the magmas, the changing composition of the magmas as they partially crystallized, and the variable depths of solidification. Geologists refer to the depth at which magma solidifies within the surrounding rocks as the ***emplacement*** depth of a pluton. Plutons can be emplaced at relatively shallow depths in the crust, less than 10 kilometers from the surface, or they could solidify in the lower continental crust, at depths of up to 50 kilometers. By examining the plutonic rocks of the batholith within the period of greatest magma generation, about 100 million years ago, it is possible to "see" several levels of the batholith exposed in different parts of the Sierra Nevada today (Fig. 8-13).

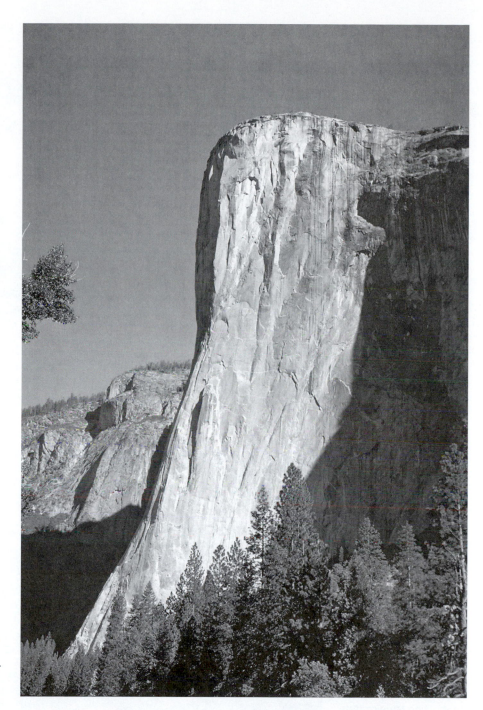

Fig. 8-8 El Capitan, Yosemite Valley, one of many famous landmarks composed of granitic rock. (Source: Sarna-Wojcicki, A. U.S. Geological Survey.)

Fig. 8-9 Close-up view of the Cathedral Peak granodiorite near Saddlebag Lake, showing large rectangular feldspar crystals.

Fig. 8-10 Close-up view of Half Dome granodiorite showing large crystals of hornblende. (Source: Huber, N.K. U.S. Geological Survey.)

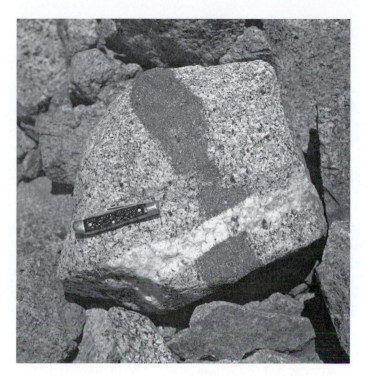

Fig. 8-11 Three distinct episodes of magma intrusion can be seen in this single rock along Highway 120 in Yosemite National Park. Following the principles of cross-cutting relations outlined in Chapter 3, the light-colored quartz vein was the last to form, because it cuts through both of the other plutonic rocks. (Source: Sarna-Wojcicki, A. U.S. Geological Survey.)

Many lines of geologic evidence indicate that the Cretaceous plutons in the central and northern Sierra Nevada, including those in Yosemite National Park, as well as those in the Peninsular Ranges Province (see Chapter 17), appear to have been emplaced in the crust at depths of less than 10 kilometers. One key used to infer the emplacement depth of plutonic rocks is the metamorphic minerals created in the surrounding rock by the magma's intrusion. The temperature and pressure conditions for the formation of some of these minerals are so restricted that geologists use these minerals as **geobarometers** and **geothermometers**. The presence of one or more of these metamorphic minerals—for example, andalusite—can be a reliable indicator of the depth at which magmas were emplaced. On a larger scale, the type of deformation of the rocks can indicate whether they were brittle or plastic when the magma was emplaced, as can the sharpness of contacts between plutons and the shapes of the plutons themselves. These physical properties also provide evidence of the depth of magma emplacement.

Cretaceous plutons in the northern and central Sierra Nevada are easily separable; that is, contacts between different bodies of magma are sharp. Geologists are able to distinguish multiple plutons, indicating that discrete batches of magma were injected from below. Because most of the plutonic rocks are rich in silica, geologists infer that magmas rose through a thick section of silica-rich continental crust, partially melting it, mixing with the melt, and thus enriching the magma in silica. Many of the intruding magmas ballooned outward as they rose within the crust, because the pressure of the overlying rocks decreases at higher levels. At the margins of the plutons, heat caused plastic deformation of the surrounding rocks. Evidence of this deformation can be seen in the rocks surrounding the Tuolumne series in Yosemite Valley.

Fig. 8-12 Proportional content of quartz and the two feldspars used to classify common plutonic rocks. Most Sierran plutonic rocks fall within the shaded area. (Source: Huber, N.K., and Wahrhaftig, C. 1987. *The Geologic Story of Yosemite National Park*. U.S. Geological Survey, Bulletin 1595.)

West   **East**

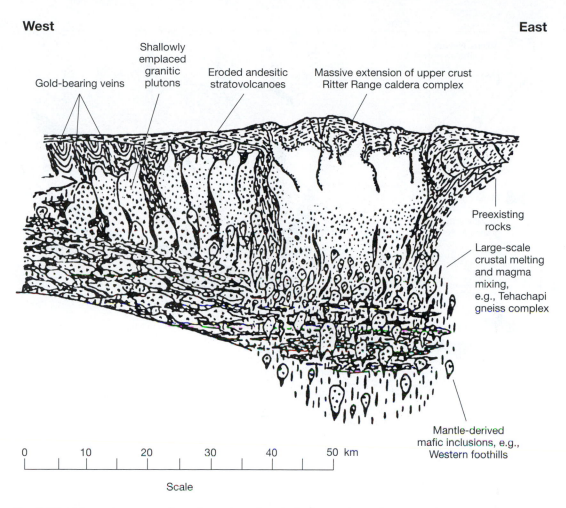

Gold-bearing veins

Shallowly emplaced granitic plutons

Eroded andesitic stratovolcanoes

Massive extension of upper crust Ritter Range caldera complex

Preexisting rocks

Large-scale crustal melting and magma mixing, e.g., Tehachapi gneiss complex

Mantle-derived mafic inclusions, e.g., Western foothills

0 10 20 30 40 50 km

Scale

Fig. 8-13 Schematic cross section of the plutonic rocks in the southern Sierra Nevada and the different levels of the Mesozoic magma chambers they appear to represent. (Source: Adapted from Saleeby, J. 1990. *Exposed Cross-Sections of the Continental Crust.* Norwell, MA: Kluwer Academic, p. 143.)

The plutonic rocks along the southern "tail" of the Sierra Nevada, where it is abruptly cut off by the Garlock fault, are surprisingly different from those seen to the north. Here, in the Tehachapi and San Emigdio Mountains, is a belt of mafic plutonic rocks and metamorphic rocks that appear to have formed at very high temperatures and pressures (Fig. 8-14). The most abundant rock in the area is ***mafic gneiss***, a banded rock composed mainly of hornblende and plagioclase feldspar, sometimes containing red garnets as large as 10 cm in diameter. Based on its composition, the most likely source rock for the gneiss was material much lower in silica than continental crust. Compared with the central and northern Sierra Nevada, the contacts between plutons are hard to define in this part of the batholith. In many areas, contact zones are gradational and the metamorphic rocks are ***ultrametamorphosed***; that is, they are difficult to distinguish from igneous rocks.

Fig. 8-14 Simplified geologic map of the southernmost Sierra Nevada. Mafic plutonic rocks and gneiss are found in dark-shaded areas, and granitic rocks in the patterned areas. (Source: Ross, D. 1992. *Geology* 13[4]:289.)

All of the characteristics of the southern Sierra Nevada batholith indicate that the plutonic rocks crystallized at a much greater depth than those seen further north. The mafic gneisses and mafic plutons in the southwestern Sierran foothills indicate that magma was emplaced in the middle crust at a depth of about 30 kilometers. Here, then, are the "roots" of the batholith (see Fig. 8-13); that is, these rocks have been brought to the surface from the deepest levels of magma emplacement. Further north, small chunks of unmelted rock similar to the mafic rocks of the southern Sierra, known as ***enclaves*** or ***inclusions***, are found in the more typical granitic plutons. It is likely that rocks similar to those seen at the batholith's tail underlie the granitic plutons further north.

About 280 kilometers to the north-northeast of the batholithic roots are the peaks of the Minarets and the Ritter Range (Figs. 8-15 and 8-16). These dark, jagged pinnacles stand in stark contrast to the smooth, rounded granitic slopes for good geologic reason: They are composed of an entirely different rock type. Naturalist John Muir called them black slate during his climbing adventures of 1873. Geologists working in the area today have recognized them as metamorphosed volcanic rocks that were originally erupted about 100 million years ago. They have thus discovered parts of the Mesozoic volcanoes that must have sat above the Sierran batholith in the

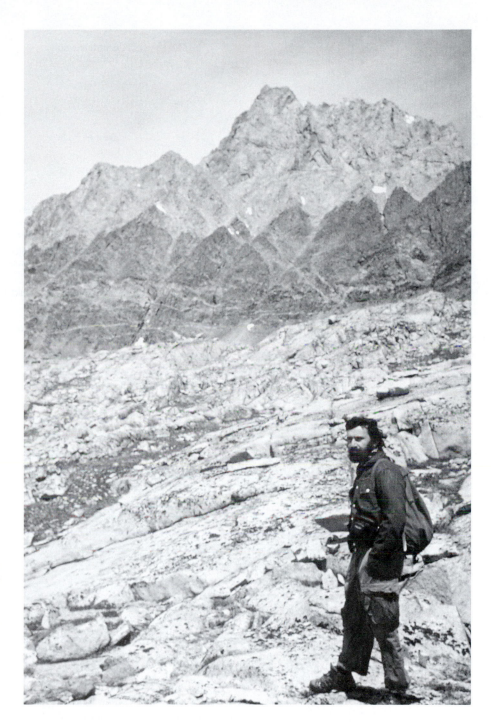

Fig. 8-15 Granitic and metavolcanic rocks near Mt. Humphreys, Sierra Nevada. The granitic rocks in the foreground are lighter colored and smoother, contrasting with the dark, jagged metavolcanic rocks behind them. (Source: Cooper, A.)

Fig. 8-16   Map (top) and cross section (bottom) showing the extent of the 100-million-year-old Minarets Caldera, reconstructed from the metavolcanic and plutonic rocks found in the area today. (Source: Fiske, R., and Tobisch, O. 1994. Geological Society of America Bulletin, 106[5].)

same fashion as the Andes Mountains today lie above a huge batholith (see Fig. 8-13). In the area near the Minarets, a large caldera covered about the same-sized area as the more modern Long Valley Caldera (see Chapter 5). Pyroclastic flows, unusual deposits formed by the collapse of huge chunks of the caldera walls, and even part of the ancient caldera rim have been recognized by careful mapping and study of the composition of these rocks (see Fig. 8-16). The recognition of the Minarets Caldera has come about only recently because it was necessary to "see through" all of the later metamorphism, intrusion of magma, and faulting that have affected the rocks in the Sierra Nevada.

ORIGIN OF THE BATHOLITH

Geologists agree that batholiths throughout the world are formed at plate margins where oceanic and continental plates converge at subduction zone boundaries. Today the Andean volcanic arc is the surface expression of an immense quantity of magma being emplaced in the South American Plate as the Nazca Plate descends beneath it to the west (see Figs. 1-1 and 1-7). During Mesozoic time, the Sierran batholith was presumably overlain by an arc of active volcanoes similar to today's Andes Mountains (Fig. 8-17). In fact, volcanic ashes found within Cretaceous sedimentary rocks in Colorado are believed to record eruptions from those now-gone volcanoes.

The generation of batholiths has long been a mystery to geologists. It is currently thought that the formation of granitic batholiths is fueled by water. Water trapped in the subducting oceanic crust is driven off as the plate descends. This water "leaks" into the mantle above it and lowers the melting point of the surrounding rocks, causing them to partially melt. Laboratory experiments have demonstrated that water effectively lowers the melting temperature of many silicate minerals. As heated basaltic magma rises from the mantle into the lower crust, some of it mixes with large amounts of silica-rich magma, generating the batholith-forming granitic magma. Some magma rises to the Earth's surface with less mixing, producing the andesitic lavas of volcanoes like the Andes or the Cascades.

One unresolved problem of batholiths is a question of space: How can such a huge volume of magma be emplaced into the crust, and what happens to the materials that were already there? One obvious way to make room for the batholith is to incorporate crustal materials into the magma, but this does not in itself create enough room. Other mechanisms include the uplift and erosion of materials above the batholith and the removal of material by volcanic eruptions. Another mechanism, increasingly the favorite of some geologists, is the creation of space by extension of the crust in a direction perpendicular to the axis of the batholith. Extension or transform faulting would favor the continued upwelling of magma into the extended crust.

PLATE MOTIONS AND THE BATHOLITH

The history of the Sierran batholith is an important record of the subduction history along the western margin of North America. The locations and ages of the Mesozoic plutons in the Sierra Nevada, as well as those further south in the Mojave Desert Province, in the San Gabriel and San Bernardino Mountains, and in the Peninsular Ranges tell us the following:

- A subduction zone existed west of the North American continent, west of where granitic rocks are found.
- The subduction zone was oriented roughly parallel to today's California coast.
- Subduction was active along the length of California during Mesozoic time.

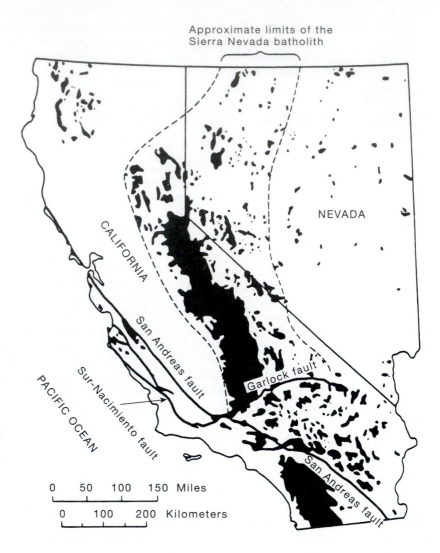

Fig. 8-17 Map showing the extent of the Sierra Nevada batholith. (Source: Norris, R., and Webb, R. 1990. *Geology of California*. New York: Wiley. From Geological Society of America.)

Plutons in the central Sierra Nevada do show some regular patterns of ages and types of rocks. The oldest plutonic rocks, about 210 million years old, are found along the eastern edge of the Sierra Nevada west of Mono Lake. Slightly younger plutons of Jurassic age are found on both the western and eastern edges of the batholith. The most voluminous plutons, accounting for more than 90 percent of the batholith, are those emplaced between about 120 and 80 million years ago during the Cretaceous period. These form the center of the batholith and include the Half Dome granodiorite, the Cathedral Peak granodiorite, and the other famous granitic rocks of the Yosemite area. During this latest phase of activity in Cretaceous time, plutonic activity moved systematically eastward (Fig. 8-18). The emplacement of magma was episodic during this period, so that today we find north-trending groups

Fig. 8-18 Diagram showing the ages of plutonic rocks in the central Sierra Nevada, demonstrating that plutonic activity migrated eastward during Cretaceous time, between crystallization of the Fine Gold and Powell plutons. (Source: Bateman, P. 1983. Geological Society of America, Memoir 159.)

of older Cretaceous plutons in the western Sierra Nevada and younger Cretaceous plutons in the eastern Sierra.

As geologists learn more about the exact timing and location of magma emplacement, they can track the motions of the oceanic plates that collided with North America during Mesozoic time. For example, the fact that the greatest volume of plutonic rock is between 150 and 80 million years old tells us that the subduction ultimately responsible for the magmas was most vigorous at that time. The episodic emplacement of magmas, which can be deduced from the clustering of ages of various plutons, probably reflects episodic periods of more rapid subduction.

Geologists speculate that about 80 million years ago, the rate of subduction of the oceanic plate to the west, known as the Farallon Plate, increased, and the angle of subduction flattened out. As a result of this change in geometry, the zone of magma

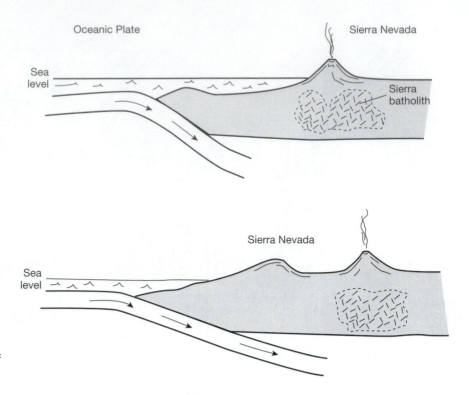

Fig. 8-19 Diagram showing how magmatic activity might shift in response to flattening of a subducting plate.

generation shifted to the east, "shutting off" the Sierran batholith and triggering magmatic activity to the east (Fig. 8-19).

METAMORPHIC ROCKS OF THE SIERRA NEVADA

Although Mesozoic granitic rocks make up most of the Sierra Nevada Province, metamorphic rocks are scattered throughout much of the batholith. In the High Sierra, blocks of metamorphic rocks are called *roof pendants* (Fig. 8-20 and Color Plate 24; see also Fig. 8-22)—where they appear to represent the rocks that were originally the top or roof of the magma chamber—or *septa*—where they form screens or dividers at the edges of plutons. A more extensive zone of metamorphic rocks, known as the western metamorphic belt, occurs along the northwestern foothills of the Sierra Nevada. The rocks here have been of particular interest to geologists because they are the host rocks for California's gold deposits.

The majority of the metamorphic rocks in the Sierra Nevada have been only moderately metamorphosed. Highly metamorphosed rocks such as gneiss (see Chapter 2) are relatively rare and are concentrated in the southern "tail" of the batholith. Many of the metamorphic rocks are *foliated*; that is, they show evidence of metamorphic layering caused by the growth of metamorphic minerals in preferred orientations (see Chapter 2 and Fig. 8-21). Some foliation is inherited from original sedimentary layering and enhanced during the process of metamorphism. In other

Fig. 8-20 Mt. Morrison roof pendant, looking north. (Source: Stevens, C. San José State University.)

Fig. 8-21 Foliated metavolcanic rocks in the western Sierra foothills, south of Folsom. Because the rocks weather along the foliation planes, they form aligned outcrops commonly called "tombstone rocks" for their resemblance to grave markers.

Fig. 8-22   Map showing roof pendants in the Sierra Nevada and the western foothills metamorphic belt. (Source: Adapted from Lahren, M., and Schwiekert, R. 1994. Geological Society of America Bulletin.)

rocks, the foliation crosses the original sedimentary layers. In some rocks, geologists can detect several directions of foliation in a metamorphic rock, suggesting that the rock was subjected to several different episodes of metamorphism. Common foliated rocks found in the Sierra Nevada include slate and schist; these have a wide variety of compositions, reflecting their diverse origins.

The most common nonfoliated metamorphic rocks are hornfels, marble, quartzite, and metachert (see Chapter 2). The presence of certain metamorphic minerals or assemblages of minerals in a metamorphic rock provides geologists with clues about the identity of the premetamorphic rock, referred to as the ***protolith***. As discussed earlier, metamorphic mineral assemblages also act as geobarometers or geothermometers to record the pressure and temperature conditions of metamorphism. Other features, including layering and even fossils, might survive metamorphism and provide additional clues about the nature of the "parents" of slates and schists.

An understanding of the metamorphic rocks of the Sierra Nevada is particularly critical to the interpretation of California's tectonic history during Mesozoic time. Unfortunately, this understanding is difficult to achieve because of the very nature of the metamorphic rocks. It is particularly challenging to identify and correlate the rocks as they were prior to metamorphism, because many of the rocks have been subjected to multiple episodes of deformation. West of the batholith, oceanic rocks were buried in subduction zones, chemically altered by hot fluids during subduction, and subjected to great stress as they were accreted to the North American margin. The arrival of each new accreted terrane generated faulting, folding, and shearing in the adjacent belts of rocks. For more than 130 million years, rocks surrounding the Sierran batholith were subjected to episodes of great heat and pressure as magma was emplaced. At depth, the intrusion of magma squeezed the rocks along the pluton walls, causing them to be folded and sheared. The heat from plutonic intrusions mobilized fluids in the adjacent rocks, causing chemical reactions and concentrating certain elements—including gold—in veins and pockets.

PIECES OF THE PUZZLE: METAMORPHIC ASSEMBLAGES OF THE SIERRA NEVADA

Metamorphic rocks of the Sierra Nevada Province can be grouped into four general categories:

1. Metasedimentary rocks formed on the Paleozoic North American Plate and correlated to sedimentary rocks in the Basin and Range Province (see Chapter 7).
2. Metavolcanic rocks of Mesozoic age erupted on the North American Plate.
3. Metavolcanic and metasedimentary rocks of Paleozoic age representing an ancient volcanic island arc terrane that was accreted to North America.
4. Metavolcanic and metasedimentary rocks, both Paleozoic and Mesozoic, originally formed on oceanic plates and accreted to North America as subduction zone complexes during Mesozoic time.

Fragments of Paleozoic North America

As outlined in Chapter 7, a stable, shallowly sloping marine platform existed in the region of today's Basin and Range Province in Paleozoic time, prior to the development of an active subduction zone. Deeper marine environments would have existed further west of that margin, in the area that is now the location of the Sierra Nevada. It is logical that sedimentary rocks would have accumulated on the deeper parts of the stable carbonate platform during Paleozoic time. With the transformation of the continental margin to a subduction zone in early Mesozoic time, however, the Sierra Nevada region became the host for the Sierran batholith. However, bits of the former stable continental margin have survived as metamorphic roof pendants.

In the eastern part of the High Sierra, roof pendants contain marble, quartzite, slate, and hornfels. The sedimentary predecessors of these rocks—limestone, dolomite, quartz sandstone, and shale—accumulated offshore of the stable western margin of North America. In some localities, fossils of early Paleozoic age have also survived metamorphism. Some of the rocks have been correlated to their unmetamorphosed equivalents further east, in the Inyo and White Mountains, and others are thought to match rocks in east central Nevada. Areas where these remnants are present include, from north to south, the Saddlebag Lake roof pendant (Fig. 8-23), parts of the Mt. Morrison roof pendant near Convict Lake (Fig. 8-24),

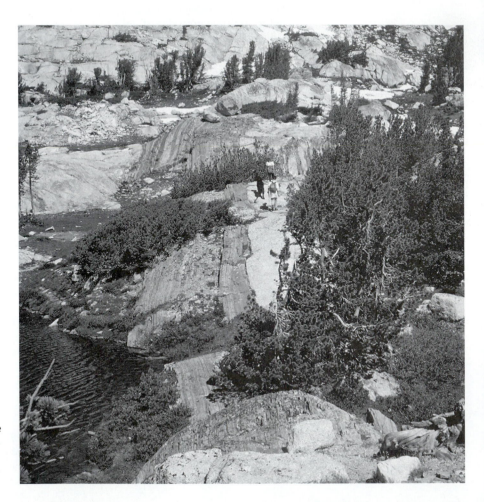

Fig. 8-23
Metasedimentary and metavolcanic rocks of the Saddlebag Lake pendant, north of Saddlebag Lake. Note the size of the dog and child for scale.

Fig. 8-24 Map showing the roof pendants of the Mt. Morrison area. (Source: Adapted from Rinehart, C.D., and Ross, D. 1964. U.S. Geological Survey, Professional Paper 385.)

the Bishop Creek roof pendant southwest of Bishop, and the Big Pine roof pendant east of Kings Canyon National Park and west of the town of Big Pine.

Paleozoic and Mesozoic Volcanic Rocks in the High Sierra

Some of the Sierran roof pendants are hornfels and schist whose protoliths were clearly of volcanic origin. Those that are of the same age as the Sierran batholith presumably represent volcanic flows and ash erupted from the same large magma bodies that formed the batholith. Rocks of the Minarets Caldera discussed earlier are included in this category. Volcanic rocks in the Mount Dana roof pendant might represent an older Mesozoic caldera.

Some volcanic roof pendants in the High Sierra predate the Sierran plutonic rocks. These volcanic rocks were deformed and intruded by younger Mesozoic intrusions of magma. Geologists hypothesize that the volcanic rocks, which are found along a narrow northwest-trending zone in the High Sierra, represent one or more earlier volcanic arcs. One example of prebatholith volcanic rocks is found in the Saddlebag Lake pendant just north of Tioga Pass, where metavolcanic rocks are found with slate and schist that were formerly marine sedimentary rocks (Fig. 8-22). Geologists have speculated that some of the prebatholith volcanic rocks were moved from their late Paleozoic volcanic centers to their present locations along later lateral faults (see Chapter 18).

Accreted Oceanic Terranes of the Sierran Foothills

During Paleozoic time, the western margin of North America stood roughly where the Mojave Desert and Basin and Range Provinces of California are today. As outlined in Chapter 7, this was a time of tectonic quiet for most of California; in fact, much of what is now California did not exist. When the active convergent plate boundary developed along western North America, the relative quiet ended. As discussed earlier in this chapter, the Sierran batholith is one striking proof of tectonic upheaval during Mesozoic time. The metamorphic rocks of the northwestern Sierra Nevada (Fig. 8-22) provide additional evidence of plate collisions and terrane accretion during Mesozoic time. The metamorphic rocks of the western foothills also document earlier collisions between the North American continent and oceanic terranes that lay to the west.

On the California geologic map (Endpaper 1), the western metamorphic belt of the Sierra Nevada appears as two extensive bands of metamorphic rock on the northwestern side of the Sierra Nevada province. On the map, Mesozoic and Paleozoic units are separated by an extensive fault system, and the faults are further accentuated by the large slices of serpentine. Although the geology of the western foothills appears relatively simple on the state map, each map unit is actually very complex. Geologists now believe that each unit contains several individual slivers of oceanic crust that were accreted to the western margin of North America during subduction (Fig. 8-25). It appears that some slivers are terranes that formed near the continental margin, whereas others formed at greater distances from North America and are more "exotic" to the Sierran region. Because all of the rocks have been metamorphosed, faulted, and folded, they have been extremely difficult to correlate and identify as separate crustal fragments. By about 140 million years ago, in the mid-Jurassic period, all of the oceanic terranes in the western foothills had been accreted, and the subduction zone moved west of the modern Great Valley.

Shoo Fly Complex.   The oldest rocks in the northern Sierra Nevada, which are as much as 500 million years old, make up the Shoo Fly Complex. The Shoo Fly Complex, named in the early 1900s for a small settlement north of Quincy, is found in the northwestern Sierra Nevada. It contains a variety of metamorphic rocks that are of oceanic origin, including both metasedimentary (Fig. 8-26) and metavolcanic rocks. Geologists have subdivided the Shoo Fly Complex into several units, and some geologists believe that the units represent different oceanic terranes. About 375 million years ago, during late Devonian time, a period of uplift and erosion is marked by an unconformity between the Shoo Fly Complex and all younger Paleozoic rocks. During this period of deformation, rocks of the Shoo Fly Complex were faulted and folded.

Fig. 8-25 Map showing the accreted terranes of the western metamorphic belt, as interpreted from studies of the metamorphic rocks. (Source: Adapted from Fagan, T., and others. 1995. *Northern Sierra Nevada Field Trip: Tectonic Belts of the Sierra Nevada.* Northern California Geological Society.)

Fig. 8-26 A rare block of dolomite marble in the Shoo Fly complex, along the North Fork of the American River. (Source: Harwood, D. U.S. Geological Survey.)

Northern Sierra Terrane: An Accreted Volcanic Arc.

East of the Shoo Fly Complex in the northern Sierra Nevada is a younger belt of Paleozoic rocks termed the Northern Sierra Terrane (Fig. 8-25). Rocks in this belt include chert and black shale, both containing radiolaria, formed in deep marine waters during the Mississipian and Permian periods. During Permian time, about 250 million years ago, large volumes of andesite erupted in the vicinity, as evidenced by the presence of Permian metavolcanic rocks associated with the sedimentary formations. The source of the eruptions was a volcanic island arc. Associated with the volcanic rocks are limestones that once formed reefs that fringed the islands. Some of the metamorphosed limestone contains fossils that are distinct and unique to the ancient reefs of the Klamath Mountains and northern Sierra Nevada (see Chapter 9). In fact, the geologic history of the northern Sierra Nevada during Paleozoic time appears to match well with that of the eastern Klamath Mountains. Geologists believe that the two areas were continuous and were part of a series of island arcs that lay west of the North American continental margin during Paleozoic time (see Chapter 9).

Late Paleozoic and Mesozoic Subduction Complexes.

West of the rocks of the northern Sierra terrane, and separated by a major fault system, are rocks that geologists interpret as belonging to an ancient subduction zone. Within this zone are complex assemblages of rocks representing many elements of a ***subduction complex***: submarine volcanic rocks erupted at spreading centers; sedimentary rocks deposited in deep-ocean trenches; and metamorphic and plutonic rocks formed at great depths

CAVES OF THE SIERRA NEVADA

Although caves are not the most famous of the Sierra Nevada's tourist attractions, visitors can tour underground caverns at several parks in the Sierra Nevada Province. These include Mercer and Moaning Caverns along Highway 4, Crystal Cave and Liburn Cave in Sequoia National Park, and Boyden Cave west of Kings Canyon National Park. Like most of the world's caverns (and unlike the lava tubes discussed in Chapter 5), the caves of the Sierra Nevada have formed by dissolution of carbonate rock due to water flowing along fractures. Over thousands of years,

water gradually dissolved the mineral calcite, opening cave passages and rooms along fractures. Within the caves, stalagmites and stalactites are created where calcite is reprecipitated from solution (Fig. 8-27).

All of the caverns in the Sierra Nevada Province have formed in the metamorphic rocks of the accreted terranes, in the metasedimentary blocks composed of marble. Mercer and Moaning Caverns lie within the Calaveras Complex, whereas the caverns of the southern Sierra Nevada are formed in rocks of the Kings Terrane.

Fig. 8-27 Solution cave in metamorphic rock of the western Sierra foothills, Calaveras County. (Source: Tinsley, J. U.S. Geological Survey.)

within and beneath the crust. This mixture of rocks forms a discontinuous belt stretching into British Columbia and including the Central metamorphic terrane of the Klamath Mountains (see Chapter 9). In the northern Sierran foothills, rocks within this belt have been named the Calaveras Complex. The types of rocks found in a subduction complex are discussed in more detail in Chapter 9 and Chapter 12.

The westernmost belt of metamorphic rocks in the Sierra Nevada is referred to as the Foothills Terrane (Fig. 8-25). Rocks found here are of middle to late Jurassic age, about 160 million years old, thus representing the youngest period of subduction and accretion in the Sierra Nevada Province. Along the central part of the Sierra foothills, the Foothills Terrane is separated from the rocks of the Calaveras Complex by the Melones fault, the easternmost branch of the Foothills fault system. The fault system runs along the boundary between blue and green units of the Sierra foothills (Endpaper 1).

GOLD!

In January 1848, James Marshall was attempting to correct a defect in his sawmill. He had convinced his boss, Captain John Sutter, to build a sawmill on the American River at Coloma and float the much-needed lumber downstream to Sutter's Fort in Sacramento. The tailrace—the stretch of channel downstream from the mill—had not been dug deep enough, and water was backing up to the wheel and preventing it from turning. Marshall blasted a deeper channel, hoping that additional sediment would be removed by the stream. On January 24, Marshall closed the gate that stopped the water flowing through the tailrace and walked down it to see whether enough sand and gravel had been removed.

Having strolled to the lower end of the race, he stood for a moment examining the mass of debris that had been washed down, and his eye caught the glitter of something that lay, lodged in a crevice, on a riffle of soft granite, some six inches under the water. He picked up the substance. It was heavy, of a peculiar color, and unlike anything he had seen in the stream before. For a few minutes he stood with it in his hand, endeavoring to recall all that he had heard or read concerning the various minerals. The weight assured him that it could not be mica. Could it be sulphurets of copper? He remembered that the mineral is brittle; he turned about, placed the specimen under a flat stone and proceeded to test it by striking it with another. The substance did not crack or flake off; it simply bent under the blow.

Marshall returned to the mill, his usually crusty face beaming, and cried:

"Boys, by God, I believe I have found a gold mine!"

He showed his nuggets as proof. His workers were not impressed. They continued about their tasks. On the morning of the fifth day [Marshall had collected enough yellow nuggets to make three ounces. He tied them into his kerchief and] started out on horseback to cover the fifty mountain miles down to the fort, sleeping that night under an oak tree. He was ostensibly searching for a wagonload of supplies which Sutter had promised.

At nine o'clock on the morning of January 28, 1848, sopping wet from a cloudburst which had enveloped him in the last eight miles, Marshall arrived at the fort. He asked Sutter where they could talk with privacy. Puzzled, Sutter took him to his bedroom-sitting room in the main building, locking the door. Marshall asked for two bowls of water, a stick of redwood and some twine and sheet copper to make a scale. Sutter told him that he had scales in the apothecary shop and went for them himself, failing to lock the bedroom door when he returned. A clerk walked in with some papers just as Marshall was about to dump the yellow nuggets onto the table. Marshall cried in consternation:

"There! Didn't I tell you we had listeners!"

Sutter quieted his overwrought partner, then gazed down at the yellow nuggets which James Marshall poured out of his kerchief onto the table. Sutter examined the specimens, pulled down a volume of his *Encyclopedia Americana*, studied it for a time, tested the nuggets with *aquafortis**, which had no effect on them, balanced them on the scales with a like amount of silver, then dipped the scales into water, the yellow nuggets quickly outweighing the silver.

Sutter turned his now wide and flashing eyes up into the face of the wildly excited Marshall.

"It's gold," Sutter said. "At least twenty-three-carat gold!"

Captain Folsom sent a flake of gold described as Marshall's first piece to Washington in 1848. This piece is now preserved in the Smithsonian Institution.

With this event, a few bits of stream sediment changed the face of California and the history of the entire world. The gold rush was slow at first, because it took months for word of Marshall's discovery to travel. The first to arrive in the summer of 1848 were Hawaiians, followed by miners from Mexico, Peru, and Chile by the end of the year. By the end of 1848, California's non–Native American population was 20,000, including 5,000 miners. Waves of Americans from the eastern states, South America, and Europe followed. At the end of 1849, the population had grown to 100,000; by 1852, 224,000 non–Native Americans lived in California, and 100,000 were miners. At that time, 24 percent of California's population was foreign-born. Most Americans knew little about mining, but among the immigrants were experienced miners with origins in Mexico, Peru, Chile, Spain, and Cornwall in England. Between 1851 and 1855, the United States accounted for 45 percent of the world's gold output.

---

* Nitric acid; "strong water" in Latin.

Source: Excerpted from Stone, I. 1956. *Men to Match My Mountains*. Garden City, NY: Doubleday & Co, pp. 105–106; with quoted materials from Marshall's biographer George F. Parsons and Sutter's biographer James P. Zollinger.

Volcanic rocks erupted from submarine spreading centers, and now metamorphosed, form a major component of this unit. Unweathered metavolcanic rocks with high iron content are commonly green, and in some areas are called **greenstone**. Metamorphosed marine sandstone and shale are also found, metamorphosed to slate and schist within the Foothills Terrane. One of the more extensive formations of this unit is the Mariposa Formation. Gabbro, formed at greater depths within the oceanic plate, and ultramafic rocks, thought to have formed in the mantle, are also present in the Foothills Terrane.

One of the most unusual and interesting metamorphic rocks in the western Sierran foothills is mariposite schist, a banded green and white rock containing the bright apple-green mineral **mariposite**. Mariposite, named for Mariposa County, where it was first discovered in 1868, is a form of mica that gets its green color from minor amounts of chromium. Mariposite forms by the alteration of serpentine by hydrothermal fluids, and it is most common along the Melones fault zone, where it is associated with massive white quartz veins. Mariposite schist can also contain gold-bearing veins. Mariposite schist has long been valued as a decorative stone. A huge boulder from Placerville was selected in 1987 to represent California in the Constitution Monument built in Philadelphia for the 200th anniversary of the Constitution and the Bill of Rights.

FORMATION OF SIERRAN GOLD DEPOSITS

Most of the gold in the Sierra Nevada foothills is pure or "native" gold, rather than gold combined with other elements to form gold-bearing minerals. The gold forms in veins of quartz 30 centimeters to 30 meters thick (Fig. 8-28). Other minerals, including calcite and pyrite, or "fool's gold," occur in some veins. Rare gold-bearing minerals such as **calaverite**, or gold **telluride** ($AuTe_2$), are also present in some veins. Calaverite was named for Calaveras County, where it was first found. Within some veins there are rich pockets of gold known as **ore shoots**.

The crystallization of magma rich in hydrothermal fluids and silica is one of the late stages of the formation of granitic plutonic bodies. In the Sierra Nevada batholith, quartz veins were injected along shear zones and faults, which provided pathways for the migration of hydrothermal fluids, because rocks are typically weaker there. Metal deposits of this type are referred to as **lode deposits**. In some areas, particularly in the southern part of the Sierran foothills, more diffuse areas of metamorphic rock are also enriched with gold. Although most of the gold has been mined from quartz veins in metamorphic rocks, veins also occur in granitic rocks.

A continuous zone of gold-bearing quartz veins and rock enriched with gold, known as the **Mother Lode**, runs for 195 kilometers along the Melones fault. A trip along California's historic Highway 49 through the Gold Rush towns and mines is also a trip along the boundary (sometimes called a suture) between accreted oceanic terranes. Hundreds of mines were excavated to follow the quartz veins within this 1.5-kilometer-wide belt. Because most of the veins are dipping, the lode deposits could extend to thousands of meters below the surface.

A

Fig. 8-28 **A,** Gold-bearing quartz vein in metamorphic rocks of the western Sierra Nevada. **B,** An early gold miner. (Source: **A,** Photo by author; **B,** California Department of Conservation, Division of Mines and Geology.)

B

EMPIRE MINE STATE PARK

The Grass Valley mining district in western Nevada County was the richest and most famous of California's gold mining districts. At least $300 million in gold was produced from its lode mines between 1852 and 1956. The Grass Valley district lies north of the Mother Lode, but along the same general trend.

Quartz veins there are found in granodiorite, rather than in metamorphic rocks. The Empire Mine alone, the richest producer of all California mines, produced $130 million in gold (Fig. 8-29). Today, visitors can see many of its original above-ground workings at Empire Mine State Park.

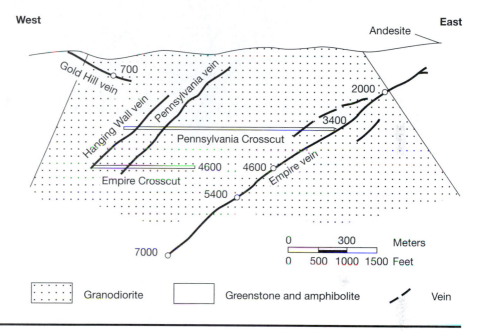

Fig. 8-29 Cross section of the Empire and Pennsylvania mines, Nevada County. The numbers show the mine levels, in feet, below the ground surface. (Source: Clark, W. 1984. Gold Mines of Grass Valley, *California Geology*, p. 49. Used with permission of the California Department of Conservation, Division of Mines and Geology.)

Although the initial finds of gold in California were nuggets in stream gravels, it did not take the early miners long to determine that gold-bearing quartz veins were the source of the stream sediments and that the veins were most common in the metamorphic rocks of the western foothills (Fig. 8-30). Geologists later made the association between lode deposits and submarine volcanic rocks from accreted terranes intruded by younger granitic rocks.

Geologists believe that the Sierran gold was originally deposited in the rocks of the accreted oceanic terranes. The migration of gold-bearing fluids into the rocks probably occurred near submarine volcanic vents and resulted in the formation of mineralized zones similar to those seen at modern submarine spreading centers (called "black smokers"). After the rocks had been accreted to North America, the heat generated during the intrusion of the Sierra Nevada batholith mobilized the gold and caused it to be concentrated in the veins where it is found today. The gold-bearing veins cut across Mesozoic metamorphic rocks, as well as some of the plutonic rocks, so the gold mineralization must postdate those rocks. By dating rocks that predate and

Fig. 8-30 Major gold deposits of the Sierra Nevada foothills. (Source: 1966. *Geology of Northern California,* Bulletin 190, p. 211. Used with permission of the California Department of Conservation, Division of Mines and Geology.)

postdate the gold-bearing veins, geologists estimate that the gold veins were emplaced between 115 and 120 million years ago.

Placer Gold

The gold discovered by James Marshall consisted of nuggets of pure gold that were part of the alluvium of the American River. Like the other rocks and minerals in the streambed, the gold did not form there. It was eroded from sources higher in the mountains, transported by the stream, and deposited with the other sediments. Here we have another example of the rock cycle in action (see Chapter 2). Because it is durable and chemically stable, gold resists chemical weathering. It is easily worn into fine particles because it is soft, but gold persists in sedimentary environments like stream channels and beaches, where less resistant materials do not survive. Sedimentary ore deposits formed in this way are known as *placer deposits*, from the Spanish word for sand bank. Other examples of valuable placer deposits include diamond placers in South America and beach placers of titanium oxide mined in various localities. Placer deposits of gold or other valuable ores can survive in sediments and sedimentary rocks long after the stream or beach where they accumulated has disappeared from the landscape.

Placer deposits accounted for more than 40 percent of California's total gold output. These include recent deposits found in and near the streams flowing from the Sierran crest to the Sacramento and San Joaquin Valleys. Sedimentary deposits of ancient, late Cenozoic rivers have also yielded rich placer gold in the northwestern Sierra. The ancient placers are the deposits of rivers that flowed westward from a previous Sierra Nevada range during Tertiary time (about 25 million years ago). These ancient deposits, the *Tertiary gravels*, were first studied by J. Whitney in the 1880s and later by Waldemar Lindgren, who reconstructed the paths of the ancient rivers (Fig. 8-31).

Placer Mining

For countless centuries the streams had been dropping flakes and chunks of the precious mineral onto sand bars, into crevices in the banks, and into "pot holes" in the beds of the rivers. Every stone along a river's course had provided an obstacle behind which fragments might lodge.

Water had been the primary agent in storing up this accumulation of wealth, and water was to be the miner's chief weapon in his attack upon it. Sometimes the miner might be fortunate enough to find crevices richly piled with comparatively pure gold dust that could be scraped out with a knife and spoon, but in most cases he had to dig down into promising-looking ground with a pickaxe and shovel. After he had thrown up mounds of "dirt," he found himself facing the problem of separating the yellow grains from the gravel and sand with which they were intermixed.*

During the early gold rush years, miners devised a number of methods for sorting gold flakes from the river sediments. All methods relied on the simple fact that gold is much denser than the other components of river alluvium—mainly silicate minerals. Because of its greater density, gold settles out more quickly from suspension in water; this is one reason for its concentration in placers. The simplest way to sort gold from other sediments was by *panning*, a method still

---

*From Rodman, P.W. 1947. *California Gold: The Beginning of Mining in the Far West.* Lincoln: University of Nebraska, p. 50.

Fig. 8-31 Map showing the ancient rivers that deposited the Tertiary gravels. (Source: 1970. *Gold Districts of California*, Bulletin 193, p. 18. After Lindgren, 1911, and Jenkins, 1935. Used with permission of the California Department of Conservation, Division of Mines and Geology.)

popular with hobbyists today. California miners used either a flat-bottomed pan made of iron or tin or a wooden bowl called a batea. The pan was filled with sediment and water, and the larger gravel was picked out. Then the miner lowered the pan into a stream or pool at a slight tilt and swirled the sediment. Lighter materials were carried away from the heavier grains or "pay dirt," which often contained heavy minerals such as magnetite and pyrite (fool's gold), as well as the gold flakes or nuggets. If the miner saw a "flash in the pan," he was hopeful, but at times the sediment didn't "pan out."

Panning is a simple and inexpensive method of placer mining, but it is a low-volume operation. Miners next developed methods of sorting larger volumes of sediment. The *rocker* or *cradle*, in use as early as 1848, was a long box mounted on rockers like a baby's cradle. Positioned on a slope with its downhill end open, the cradle was fed at the top end by a hopper with a sieve on its base. Water was continuously poured into the hopper to create a current, and the gold tended to drop behind in the rocker box, where it was trapped by cleats. The operation of the cradle enabled three or more men to work together to wash a larger volume of sediment. Many of California's richest and most accessible placer deposits were mined using the pan and the cradle.

In late 1849 miners began to move vast quantities of sediment, diverting rivers to reach the gold that lay within the streambeds. Large mining associations formed to handle the cost of these operations, and new methods for moving vast tonnages of sediments came about. A 12-foot-long trough called a *long tom* operated in the same fashion as a cradle. Miners improved the design by adding a chain of riffle boxes at the lower end of the long tom to catch more of the gold with each washing. The end result of these modifications was the *sluice box*, an open wooden trough with a series of riffle boxes, fitted with cleats to catch the heavy sediment. By 1850, groups of miners were constructing ditches and flumes to move enough water to run their sluices (Fig. 8-32). These water diversions had lasting effects on California's water laws (see Chapter 10).

One of the problems with gold washing is that the process does not completely separate the gold from other heavy minerals. Miners from South America purified their washings by first drying the mixture and then tossing it into the air or blowing on it. The process winnowed the heavier gold from the other grains. However, this was an inefficient if nonpolluting process, and it was soon replaced by the use of mercury. When gold-rich sediment is mixed with mercury, the mercury binds the gold into an amalgam. Subsequent heating of the mercury vaporizes it, leaving behind pure gold. This technique, which had been practiced by medieval alchemists, was used by late 1849 and is still used today. The demand for mercury fueled the development of mercury mines in the Coast Ranges (see Chapter 12).

Fig. 8-32 An early drawing of California placer miners. (Source: Paul, R. 1947. *California Gold*. Lincoln: University of Nebraska Press.)

HOW THE CALIFORNIA MINES ARE WORKED.

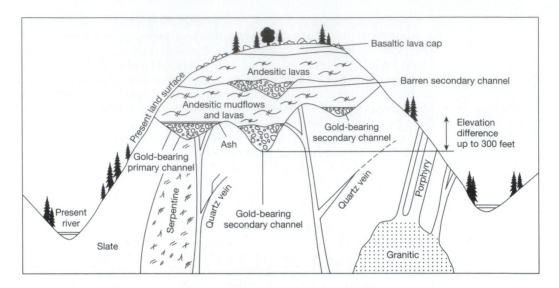

Fig. 8-33 Location of Tertiary placer deposits. (Source: Chakarun, J. *California Geology,* 1988.)

By the early 1850s, much of the easily mined placer gold had been extracted from the riverbeds of the western foothills. The entire beds of most stream channels had been overturned, some more than once, and the rivers themselves had been diverted and dammed to get to the placer gold. Even today, piles of sediment persist throughout the foothills, looking like the diggings of countless large gophers.

Miners next turned their attention to older placer deposits, those formed by streams that flowed westward from an older Sierra Nevada mountain range (Fig. 8-31). Cornish miners had been the first to look for older, buried gravel deposits, which have long been referred to as the Tertiary gravels by geologists. Because of uplift of the Sierra Nevada, Tertiary gravels are now located above the modern river channels (Fig. 8-33). Consequently, it was necessary for the miners to find a way of washing these gravels to get to the gold that was concentrated near the bottom layers of sediment. In some areas, miners tunneled under the Table Mountain volcanic flows to reach the ancient gravels beneath them. One example of successful tunneling was the Omega Mine near Jamestown.

In 1853, ingenious miners came up with the idea of attaching a nozzle to a hose and forcefully washing the sediments down from the hillsides. This marked the birth of **hydraulic mining**, a practice responsible for changing the landscape from the Sierran foothills to San Francisco Bay. Instead of being able to wash 1 cubic yard of sediment in a day, or 4 cubic yards with a sluice, a miner could wash 50 to 100 cubic yards per day using a hydraulic system together with a sluice. Soon so much sediment had washed from the hillsides into the streams of the Sacramento Valley and into the San Francisco–San Joaquin delta that navigation and agriculture were affected. The Sawyer Decision of 1884 prohibited the dumping of mining-related sediment into the Sacramento and San Joaquin Rivers and their tributaries. Sediment from hydraulic

Fig. 8-34 A stereogram pair of dredge tailings along the lower Yuba River near Marysville in 1932. To view this figure three-dimensionally, use a pocket stereoscope. (Source: U.S. Department of Agriculture.)

mining in the late 1800s can still be identified in the eastern part of San Francisco Bay more than a century later (see Chapter 12).

During the 20th century, sediments have been extracted from riverbeds by *dredging*. Floating suction dredges are able to remove sediment for gold washing while the rivers are flowing, thus eliminating the need for diversion. Dredging operations were active in some Sacramento River tributaries until the early 1960s. Throughout the western foothills, great piles of sediment known as *dredge fields* can still be found (Fig. 8-34). Visitors can see an excellent example of the hydraulic mine and dredge fields at Malakoff Diggins State Historic Park in North Bloomfield (Fig. 8-35).

Fig. 8-35 The hydraulic pits at Malakoff Diggins State Park, North Bloom-field, Sierra County, more than a century after hy-draulic mining ceased. (Source: Photo by author.)

Mining the Lode Deposits

California's lode deposits are much less accessible than most of the placer deposits because they are exposed only where quartz veins come to the surface. Underground mining presented many more challenges than placer mining, so the lode deposits were not extensively mined until 1851, after the easy pickings offered by the rivers had dwindled. Early miners dug shafts and tunnels to follow the quartz veins, developing new technology for underground extraction as they went (Fig. 8-36). The problem of separating the gold from the vein quartz was handled by the ***stamp mill***, a device that pulverized the quartz with a heavy iron pounder. After crushing, the gold was sorted using mercury and a variety of sluices, in a manner similar to that used in placer mining. As placer mining declined, lode mining continued as new and deeper veins were discovered. Because of the high cost of underground mining, companies were formed and stock sold to finance the cost of operations.

Underground mining continued in the Sierran foothills until the 1940s. The Jackson-Plymouth mining district in western Amador County opened its first lode mines in 1850 and continued to produce gold until 1942. One of those mines was the Argonaut, a mile-deep mine with miles of tunnels. Two of the hazards of underground mining—poor ventilation and wooden timbers—combined with deadly results in 1922, when a disastrous fire at the Argonaut killed 47 miners. The fire began at the 3,350-foot level of the mine, trapping an entire shift of men who were working at levels beneath the fire.

Fig. 8-36 Typical features of an underground mine. (Source: Peters, W.C. 1978. *Exploration and Mining Geology*. NY: Wiley.)

Gold Production Since 1848

Two famous gold finds in California were the 195-pound mass found at Carson Hill in 1854, and a 54-pound, pure gold nugget found at Magalia in 1854, which was melted down shortly after its discovery. Unfortunately, masses of this size are exceptional. Most Sierran gold has been mined from deposits containing between 1/4 and 1/2 ounce of gold per ton of *ore*. An ore is any accumulation of a particular element, including metals like gold, silver, or mercury, that has been sufficiently concentrated to be profitably mined. It is easy to see from this definition that economics, accessibility, and technology define ores as much as geologic circumstances. What was discarded as waste rock during one period might be re mined profitably in later years if the price of the commodity rises or if new mining methods are developed. California's gold production has fluctuated with changes in all of these conditions (Fig. 8-37). Mining ventures and investments in precious metals like gold are usually financially risky, partly because of the complexities of market conditions, and partly because of the geologic uncertainties associated with ore deposits.

In the 1980s, a rise in the price of gold, combined with a new method of extracting it, produced a new wave of mining at abandoned sites in the Sierra Nevada. Today, mining companies are using a chemical process called ***heap leaching*** to extract gold from deposits that have not been profitable in the past. Heap leaching, which uses cyanide to dissolve gold from the quartz veins, increased gold recovery by 100 times over other methods of mining lode deposits. It is currently being used at several mines in California and Nevada. In 2000 and 2001, 16 gold mines were active in California.

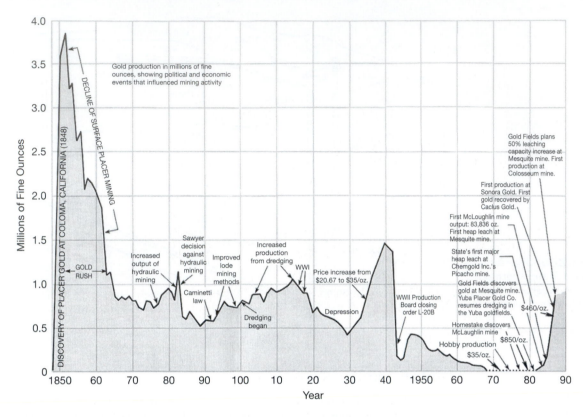

Fig. 8-37   Diagram showing California's gold production between 1848 and 1988. (Source: Fuller, W. California gold production 1848-1988. *California Geology* 41:180-181. Used with permission of the California Department of Conservation, Division of Mines and Geology.)

CENOZOIC VOLCANIC ROCKS OF THE SIERRA NEVADA

During middle Cenozoic time, between about 28 and 22 million years ago, huge ash flows erupted from the Great Basin, from what is now Nevada. These volcanic flows covered Mesozoic rocks in parts of the Sierran Province, and the ash flows must have flowed downhill from Nevada to deposit the Valley Springs Formation, found today in the northwestern Sierran foothills. Younger andesitic volcanoes erupted about 9 to 11 million years ago from the high Sierra Nevada to deposit the distinctive Mehrton Formation in the western Foothills west of Jamestown, the Table Mountain flows (discussed earlier), and widespread volcanic flows in other areas (Fig. 8-38).

GLACIERS

Yosemite Valley, with its sheer granite walls, domes, and waterfalls, is probably the most famous scenery in the United States (Figs. 8-8, 8-39, and Color Plates 22 and 25. Without the action of glaciers, Yosemite Valley would look quite different.

Fig. 8-38 Outcrops of the Mehrton Formation west of Jamestown along Highway 108. (Source: Sarna-Wojcicki, A. U.S. Geological Survey.)

Glaciers are responsible for much of the landscape of the High Sierra, and although most of the ice has melted during the past 10,000 years, it has left a record of the Sierra Nevada's past climate for geologists to study.

A glacier is a body of ice that moves over a land surface. Glaciers form by the accumulation of snow over many years. As older snow is buried by new snowfall, it undergoes a transformation not unlike metamorphism. Snowflakes are compacted, their edges are rounded off, and air is driven from the snow. The ice recrystallizes, forming larger, more compact crystals. During its transformation to glacier ice, snow increases its density about 9 times and its strength about 50 times. Ice fields form by the continual accumulation of snow and ice. When the ice is sufficiently thick, it begins to flow under its own weight. As it flows, it carries rock and sediment with it— on the ice, within it, and along the sides and bottom of the glacier. The bits of rock actually accomplish the grinding and scraping of the rocks they override. Although harder than snow, glacier ice is still much softer than most of the minerals found in the Sierra Nevada.

For a glacier to form and to grow, winter snowfall must exceed summer melting over a number of years. A glacier is thus a balanced system that responds to fluctuations in both temperature and precipitation. Favorable conditions for the formation and growth of a glacier are maximized when winter snowfall is high and summer temperatures are low. If more snow melts in summer than accumulates in winter, the glacier will shrink. This is often referred to as the *retreat* of the glacier, although, of course, the ice does not actually move back up the valley. A mountain glacier is also balanced between the *accumulation area* at the highest elevations and the *ablation area* at the lowest points, because temperatures are lower at higher elevations (Fig. 8-40).

Fig. 8-39 Yosemite Valley and Half Dome, viewed from Olmstead Point along Highway 120. (Source: Sarna-Wojcicki, A. U.S. Geological Survey.)

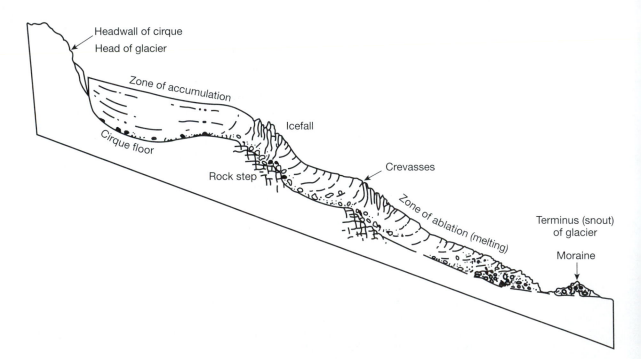

Fig. 8-40 Diagram along the length of a valley glacier. (Source: Hill, M. 1974. Glaciers: A picture story: *California Geology* 27:28. Used with permission of the California Department of Conservation, Division of Mines and Geology.)

THE EARTH'S CLIMATIC CYCLES

Geologists have long been aware that the Earth's climate has experienced fluctuations far grander than the variability we see by examining the weather records of the past few centuries. Since the early 1800s, naturalists have recognized that great masses of ice had once extended far south from the Arctic Circle to cover much of northern Europe and North America. This has occurred several times during the past 2 million years, a geologic time period recognized as the great Ice Age, or the Pleistocene. Geologists also recognized that, during the Pleistocene, mountain ranges in temperate latitudes, including the Sierra Nevada, the Rocky Mountains, and the Alps of Europe, were covered by *alpine glaciers*, also called mountain glaciers.

The Late Cenozoic period, about the last 10 million years, has been a time when glaciers have periodically built up throughout the world. One of the requirements for the occurrence of worldwide glaciation is to have continents near the Earth's poles, where temperatures are cold enough for glaciers to form. Another requirement is the uplifting of young mountain chains to elevations sufficient to sustain alpine glaciers. Because of plate tectonics, the Late Cenozoic period has been favorable for glacial conditions. The Sierra Nevada has been uplifted high enough to sustain alpine glaciers for at least the past 3 million years.

Astronomic factors are thought to trigger the episodic global cooling that results in glacial periods.

Variations in the tilt of the Earth on its axis, the precession in its elliptical orbit, and the solar sunspot cycle affect the amount of solar radiation that strikes Earth. About every 100,000 years, these variables combine to minimize the amount of solar radiation, and cooling occurs. If the continents and mountain ranges are in the correct positions, then glaciers can form and grow.

In the western United States, large alpine glaciers occupied the higher valleys of the higher mountain ranges during periods of global cooling. Large valley glaciers were present in the Rocky Mountains, the Cascade Range, and some of the higher ranges of the Basin and Range Province. In California, geologists have found evidence of Pleistocene glaciation in much of the Sierra Nevada, in the higher parts of the Klamath Mountains, on the flanks of Mt. Shasta and Lassen Peak, and even in the high parts of the northern Coast Ranges, including high peaks of the Yolla Bolly Wilderness west of Redding.

During glacial periods, more of the Earth's water is held in the great ice sheets on land than during warm periods. As a result, worldwide sea level is lower during times of glaciation. Conversely, when the Earth warms and ice sheets melt, water is added back to the oceans, causing a rise in sea level. As we will see in Chapter 12, climatically caused changes in sea level had dramatic effects on San Francisco Bay and other coastal environments.

Sierran Glaciers

In the Sierra Nevada, three factors influenced the distribution of Pleistocene glaciers. Because the range is lower in the north than in the southern and central Sierra, glaciers were smaller there. On the other hand, the southern Sierra is warmer and drier than the northern latitudes, so glaciers were restricted in the southern part of the range. Because of the rain shadow effect (see Chapter 6), snowfall was far greater on the western slopes, causing much longer valley glaciers to form there compared with the eastern side. The largest ice fields in the Sierra Nevada during Pleistocene glaciation were in the vicinity of the upper Tuolumne, Merced, and San Joaquin Rivers (Fig. 8-41). Valley glaciers flowed down from the Sierran crest for as much as 100 kilometers on the western slope. Using the evidence left behind by the most recent (Tioga) glaciers, geologists have reconstructed the glaciers that once filled the Yosemite area (Fig. 8-42).

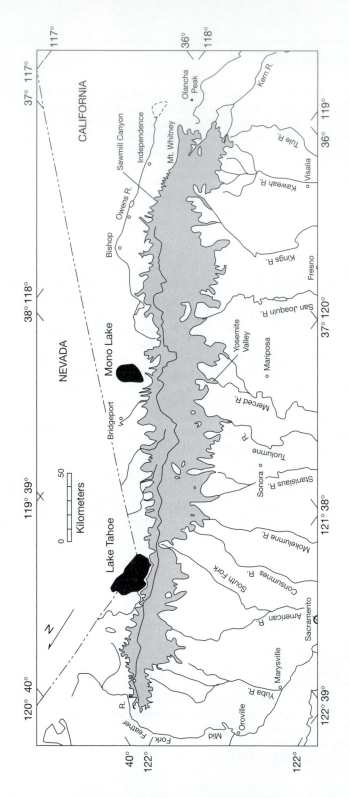

Fig. 8-41 This map shows the approximate extent of ice in the Sierra Nevada 20,000 years ago during the Tioga glaciation. (Source: Bailey, E.H., ed. 1966. *Geology of Northern California*. Bulletin 190, p. 158. Used with permission of the California Department of Conservation, Division of Mines and Geology.)

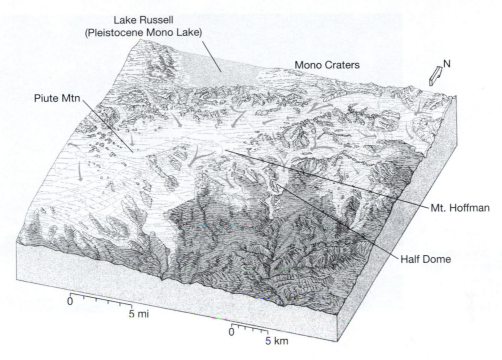

Fig. 8-42   Oblique view across the Sierra Nevada, showing the glaciers in the Yosemite area 20,000 years ago. (Source: Alpha, T.R. U.S. Geological Survey.)

Today more than 100 glaciers remain in the Sierra Nevada, most of them on the sun-sheltered northern sides of peaks (Fig. 8-43). The largest, Palisade glacier, covers less than 1 square kilometer. Compared with the more than 500 square kilometers covered by the largest glaciers about 20,000 years ago, today's glaciers are extremely small. During this century, the Sierran glaciers have shrunk considerably, as indicated by comparison of early photographs with those taken more recently.

The Work of Glaciers

John Muir, one of California's most famous naturalists, spent many years exploring the Sierra Nevada during the late 1800s and early 1900s. Muir was one of the first to observe the glaciers of the Sierra Nevada, and he firmly believed that glacial ice had been the main creator of Yosemite:

> In the beginning of the long glacial winter, the lofty Sierra seems to have consisted of one vast undulated wave, in which a thousand separate mountains, with their domes and spires, their innumerable canyons and lake basins, lay concealed. In the development of these, the Master Builder chose for a tool, not the earthquake nor lightning to rend asunder, not the stormy torrent nor eroding rain, but the tender snow-flowers, noiselessly falling through unnumbered seasons, the offspring of the sun and sea.*

---

*Excerpted from Gunsky, F.R., ed. 1968. *South of Yosemite: Selected Writings of John Muir.* Garden City, NY: Natural History Press, p. 269.

Fig. 8-43 Dana glacier in 1975. Note the crevasses in the glacier ice. (Source: Clark, M. U.S. Geological Survey.)

Glacial Erosion

In the High Sierra, most of the evidence of past glacial activity is in the form of erosional features unique to glaciers. As the moving glaciers dragged rock fragments over vast areas of bedrock, they created scratches called glacial *striations*, grooves, crescent-shaped gouge marks, and a very shiny *glacial polish*. Granitic surfaces along the Tioga Pass Road are polished as if they had been subjected to a giant lapidary wheel (Fig. 8-44). On a larger scale, granitic mounds called *roche moutoneés* were streamlined underneath the ice, leaving asymmetrical domes that point in the direction of ice flow (Fig. 8-45).

Another important erosional feature of alpine glaciers are *cirques*, bowl-shaped depressions that the ice has carved at the head of glacial valleys (see Fig. 8-3). It is the elevations of the cirques that give geologists an estimate of the ancient position of the snow line during glacial periods. The cirque elevations show that during the most recent glacial period in the Sierra Nevada, the snow line was about 750 meters lower than it is today. Today, many of the cirques contain small lakes called *tarns*, which are a favorite destination of the Sierra Nevada's backpackers.

Valleys eroded by glaciers have a characteristic U shape (see Fig. 8-39)—steep-sided and broad-bottomed—that contrasts with the normal V shape of a stream-eroded valley. The carving of glacial valleys also created another erosional feature—waterfalls. The great waterfalls of Yosemite are spectacular examples of *hanging valleys*. When ice filled the Merced River canyon, tributary streams flowed into the glacier at the level of the ice. When the ice later melted, the Merced River flowed near its much lower, current elevation, and the tributary streams were left "hanging" hundreds of meters above the river (Fig. 8-46).

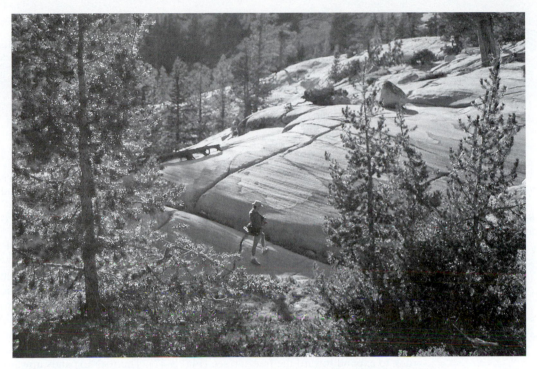

Fig. 8-44   Glacial striations and polish on granodiorite at Olmstead Point, south of Highway 120 in Yosemite National Park. (Source: Photo by author.)

Fig. 8-45   Pothole Dome, a roche moutonée in Tuolumne Meadows, Yosemite National Park. (Source: Huber, N.K. U.S. Geological Survey.)

This V-shaped valley . . .   modified by water . . .   then modified by ice . . .   results in this U-shaped valley.

Fig. 8-46 The evolution of U-shaped and hanging valleys. (Source: Huber, N.K., and Wahrhaftig, C. 1987. *The Geologic Story of Yosemite National Park*. U.S. Geological Survey, Bulletin 1595.)

Some of the Sierra Nevada's spectacular landforms are only partly created by glacial erosion. The structure of the granitic rock itself—the Half Dome granodiorite—is partly responsible for the domes of Yosemite, including Half Dome. This particular plutonic rock is unusually massive, containing few fractures or joints. As the rocks were uplifted, they were unburdened of the huge weight of rock that once covered them. The release of that pressure, plus the removal of the glacial ice, causes the rock to expand toward the free surface (Fig. 8-47). The rock expands upward or toward canyon walls, and the outside layers then flake off as uncovering proceeds. This process is called *exfoliation*, and it is responsible for dome formation in the Sierra Nevada. In areas where the rocks already have more typical, linear joint systems, the "release stress" is relieved by those fractures, and exfoliation domes do not develop.

Glacial Deposits

In the balanced system of alpine glaciation, it is logical to assume that the rock removed by glacial erosion must be deposited at some lower elevation point. Glacial

Fig. 8-47 The development of exfoliation on massive granitic surfaces. (Source: Huber, N.K., and Wahrhaftig, C. 1987. *The Geologic Story of Yosemite National Park*. U.S. Geological Survey, Bulletin 1595.)

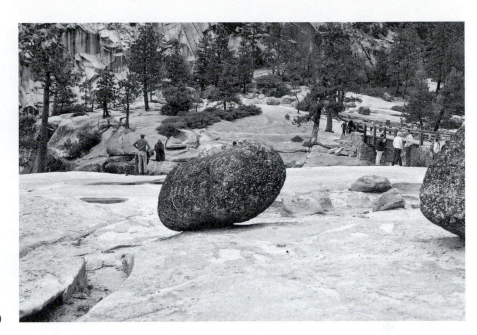

Fig. 8-48 Glacial erratic in Yosemite Valley. (Source: Sarna-Wojcicki, A. U.S. Geological Survey.)

deposits are recognized throughout the Sierra Nevada, most of them at the melting areas of the former glaciers. However, ice-transported boulders known as *erratics* have been dumped on glacially polished surfaces in the highest areas (Fig. 8-48). Glacial erratics are identified by their positions in places where no stream, avalanche, or other process could have deposited them. Rocks matching the composition of the erratic boulders are usually found higher in the ranges, where the boulders were originally picked up by the glaciers.

Along the sides and at the foot of a valley glacier, ridges of sediment known as *moraines* pile up. At the ends of glaciated valleys, moraines form characteristic loops (Fig. 8-49); these might enclose a lake or a meadow if streams have not cut a notch through them. Moraines are composed of the material pushed along by the glacier and left behind after the ice melted. This poorly sorted material can contain huge boulders and is called glacial *till*. Even when the glacial moraines have been eroded, older deposits of till can be recognized along the slopes of the Sierra Nevada by their characteristically poor sorting and large size of the largest boulders.

By identifying and mapping moraines and till, geologists are able to reconstruct the extent of glaciers that have long since melted. If the deposits can be dated, then the timing of glacial periods can also be established. Unfortunately, Sierran glacial deposits are very difficult to date because they are not completely preserved, and because they contain very little datable organic material. Most of the glacial deposits on the western slope of the Sierra Nevada have been eroded by the powerful rivers there. Only on the dry eastern side, where stream erosion is minimal because of the rain shadow effect, is a partial record of glaciation preserved in the form of moraines (Fig. 8-49).

Fig. 8-49 Aerial view of glacial moraines along the eastern side of the Sierra Nevada. Grant Lake appears at the left edge of the photo. (Source: Gilliam, M. University of Colorado.)

The History of Sierran Glaciation

Geologists are able to recognize two sets of glacial moraines in the canyons of the eastern Sierra Nevada, as well as older glacial deposits left by two or more earlier periods of glaciation. The most recent moraines are, of course, the least eroded and easiest to recognize. They were deposited during the Sierran glacial period named the ***Tioga glaciation*** by geologist Eliot Blackwelder. Tioga glaciers filled the valleys of the Sierra Nevada, between about 20,000 and 100,000 years ago, during the most recent episode of worldwide glaciation.

Before the Tioga glaciation, more extensive glaciers had occupied the Sierra Nevada. Moraines of these glaciers, called the ***Tahoe glaciers***, are found further down the valleys than the Tioga moraines, and they form much larger ridges (Fig. 8-52). In fact, it is only because the Tahoe glacial period came before the less extensive Tioga glaciation that any evidence of the Tioga period has been preserved. The Tahoe glaciers would have completely destroyed any moraines that lay in their path. Geologists have disagreed vigorously about the age of the Tahoe glacial period, but many believe that it correlates with the worldwide glacial period that occurred between about 130,000 and 200,000 years ago.

Evidence for glacial periods before the Tahoe glaciation is found at several Sierran localities. The evidence consists mainly of older glacial tills that no longer preserve any moraine-like topography. One locality where older Sierran till has been preserved is world-renowned to geologists. Along State Highway 395 near Sherwin Summit, glacial till is buried by thick deposits of the Bishop Ash, erupted 758,000 years ago from the Long Valley Caldera (Fig. 8-53). The glacial period

LAKE TAHOE

Lake Tahoe, the second deepest lake in the United States (after Oregon's Crater Lake) is one of the Sierra Nevada's jewels. The water in Lake Tahoe would cover a flat California-sized area with water 36 centimeters deep. Its maximum depth is 501 meters, and it lies at an elevation of 1,897 meters above sea level, making it the highest lake of its size in the country (Fig. 8-50).

The lake itself lies within a down-dropped basin created by late Cenozoic faulting, and Pleistocene glaciers further deepened the basin. The dam that blocked the Truckee River to create Lake Tahoe was created partly by moraines. Recent surveys of the lake's bottom discovered debris from huge pyroclastic flows along the lake's western edge (Fig. 8-51).

The Truckee River flows out of the northern end of the lake to the northeast, past Reno, Nevada, and into Pyramid Lake. About 40 percent of the precipitation falling into the Tahoe basin falls directly into the lake. The rest enters the lake from streams and slopes flowing through mostly granitic rocks. Sediment produced from these areas is sand-sized and lacking in nutrients. As a result, Lake Tahoe has naturally clear water: One can see about 30 meters below the water surface. However, the clarity of the water is being reduced by the growth of algae in the lake, and scientists are working on ways to identify and reduce sources of fine sediments that provide nutrients for the algae. Increasing development in the Tahoe basin, including recreational homes, ski areas, and commercial areas, poses a continued threat to the lake's clarity.

Fig. 8-50 Map of Lake Tahoe. (Source: U.S. Geological Survey.)

Fig. 8-51 Map of the floor of Lake Tahoe showing debris along the western edge. (Source: U.S. Geological Survey.)

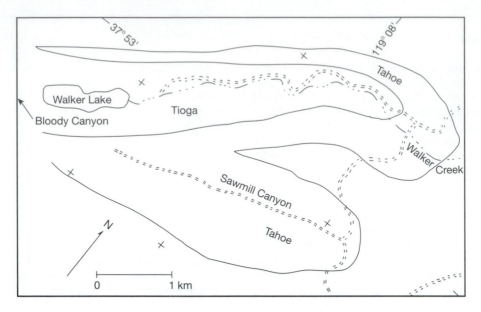

Fig. 8-52 Sketch of glacial moraines in Sawmill and Bloody Canyons on the eastern side of the Sierra Nevada. These moraines can be seen in Fig. 8-49. (Source: Burke, R.M., and Birkeland, P. 1979. *Friends of the Pleistocene Field Trip Guidebook*.)

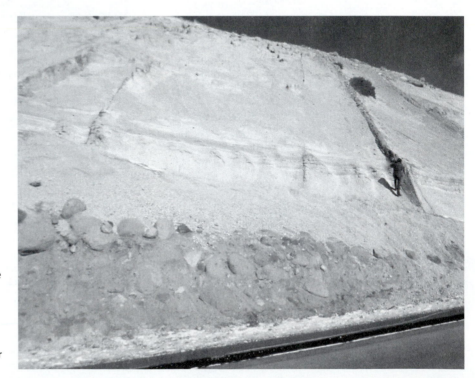

Fig. 8-53 Gravels of the Sherwin till (bottom of outcrop) were buried by the Bishop Ash when it erupted about 760,000 years ago. Photo taken on U.S. Highway 395 near Sherwin summit.

recorded by the older till is appropriately known as the Sherwin glaciation. Older glacial deposits can also be seen further north along Highway 395 near Bridgeport. Geologic evidence indicates that once the Sierra Nevada had been uplifted to a high enough elevation, it was subjected to periodic episodes of glaciation during the past few million years.

Rivers below the alpine glaciers of the Sierra Nevada were also dramatically affected by glacial episodes. Fed by enormous volumes of summer meltwater and glacial sediment, these rivers carried huge amounts of *glacial outwash*. Some of these sediments were deposited in the Central Valley, where they can be seen today as alluvial deposits of at least four different ages (see Chapter 11). In the Basin and Range Province, large lakes formed during the cooler glacial periods (see Chapter 6). Next to plate tectonics, recent climatic changes might be the most important natural factor that controls California's landscape.

SUMMARY

1. During Paleozoic time, the western edge of the North American continent was a passive margin, far from an active plate boundary. Only the southeastern part of today's Sierra Nevada province was part of Paleozoic North America (see Chapter 7 for more discussion of the passive margin). Also during the Paleozoic, rocks that would later become part of North America were forming on oceanic plates far from North America.

2. At the beginning of Mesozoic time, about 230 million years ago, a subduction zone formed along the western edge of North America, drastically changing California's plate tectonic setting. As a result, granitic magma crystallized within the North American plate, forming the oldest plutons of the Sierra Nevada Province about 210 million years ago.

3. As subduction continued during the Mesozoic, pieces of oceanic plates were accreted to the western edge of North America. Although subduction and accretion occurred during Mesozoic time, the eastern accreted terranes contain Paleozoic rocks. Those found today in the western part of the Sierra Nevada foothills are made up exclusively of Mesozoic rocks.

4. About 100 million years ago, the subduction zone moved westward to the location of today's Coast Ranges. At this time, most of the plutons in the Sierran batholith crystallized from magma. From 100 to 80 million years ago, giant andesitic volcanoes (similar to today's Andes Mountains) formed a towering mountain chain in the position of today's Sierra Nevada. Emplacement of the Sierran batholith metamorphosed both the rocks of Paleozoic North America and the rocks of the accreted terranes.

5. At the end of the crystallization of the batholith, gold was concentrated in quartz veins emplaced in the rocks of the western Sierran foothills.

6. During late Cenozoic time, about 9 to 10 million years ago, major volcanic eruptions from sources close to and east of today's Sierra Nevada crest produced lava flows and pyroclastic materials that flowed into east-flowing rivers. These lava flows buried ancient river gravels, some of which contained placer gold deposits.

7. During the worldwide glacial periods of the past million years, the high parts of the Sierra Nevada have experienced several periods of glaciation. The most recent glacial periods, the Tahoe and Tioga glaciations, reached their maximum extent about 160,000 and 20,000 years ago, respectively. Glacial erosion is responsible for much of the spectacular landscape of the High Sierra.

8. Following the discovery of gold in 1848, miners from all over the world swarmed into the Sierran foothills. Hydraulic mining washed millions of tons of sediment into the Sacramento River between the 1850s and 1883, when it was banned. Thus the gold rush changed both the natural and human landscape of California.

9. Today the Sierra Nevada mountains are California's most important source of water for irrigation and municipal use (see Chapter 10). They are also home to some of the world's most famous scenery.

REFERENCES

CLARK, W.B. 1970. *Gold Districts of California*. California Division of Mines and Geology, Bulletin 193, 186 pp.

FARQUAR, F. 1965. *The History of the Sierra Nevada*. Berekeley: University of California Press, 262 pp.

GUNSKY, F.R., ed. 1968. *South of Yosemite: Selected Writings of John Muir*. Garden City, NY: Natural History Press, 269 pp.

GUYTON, B. 1998. *Glaciers of California: Modern Glaciers, Ice Age Glaciers, the Origin of Yosemite Valley, and a Glacier Tour in the Sierra Nevada*. Berkeley: University of California Press, California Natural History Guides No. 59.

HILL, M. 1974. Glaciers: A picture story. *California Geology* 27(2):23-44.

_____ 1975. *Geology of the Sierra Nevada*. Berkeley: University of California Press, California Natural History Guide 37, 232 pp.

HOLLIDAY, J.S. 1999. *Rush for Riches: Gold Fever and the Making of California*. Berkeley: University of California Press, 355 pp.

HUBER, N.K., BATEMAN, P.C., AND WAHRHAFTIG, C. 1989. *Geologic Map of Yosemite National Park and Vicinity, California*. U.S. Geological Survey, Miscellaneous Investigations Map I-1874, scale 1:125,000.

HUBER, N.K., AND WAHRHAFTIG, C. 1987. *The Geologic Story of Yosemite National Park*. U.S. Geological Survey Bulletin 1595, 64 pp.

MOORE, J.G. 2000. *Exploring the Highest Sierra*. Stanford, CA: Stanford University Press, 427 pp.

PAUL, R.W. 1947. *California Gold: The Beginning of Mining in the Far West*. Lincoln: University of Nebraska Press, 379 pp.

Web Sites and Media Resources

National Park Service, Yosemite National Park *http://www.nps.gov/yose*

National Park Service, Kings Canyon-Sequoia National Parks *http://www.nps.gov/ seki*

National Snow and Ice Data Center: All About Glaciers *http://nsidc.org/glaciers*

U.S. Geological Survey, Lake Tahoe *http://tahoe.usgs.gov*

U.S. Geological Survey, Geology in the National Parks *http://www2.nature.nps.gov/ grd/usgsnps/project/home.html*

California Geological Survey, Gold *http://www.consrv.ca.gov/CGS/geologic_resources/gold*

California Geological Survey, Online Geologic Field Trip Guides of California *http://www.consrv.ca.gov/cgs/geotour*

Oakland Museum of California: California Gold Rush *http://www.museumca.org/ goldrush*

The Klamath Mountains
Accreted Terranes and a View of the Mantle

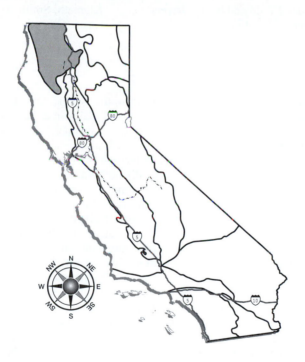

Fig. 9-1 Map showing the location of the Klamath Mountains Province.

Klamath Mountain Highlights

Major geographic feature:	Klamath Mountains
Named mountain ranges:	Marble, Salmon, Scott, Siskiyou, Trinity Alps
Other major landscape elements:	Klamath, Trinity, Scott, and Salmon Rivers
Selected high points:	Thompson Peak: 2744 m (9,002 ft.)
	China Mt.: 2604 m (8,542 ft.)
	Mt. Eddy: 2751 m (9,025 ft.)
	Preston Peak: 2228 m (7,309 ft.)
Predominant rock types:	Paleozoic marine sedimentary, Mesozoic marine sedimentary and volcanic, Mesozoic ultramafic
Plate tectonic events:	Mesozoic subduction, accretion, and intrusion of plutons
Resources:	Gold, chromium, cobalt, nickel, water
Wilderness Areas:	Marble Mountains, Salmon-Trinity Alps, Siskiyou Mountains
Federal Wild and Scenic Rivers:	Klamath, Smith, Trinity
State Parks of interest:	Castle Crags

THE LANDSCAPE

When viewed on a relief map or from the air, the Klamath Mountains Province appears quite different from other areas of California. Several mountain ranges make up the Klamath Mountains: the Trinity Alps, the Marble Mountains (Fig. 9-2), the Salmon Mountains, and the Siskiyou Mountains are the major ones. These mountain chains are lower than the Sierra Nevada, and not as parallel as those of the Coast Ranges, although most run generally north–south. The mountain summits of the Klamath Province all have about the same elevations, between 1,500 and 2,100 meters, with the highest peaks standing above that level. The consistency in the elevation of the ridges, even though they are separated by deep canyons that wind between them, gives the impression of a carved-up plateau. The Klamath Mountains Province continues across the California border into southern Oregon.

Early geologists working in northwestern California suggested that the plateau-like surface of the Klamath Mountains Province was once a much lower plain. As the ranges were uplifted during the past few million years, the rivers carved impressive gorges through this formerly low-lying surface. Today the Klamath River and its largest tributary, the Trinity River, flow through the province in canyons cut more than 1,000 meters into the hard metamorphic rocks of the province. The high surfaces of the province remain as evidence of a time when the ranges were much lower.

ACCRETED TERRANES OF THE KLAMATH MOUNTAINS

A look at the California geologic map (Endpaper 1) gives the impression that the geology of the Klamath Mountains Province is similar to that of the northwestern Sierra Nevada, and in general, this impression is accurate. Like the metamorphic belts of the northern Sierra, the Klamath Mountains are a composite of accreted oceanic terranes. In fact, the concept of accreted terranes was first developed to explain the geology of

Fig. 9-2 Marble Mountains Wilderness Area. (Source: Harwood, D. U.S. Geological Survey.)

the Klamath Mountains Province. After careful study of the rocks and fossils of the Klamath Mountains and northern Sierra Nevada, geologists have concluded that some of the rock units in each province are in fact fragments of the same oceanic plates.

As mapping and other studies allowed geologists to sort out the complexities of the Klamath Mountains, they have divided the rocks of the province into an increasing number of distinct terranes. At present, five major terranes are recognized, and several of these are subdivided into two or more units (Fig. 9-3). The terranes are generally younger from east to west, reflecting the succession of plate collisions that added each oceanic slice to the North American continent. Each terrane is bordered by major faults that represent the lines or sutures where plate fragments are joined. By dating the movement on the faults and the metamorphism that accompanied periods of deformation (see Chapter 8), geologists have confirmed that terranes were accreted in a westward succession.

All except the easternmost terranes of the Klamath Mountains are thought to be the remnants of far-traveled oceanic plates. All contain metachert, oceanic metavolcanic rocks (Fig. 9-4), and fragments of rock from the mantle beneath the plates (discussed later in this chapter). The youngest accreted terrane, the Western Klamath terrane, includes the Galice Formation, a long belt of metasedimentary rocks formed about 150 million years ago, during Jurassic time. The Galice Formation was deposited on the Josephine Ophioloite, which is discussed later in this chapter. To the west, the Western Klamath terrane is separated from rocks of the Coast Ranges by a major fault, the South Fork Mountain thrust.

Traveling eastward, and back in time, one next encounters the late Triassic to early Jurassic Rattlesnake Creek terrane, a complex tectonic mixture containing metavolcanic rocks and chert. The Sawyers Bar terrane (see Fig. 9-3) includes several blocks of Jurassic oceanic metavolcanic rocks and metachert. The Stuart Fork terrane, a still-older belt of rocks metamorphosed at high pressures and relatively low temperatures during late Triassic time, has been correlated with one of the subduction

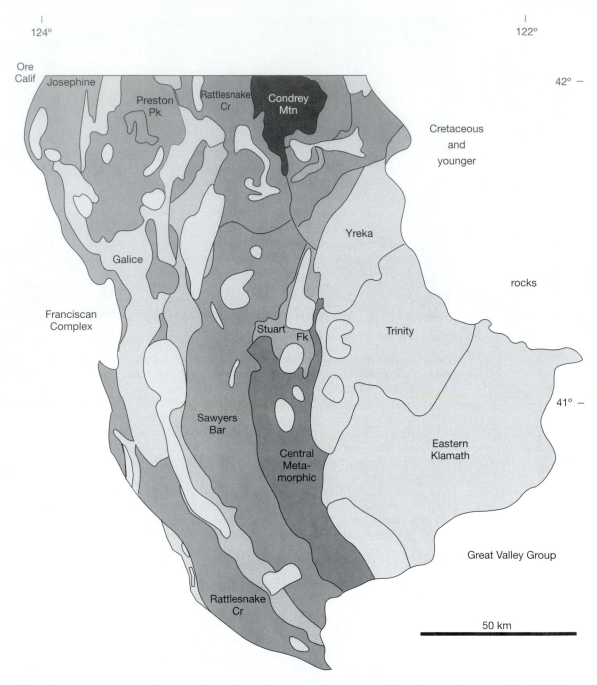

Fig. 9-3 Map showing the major accreted terranes of the Klamath Mountains province. (Source: Adapted from Hacker, B.R., and others. 1995. Timescales of orogeny: Jurassic construction of the Klamath Mountains. *Tectonics* 14:681.)

Fig. 9-4 Metavolcanic rocks of the Klamath Mountains at Condrey Mountain, showing well-preserved pillows. The matchbook is shown here for scale. (Source: Hotz, P. U.S. Geological Survey.)

complexes of the Sierra Nevada and contains rocks. East of the Stuart Fork terrane, the central metamorphic belt contains rocks very similar to the Feather River terrane, another subduction zone complex in the Sierra Nevada. Based on results of radiometric dating, geologists have determined that the metamorphic minerals in the central metamorphic belt formed during deformation in Devonian time. The original rocks of this terrane were oceanic sedimentary and basaltic volcanic rocks.

Like the rocks of the western metamorphic belt in the Sierra Nevada, the rocks of the Klamath Mountains were intruded by Mesozoic plutons. During Jurassic time, magma was generated as a result of subduction, forming volcanic arcs and, beneath them, plutons that intruded into the rocks that had already been accreted. Dating these plutons, which intrude across terrane boundaries, allows geologists to date tectonic events by applying the principles of cross-cutting relations (see Chapter 3). In the Klamath Mountains Province, magmatic activity was most active between about 200 and 155 million years ago, with minor volcanism and intrusion continuing until about 135 million years ago.

Along the eastern edge of the province is the eastern Klamath Belt, which includes the Eastern Klamath terrane, the Trinity ultramafic belt, and the Yreka terrane (see Fig. 9-3).

The oldest rocks in the Klamath Mountains Province are those of the Eastern Klamath terrane. These rocks, which range from Cambrian to Triassic in age, are thought to match the Northern Sierra terrane discussed in the previous chapter. Both terranes consist of metamorphosed andesitic volcanic rocks erupted from an oceanic arc, both associated with metasedimentary rocks that originated on a carbonate platform. The similarities between these Paleozoic formations in the Northern Sierra terrane and the Eastern Klamath terrane, especially the ages and geometric relations of individual rock units, have led geologists to conclude that both originated as parts of the same offshore volcanic arc.

THE McCLOUD FAUNA

Carbonate rocks of Permian age are found adjacent to and overlying the fractured metavolcanic rocks of the Eastern Klamath and Northern Sierra terranes. The Mc-Cloud limestone, named for the town of McCloud in Siskiyou County, is an important Permian formation within the Eastern Klamath terrane. The limestone, which can be viewed in Shasta Caverns in Shasta Lake Recreation Area, contains abundant corals (Fig. 9-5), brachiopods, and fusulinids Geologists interpret its original environment of deposition as carbonate reefs fringing the eastern Klamath volcanic arc. These ancient reefs might have appeared similar to coral reefs found today near Indonesia (Fig 9-6).

Paleontologists now recognize that the Permian-aged fossils preserved in the McCloud limestone are a unique assemblage of marine organisms. They have named this assemblage the McCloud fauna because its distinctive corals and fusulinids are best seen in the rocks of the eastern Klamath Mountains near McCloud. The same distinctive assemblage of fossils is found in a belt of Permian rocks stretching from the Northern Sierra terrane in California north to British Columbia (Fig. 9-7). The fossils of the McCloud fauna are unlike those found in other accreted terranes of western North America.

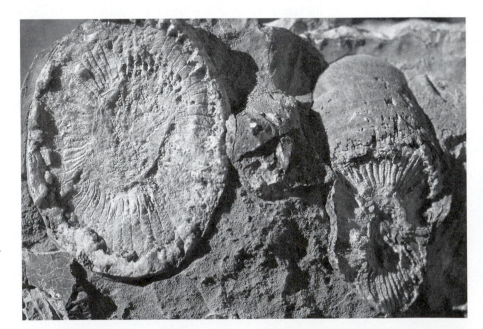

Fig. 9-5 Large fossil corals of the Permian Mc-Cloud fauna. The sample is from Quinn River Crossing, Nevada. The larger coral measures about 4 centimeters across. (Source: Stevens, C. San José State University.)

Fig. 9-6 Modern corals from a fringing reef.

The carbonate rocks in the western Klamath Mountains contain a completely different community of marine organisms. The assemblage of corals, fusulinids, and other species found in those rocks are representative of the Tethyan fauna, a group of Permian organisms represented in rocks found in Japan, China, and southeast Asia. The distinct differences between the McCloud and Tethyan faunal assemblages, even though the rocks are the same age, support the hypothesis that some of the terranes of the Klamath Mountains and the Sierra Nevada originated at great distances from Paleozoic North America. However, geologists are uncertain about the original position of the island arcs of the northern Sierra and eastern Klamath Mountains. The community of reef organisms found in the McCloud fauna is also distinctly different from the community that inhabited the shallow ocean immediately adjacent to North America, as represented by Permian fossils found in the Basin and Range Province a few hundred kilometers to the east (see Chapter 7). This would suggest that the Permian island arc was distant enough from North America that the populations were isolated from each other. On the other hand, many geologists believe that the volcanic arc was relatively close to North America because of correlations with the tectonic events recorded by Paleozoic North American rocks. Studies of modern coral populations and their ability to mix suggest that the Eastern Klamath terrane must have been several thousand kilometers offshore from North America during Permian time to prevent the populations from intermixing.

OPHIOLITE SUITES: REMNANTS OF OCEANIC BASEMENT

The sedimentary and volcanic rocks of the Eastern Klamath terrane rest on older ultramafic rocks (see Chapter 2) known as the Trinity peridotite, the largest exposed body of ultramafic rock in North America. Further west, belts of ultramafic rock, including the Josephine Ophiolite (see Fig. 9-3), form the base of several of the younger

ST - Stikine
CC - Cache Creek
CA - Cache Creek Affinity
EK - Eastern Klamath
NS - Northern Sierra

Fig. 9-7 Map showing the distribution of terranes containing fossils of the Permian McCloud fauna and terranes with Permian Tethyan fauna. (Source: Miller, E. 1992. Late Paleozoic paleogeographic and tectonic evolution of the western U.S. Cordillera. Geological Society of America. *Decade of North American Geology*, Vol. G-3, p. 83.)

terranes of the Klamath Mountains. It is these rare, dark-colored ultramafic rocks that are responsible for the dark alpine peaks like Preston Peak that are so unusual to those accustomed to light-colored granitic alpine scenery.

Geologists now recognize a characteristic assemblage of unusual rocks as the components of an *ophiolite* or *ophiolite suite*. At most localities, the rocks in an ophiolite suite are highly disrupted by faulting, folding, and metamorphism, but geologists have been able to reconstruct their original geometry by detailed mapping and other studies. At the base of a reconstructed, or idealized, ophiolite are rocks from the upper mantle beneath the oceanic crust. These are ultramafic plutonic rocks rich in iron and magnesium and low in silica (see Chapter 2). Ultramafic rocks include *dunite*, composed entirely of the mineral olivine and *peridotite*, composed mostly of olivine with some plagioclase feldspar and pyroxene (Fig. 9-8). (The fact that gem-quality olivine is known as peridot gives some commonality to the terminology of ultramafic rocks.) After they have been displaced from the mantle into the crust, most ultramafic rocks are altered by hot (*hydrothermal*) fluids to serpentinite, California's state rock.

Above the ultramafic rocks in an ophiolite suite is a layer of mafic plutonic rock, mostly gabbro (Fig. 9-9). The lower section within the gabbro typically displays a layering of crystals. The layering is caused by the sinking of the first-formed, dense minerals to the bottom of the magma chamber, followed by the accumulation of minerals crystallized later (Fig. 9-10). The gabbro layer is in turn overlain by a complex of vertical *sheeted dikes*, mafic plutonic rocks that crystallized at shallower depths

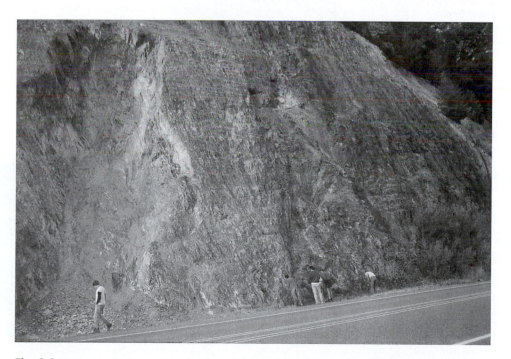

Fig. 9-8 Outcrop of ultramafic rocks along Highway 70 in the northern Sierra Nevada. (Photo by author.)

Unconsolidated
sediments

Partially consolidated sediments
and sedimentary rock

Interlayered baked sediment
and basalt flows

Pillow basalts

Sheeted basalt dikes

Gabbro

Ultramafic rocks
(dunite, peridotite)

Fig. 9-9 A section through the ocean floor showing the rock assemblage that consitutes an ophiolite suite. (Source: Coch, N., and Ludman, A. 1991. *Physical Geology.* New York: Macmillan, p. 434.)

Fig. 9-10 Accumulation of early-formed crystals at the bottom of a magma chamber. The crystals have higher densities than the surrounding magma and **A,** sink to the floor of the magma chamber. **B,** The early-formed crystals accumulate and are segregated to form an ultramafic rock. (Source: Coch, N., and Ludman, A. 1991. *Physical Geology.* New York: Macmillan, p. 434.)

beneath the ocean floor. On the top of this igneous pile is a section of mafic volcanic rocks, typically pillow basalt, erupted on the ocean floor. Geologists believe that the entire assemblage represents a cross section from the ocean floor crust into the upper mantle beneath it. Sedimentary rocks formed in the deep ocean, including chert and shale, are usually found at the top of such an ophiolite suite.

Direct observation and sampling of the ocean floor has enabled researchers to verify that oceanic crust is composed of basalt, both pillow lavas and sheet flows. Deep drilling has penetrated into the upper zone of sheeted basalt dikes, but no direct sampling has reached below about 1 kilometer. Indirect studies using seismic techniques confirm the basic three-layer model of ultramafic upper mantle, gabbro, and basaltic volcanic rocks. However, knowledge about the nature of oceanic crust and mantle relies heavily on studies of ophiolites that we interpret as on-land samples of oceanic crust and upper mantle. In part because they are commonly associated with volcanic rocks produced by explosive eruptions, rather than those produced at mid-ocean ridges, California's ophiolites are thought to have formed beneath island arcs or basins fringing the arcs.

On the California geologic map (Endpaper 1), the largest ophiolite suites can be identified by the purple slivers and pods representing areas underlain by ultramafic rocks. The Trinity and Josephine ophiolites are the largest bodies in the Klamath Mountains Province. The Feather River peridotite, the Smartville Complex, and the Kings-Kaweah ophiolite are examples of major bodies in the Sierran metamorphic terranes (Fig. 9-11). The Coast Range ophiolite extends along the western margin of the Great Valley Province (see Chapter 11). Ophiolitic rocks are also found within the Franciscan Assemblage in the Coast Ranges; these are probably displaced fragments of the Coast Range ophiolite (see Chapter 12). Considering California's long history as a convergent plate margin, it is not surprising that many fragments of oceanic basement are present in the accreted terranes of these three provinces.

RICHES FROM THE BOTTOM OF A MAGMA CHAMBER

Found within many of California's ultramafic rocks are rich pods and slivers of *chromite*, an oxide mineral containing chromium and iron. Chromium is used as a steel alloy and for plating metal surfaces. Chromite has been mined from the Josephine and Trinity ophiolites, as well as from smaller ultramafic bodies in the Klamath Mountains, the western foothills of the Sierra Nevada, and the Coast Ranges. The most intense chromite mining in California occurred during World War II, when supplies from overseas were cut off. California has been the second largest chromium supplier in the United States, although no commercial mining has occurred since the mid-1960s.

Geologists believe that chromite concentrations developed during the original crystallization of the ultramafic plutonic rocks. As magma crystallization began, the early crystals sank to the floor of the magma chamber because they were denser than the surrounding magma. Settled at the floor of the magma chamber, the early-formed minerals were isolated from further reaction with the magma (see Fig. 9-10). Early-formed silicate crystals, those with the highest temperature of crystallization, include the mafic minerals olivine and pyroxene. Geologists believe that many ultramafic plutonic rocks

1. Copper Creek
    Mountain View
2. High Plateau
3. French Hill
4. Rattlesnake
5. Holliday
6. Cyclone Gap
7. Seiad Creek
8. Fairview
9. Croggins
10. Little Castle Creek
11. Lambert
12. Zenith
13. Red Ledge
14. Boiler Pit
15. Bunker
16. Shelly
17. Pillikin
18. Murphy
19. Ward and Lyons
20. Madrid
21. McCormick
22. Lacey
23. Clara H
24. Jack Sprat group
25. Holston
26. Grau
27. Gray Eagle
    Black Diamond
28. Lucky Strike
29. White Angel
30. Newman
31. No. 5
32. Mistake
33. Norcross
    Sweetwater
    Victory No. 3
34. Castro
    Colorado
    Estrella
35. Froom Ranch
36. London
    Trinidad
37. Pick & Shovel

Fig. 9-11 The distribution of ultramafic rocks and chromite mines (shown by numbers) in California. (Source, Wight, L.A., ed. 1957. California Department of Conservation, Division of Mines and Geology, Bulletin 176, p. 124.)

crystallize by this process of *magma differentiation*. Minerals that crystallize later might also settle out from the magma, ultimately producing a layered plutonic rock.

Chromium is one of the Earth's elements that is concentrated in the mantle and thus found associated with mantle rocks. Many of the ultramafic plutonic rocks in ophiolites have been altered to serpentinite, but the chromite remains mostly unaltered. Some of the richest concentrations yielded as much as 20,000 tons of chromite from a single pod deposit. Nickel, another valuable element that is rare in crustal rocks, is also associated with the ultramafic rocks of the Klamath Mountains Province, but has not been mined at this time. The ultramafic rocks are also the probable source of reported platinum found in beach deposits near Crescent City.

Like the Sierra Nevada, the Klamath Mountains were a major destination of gold seekers during the 1850s. Early miners recognized the geologic similarities between the two provinces and were quick to discover both lode and placer deposits. Today, visitors can see the remains of gold-mining operations near Weaverville, and dredge operations are still active in a few areas.

SUMMARY

1. The Klamath Mountains Province consists of five major Paleozoic and Mesozoic oceanic terranes, some of which are subdivided. The Eastern Klamath terrane is the oldest, representing a continental platform and island arc, similar to the Northern Sierra terrane of the Sierra Nevada Province. The other terranes represent oceanic plate fragments.
2. Like the Sierra Nevada terranes, the terranes of the Klamath Mountains were intruded by Mesozoic plutons.
3. Ophiolite suites of rocks, representing slices of the upper mantle beneath the ocean floor, are extensive in the Klamath Mountains. These include the Josephine ophiolite and the Trinity peridotite, which are dominated by ultramafic plutonic rocks.
4. Chromite and nickel have been mined from ultramafic rocks of the Klamath Mountains Province, and gold is associated with the Mesozoic granitic intrusions.

REFERENCES

HACKER, B.R., DONATO, M.M., BARNES, C.G., MCWILLIAMS, M.O., and ERNST, W.G. 1995. Timescales of orogeny: Jurassic construction of the Klamath Mountains. *Tectonics* 14(3):677–703.

HARWOOD, D.S., and MILLER, M.M. 1990. *Paleozoic and Early Mesozoic Paleogeographic Relations; Sierra Nevada, Klamath Mountains, and Related Terranes.* Geological Society of America, Special Paper 255, 422 pp.

Web Sites and Media Resources

National Park Service, Wild and Scenic Rivers: *http://www.nps.gov/rivers/*
Salmon River Restoration Council and Library: *http://srrc.org/*

CHAPTER

10

Water in California

Water issues have dominated California's politics, history, and economy since the state's early years. Nowhere else is so much water stored, diverted, and transported to overcome four of nature's water-related limitations to human activities:

1. Most of the state is a semidesert naturally unsuitable for agriculture or major cities.
2. Precipitation is seasonal, falling almost entirely between October and April.
3. Years of drought and years of flood are part of the normal, natural pattern.
4. The water in California is not where most of the people are.

Californians used more than 11 trillion gallons of water in 1985, yet too few people know where their water comes from or where their wastewater goes. As concern for the availability and safety of water grows, it becomes increasingly important to understand the fundamental principles of *hydrology*, the study of water, and the basic elements of California's hydrologic systems.

PRECIPITATION

Much of California has a **Mediterranean climate**, with warm, dry summers and mild, wet winters. The Pacific Ocean moderates the temperatures along the coast, but in the Sierra Nevada, winters are cold, and winter precipitation falls as snow that is stored until the spring snowmelt. The average annual precipitation in California is about 58 centimeters (22.8 inches), with about two thirds falling in the northern third of the state (Fig. 10-1). Annual precipitation ranges from over 254 centimeters in the northwestern Coast Ranges to less than 25 centimeters in the southern San Joaquin Valley and most of southeastern California. Most of the state is therefore semidesert or desert, and only great human effort has made it possible for 35 million people to inhabit California.

As discussed in Chapter 6, California's mountain ranges create a rain shadow effect for the slopes and valleys east of them (see Fig. 6-3). The effect of the rain shadow can be seen in the distribution of precipitation within the state, particularly on the dry eastern side of the Sierra Nevada. The winter storms that bring most of

Fig. 10-1 Average yearly precipitation in California, in inches. To convert the values on the map to centimeters, multiply values by 2.54. (Source: California Department of Resources. 1987. *California Water: Looking to the Future,* Bulletin 160-87.)

Legend (in inches)

- Over 80
- 60–80
- 40–60
- 20–40
- 10–20
- 5–10
- Under 5

California's water originates in the northern Pacific Ocean and travels eastward. Storms from in and around the Gulf of California can also bring precipitation to southern California in late summer, and local thunderstorms provide an additional moisture source in the mountains during summer. However, in most of California, little or no precipitation falls between May and October. The lack of reliable rainfall during the main growing season creates one of California's major reasons for water storage and diversion.

Another important and often overlooked aspect of California's climate is the normal cyclic occurrence of both flood and drought. Prolonged, severe droughts are part of this recurrent, variable pattern. Between 1928 and 1934, 1976 and 1977, and 1987 and 1992, insufficient supplies stressed the state's ability to meet the demand for water. In contrast, damaging floods produced by extremely high rainfall occur somewhere in the state at least once every decade (Fig. 10-2).

A

B   C

Fig. 10-2 A, Effects of the 1964 flood on the town of Orick, Humboldt County. **B**, Flooding on the Russian River, March 1998. **C**, Lakeport, along the shores of Clear Lake, February 1998. (Source: **A**, Janda, R. U.S. Geological Survey; **B** and **C**, Gatley, D. Federal Emergency Management Agency.)

THE HYDROLOGIC CYCLE AND DRAINAGE BASINS

Of all the precipitation that falls over California, only about 35 percent becomes *runoff* to streams and lakes. Because of California's relatively high temperatures, *evaporation* from the land surface takes up much of the precipitation, and *transpiration* by the vegetation uses another substantial portion (Fig. 10-3). Unless the ground surface is completely impervious to water, a portion of the precipitation enters the soil by *infiltration*. Only the water that does not evaporate, is not used by the plants, and does not infiltrate into the subsurface becomes runoff. In California, the mean annual runoff is only about 200 centimeters, and many of the hot, dry deserts experience years without any runoff.

Over thousands of years, the typical amount of runoff that occurs in different parts of the California landscape is extremely variable. However, in all landscapes, running water is the most effective agent of erosion, even in desert areas that experience runoff only once or twice in a decade. Runoff from the land surface flows downhill under the influence of gravity, and water collects to carve *stream channels*. Larger channels naturally are required to carry larger amounts of water that collect as small channels join.

It is not surprising that the network of stream channels in a given area would be adjusted to accommodate the typical runoff conditions. The natural stream network that develops is a system of *drainage basins* or *watersheds* (Fig. 10-4). Each stream channel, regardless of size, collects runoff from an area defined as its drainage basin. All of the precipitation that falls within the boundaries of that basin and becomes runoff will eventually pass through the mouth of the channel. All runoff outside of the boundaries flows into the channels of other drainage basins. Because the mouth of the channel is the lowest point in the drainage basin, it is often called the *base level* for that stream. The boundary of a drainage basin is termed its *divide*, and high areas near the divide are often termed the *headwaters* of a stream. Drainage

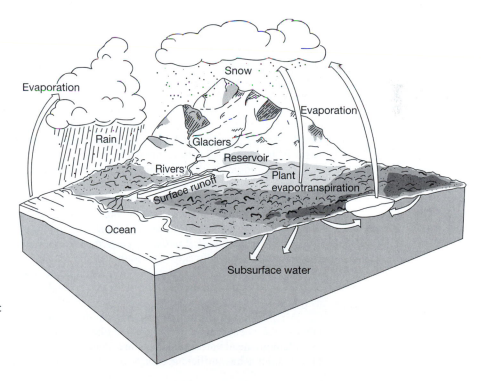

Fig. 10-3 The hydrologic cycle. (Source: Coch, N., and Ludman, A. 1991. *Physical Geology.* New York: Macmillan.)

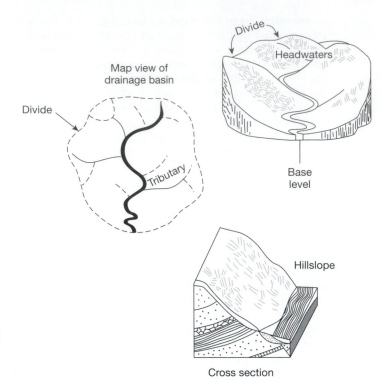

Fig. 10-4 Elements of a drainage basin. (Source: Ryan, D. 1976–77. *The Co-Evolution Quarterly.* Sausalito, CA.)

basins could have areas much less than a few acres, or they might encompass huge portions of continents (see Table 10-1). The drainage basin of a large river such as the Sacramento River contains many smaller *tributary* drainage basins, such as the American and Yuba Rivers, and these tributaries collect water from even smaller tributaries (see Endpaper 3). The ultimate base level for most of California's streams is sea level, because most runoff eventually flows to the Pacific Ocean. However, in eastern California, streams commonly flow into closed basins, some of which contain lakes (see Chapter 6). In these cases, the lake or the dry floor of the basin is the base level for the stream.

The geologic work accomplished by streams is to carry materials from high areas of the landscape to low-lying areas, thus eroding the mountains and depositing sediments in deltas, lakes, dry basins, or along the continental shelf. The material carried by a stream is its *load,* which consists of the *dissolved load* of molecules and ions in solution and the *mechanical load* of solid particles. The mechanical load consists of the *suspended sediment* carried in the flowing water and the *bed load* of larger particles that roll, skip, and bounce along the bed of the stream. During floods, streams in California can carry very high sediment loads; for example, during the December 1964 flood, the Eel River carried as much suspended sediment in one day as it had during the previous 23 years.

A typical large stream in California changes radically as it flows from the mountains to the ocean or into a large river like the Sacramento or San Joaquin. In the mountainous headwaters, channels are steep, narrow, and irregular, and the steeply falling stream has sufficient energy to transport boulders and cobbles (Fig. 10-5).

TABLE 10-1 DATA FOR SELECTED RIVERS IN CALIFORNIA AND ELSEWHERE

River	Drainage Area, mi² (km²)	Average Discharge, cfs (m³/sec)	Average Annual Runoff (in thousands of acre-feet)
Klamath	12,000 (31,080)	17,785 (504)	12,900
Eel	3,113 (8,063)	7,410 (210)	5,379
Pit	4,711 (12,200)	3,880 (110)	2,819
Sacramento	8,900 (23,050)	24,670 (699)	17,870
Russian	793 (2,054)	2,365 (67)	1,712
San Joaquin	1,676 (4,341)	4,375 (124)	3,179
Kings	1,545 (4,001)	2,225 (63)	1,624
Colorado	242,000 (626,780)	17,750 (503)	12,860
Mojave	2,121 (5,493)	6 (0.17)	4.3
Los Angeles	827 (2,142)	215 (6)	156
Mississippi	1,250,965 (3,240,000)	635,200 (18,000)	460,360
Amazon	2,374,500 (6,150,000)	7,100,000 (200,000)	5,115,000

cfs = cubic feet per second.
Sources: California Water Atlas, U.S. Geological Survey Water-Resources Data for California.

Rapids and riffles with pools separating them are common. As the stream collects runoff from tributaries, its channel becomes larger, and as it flows to more gently sloping terrain, its channel becomes less steep. In California's valleys, a typical stream transports sand and finer sediment and flows in a broad and more gently sloping channel (Fig. 10-6).

Fig. 10-5 Large boulders and rapids of a typical steep mountain stream; this is the Truckee River. Note the raft for scale. (Source: Photo by author.)

Fig. 10-6 The lower channel and floodplain of Redwood Creek, Humboldt County. The town of Orick is built on the floodplain on the right side of the photograph. The headquarters of Redwood National Park are just off the left side of the photograph. Following the 1964 flood, levees were constructed along the banks of lower Redwood Creek; compare with Fig. 10-2A to see 1964 flooding and levee modifications. (Source: Photo by author.)

A Balanced System

The network of stream channels on the landscape can be thought of as a natural system of canals built over many thousands of years to be perfectly designed to carry the natural runoff that is supplied. Within a given watershed, the channels of tributaries are adjusted to each other and to the main stream. A drainage system reflects the prevailing conditions of climate, vegetation, topography, and geologic materials that determine runoff. Changes in runoff, base level, or topography are accommodated by changes in the stream system, so that a condition of ***dynamic equilibrium*** exists in the drainage basin. The streams in the drainage basin have just the gradient needed to carry their sediment loads, without becoming either more or less erosive as they flow downstream.

The dynamic equilibrium of a drainage basin can be disturbed by a variety of events, both natural and human induced. Some examples of natural causes of disturbance are climate changes, which can change the runoff available in the system, the sediment load, or both; rapid uplift or down-dropping, resulting in base-level changes, which cause the stream to cut down or fill with sediment (Fig. 10-7); and major landslides, which can overwhelm the system with sediment. Examples of human disturbances are urbanization, deforestation, irrigation, hydraulic mining, and water diversions and dams (see Fig. 8-34).

Fig. 10-7 River terraces along Colorada Baranca near Ventura. Tectonic uplift and base-level changes caused this stream to fill with alluvium and subsequently cut a new channel. (Source: Sarna-Wojcicki, A. U.S. Geological Survey.)

STREAMFLOW IN CALIFORNIA

Given the enormous demand for and sometimes unreliable supplies of water in California, it is important for managers to have exact measurements of the volume of runoff in the state's rivers. Long-term records also provide an important measure of the potential for drought and flood. Streamflow measurements are made regularly on dozens of streams by the U.S. Geological Survey, the California Department of Water Resources, irrigation companies, municipal water districts, flood control agencies, and others.

The volume of water flowing in a stream at any given time is the stream's *discharge*. The discharge is the volume of water flowing past a certain point in a given unit of time. The elements of a river's discharge are the average velocity of flow, the width of the channel, and the water depth. Discharge is usually expressed in cubic feet per second (cfs) or cubic meters per second (m^3/s; see Fig. 10-8). Table 10-1 and Fig. 10-9 illustrate the average flows for selected streams.

To measure discharge, hydrologists measure the width and depth of the flowing water and use a velocity meter to record velocity. At many stations, discharge is estimated from the water depth, which is continuously recorded by a *stage recorder*. At flood warning stations, alarms are automatically triggered when the flow exceeds a certain height. Records of total discharge, average annual discharge, peak

Fig. 10-8 Cross section of a stream channel showing important stream elements and the parameters used to calculate streamflow. At the present discharge, shown by the water:

Discharge = Average width × Average Depth × Average Velocity
= Cross-sectional area × Velocity
ft³/sec = ft² × ft/sec

(Source: Adapted from Coch, N., and Ludman, A. 1991. *Physical Geology.* New York: Macmillan.)

flood discharge, and record low flows are available for many of California's streams (see *http://water.usgs.gov*).

Because water managers in California need to know the available volume of water at a given moment in time, they commonly use the unit of ***acre-feet*** to express annual runoff or volumes of water stored in a reservoir. One acre-foot is the amount of water that covers an acre of land to a depth of 1 foot, and it is equivalent to about 43,560 cubic feet of water, or about 325,851 gallons.

Water Units
1 cubic foot per second (cfs) = 677,376 gallons per day
43,560 cfs = 1 acre-foot per second
3.07 acre-feet = 1 million gallons
34.2 million acre-feet (used in California in 1985) = 11,144,104,000,000 gallons

Because they have been carved by the streams, stream channels are sufficiently wide and deep to contain their average flows. Geologists have long debated about which "typical" flow determines channel size, but most agree that it is a discharge

A

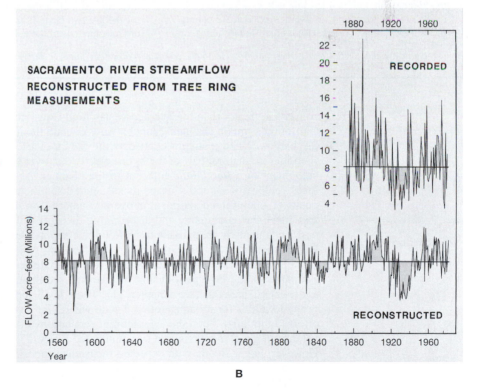

Fig. 10-9 A, Core taken from a Douglas fir tree at Muir Woods, Marin County. Drought years are recorded by very small growth rings. **B**, The long-term stream-flow record for the Sacramento River, reconstructed from tree-ring records. (Source: **A**, Sarna-Wojcicki, A. U.S. Geological Survey; **B**, California Department of Resources. 1987. *California Water: Looking to the Future*, Bulletin 160-87.)

B

THE GREAT FLOODS OF 1862

In 1860, naturalist William H. Brewer began his assignment as a member of the State Geological Survey of California under the direction of State Geologist Josiah Whitney. Because his wife and only child had just died, Brewer welcomed the opportunity to leave the east for California. During his 4 years with the Survey, he meticulously recorded his observations of 19th-century California, including details on everything from life at the mines and missions to earthquakes and fossil localities. His priceless collection of diaries, letters, and field notebooks are an important resource for historians, geologists, and naturalists. From *Up and Down California in 1860-1864, The Journal of William H. Brewer*, we get a vivid picture of the floods of January 1862:

January 19, 1862

The great central valley of the state is under water—the Sacramento and San Joaquin valleys—a region 250 to 300 miles long and an average of at least twenty miles wide, a district of 5,000 or 6,000 square miles, or probably three to three and a half millions of acres! Although much of it is not cultivated, yet a part of it is the garden of the state. Thousands of farms are entirely under water—cattle starving and drowning.

January 31, 1862

In this city [San Francisco], 37 inches of water has fallen, and at Sonora, in Tuolumne, 102 inches, or $8\frac{1}{2}$ feet, at the last dates. These last floods have extended over this whole coast. At Los Angeles it rained incessantly for 28 days; immense damage was done—one whole village destroyed. It is supposed that over one-fourth of all the taxable property of the state has been destroyed. The legislature has left the capital and has come here, that city being under water. . . .

All the roads in the middle of the state are impassable, so all mails are cut off. We have had no "Overland" for some weeks, so I can report no new arrivals. The telegraph also does not work clear through, but news has been coming for the last two days. In the Sacramento Valley for some distance the tops of the poles are under water!"

Source: Farquar, F., ed. 1966. *Up and Down California in 1860–1864: The Journal of William H. Brewer.* Berkeley: University of California Press, 583 pp.

somewhat greater than the average yearly high flow. However, it is very clear that California's stream channels cannot always contain the infrequent, but normal, peak discharges. When a stream overflows the banks of its channels, flooding occurs. Flooding is a natural part of the hydrologic cycle; over time, it can result in the development of a ***floodplain*** adjacent to the channel. As the water spreads out and loses velocity, it naturally drops its suspended sediment load, which is left as deposits of fine-grained sediment above and adjacent to the channel (see Figs. 10-2 and 10-6). By recognizing and determining the ages of floodplain sediments, geologists can identify areas likely to be flooded every 10 years, 50 years, 100 years, or less frequently.

Floodplains are highly desirable agricultural areas because they are relatively flat surfaces and they periodically receive fine sediment with nutrient-rich organic and mineral material (see Chapter 11). Their level topography also makes them prime localities for urbanization in a state where flat land is relatively scarce. In an effort to minimize flood damage, current regulations in some counties prohibit development on floodplains that are inundated every 25 years or more frequently. However, larger floods with expected recurrence times of 50, 100, or even several hundred years, are a predictable, damaging occurrence in low-lying areas in the state (see Table 10-2).

The lives of Californians continue to be affected by floods, sometimes with catastrophic results. The normally dry canyons of southern California can be transformed into rivers of death, as evidenced by the narrative in the box that follows.

DEBRIS FLOW AND A FAMILY'S MIRACULOUS SURVIVAL IN THE SAN GABRIEL MOUNTAINS

On a February night some years ago, the Genofiles were awakened by a crash of thunder—lightning striking the mountain front. Ordinarily, in their quiet neighborhood, only the creek beside them was likely to make much sound, dropping steeply out of Shields Canyon on its way to the Los Angeles River. The creek, like every component of all the river systems across the city from mountains to ocean, had not been left to nature. Its banks were concrete. Its bed was concrete. When boulders were running there, they sounded like a rolling freight. On a night like this, the boulders should have been running. The creek should have been a torrent. Its unnatural sound was unnaturally absent. There was, and had been, a lot of rain.

The Genofiles had two teenage children, whose rooms were on the uphill side of the one-story house. The window in Scott's room looked straight up Pine Cone Road, a cul-de-sac, which, with hundreds like it, defined the northern limit of the city, the confrontation of the urban and the wild. Los Angeles is overmatched on one side by the Pacific Ocean and on the other by very high mountains. With respect to these principal boundaries, Los Angeles is done sprawling. The San Gabriels, in their state of tectonic youth, are rising as rapidly as any range on earth. Their loose inimical slopes flout the tolerance of the angle of repose. Rising straight up out of the megalopolis, they stand ten thousand feet above the nearby sea, and they are not kidding with this city. Shedding, spalling, self-destructing, they are disintegrating at a rate that is also among the fastest in the world. The phalanxed communities of Los Angeles have pushed themselves hard against these mountains, an aggression that requires a deep defense budget to contend with the results. Kimberlee Genofile called to her mother, who joined her in Scott's room as they looked up the street. From its high turnaround, Pine Cone Road plunges downhill like a ski run, bending left and then right and then left and then right in steep christiania turns for half a mile above a three-hundred-foot straightaway that aims directly at the Genofiles' house. Not far below the turnaround, Shields Creek passes under the street, and there a kink in its concrete profile had been plugged by a 6-foot boulder. Hence the silence of the creek. The water was now spreading over the street. It descended in heavy sheets. As the young Genofiles and their mother glimpsed it in the all but total darkness, the scene was suddenly illuminated by a blue electrical flash. In the blue light they saw a massive blackness, moving. It was not a landslide, not a mudslide, not a rock avalanche; nor by any means was it the front of a conventional flood. In Jackie's words, "It was just one big black hill coming toward us."

In geology, it would be known as a ***debris flow.*** Debris flows amass in stream valleys and more or less resemble fresh concrete. They consist of water mixed with a good deal of solid material, most of which is above sand size. Some of it is Chevrolet size. Boulders bigger than cars ride long distances in debris flows. Boulders grouped like fish eggs pour downhill in debris flows. The dark material coming toward the Genofiles was not only full of boulders—it was so full of automobiles that it was like bread dough mixed with raisins. On its way down Pine Cone Road, it plucked up cars from driveways and the street. When it crashed into the Genofiles' house, the shattering of safety glass made terrific explosive sounds. A door burst open. Mud and boulders poured into the hall. We're going to go, Jackie thought. Oh, my God, what a hell of a way for the four of us to die together.

The parents' bedroom was on the far side of the house. Bob Genofile was in there kicking through white satin draperies at the panelled glass, smashing it to provide an outlet for water, when the three others ran in to join him. The walls of the house neither moved nor shook. As a general contractor, Bob had built dams, department stores, hospitals, six schools, seven churches, and this house. It was made of concrete block with steel reinforcement, sixteen inches on center. His wife had said it was stronger than any dam in California. His crew had called it "the fort." In those days, twenty years before, the Genofiles' acre was close by the edge of the mountain brush, but a developer had come along since then and knocked down thousands of trees and put Pine Cone Road up the slope. Now Bob Genofile was thinking, I hope the roof holds. I hope the roof is strong enough to hold. Debris was flowing over it. He told Scott to shut the bedroom door. No sooner was the door closed than it was battered down and fell into the room. Mud, rock, water poured in. It pushed everybody against the far wall. "Jump on the bed," Bob said. The bed began to rise. Kneeling on it, they could soon press their palms against the ceiling. The bed also moved toward the glass wall. The two teenagers got off to try to control the motion and were pinned between the bed's brass railing and the wall. Boulders went up against the railing, pressed it into their legs, and held them fast. Bob dived into the muck to try to move the boulders, but he failed. The debris flow, entering through windows as well as doors, continued to rise. Escape was still possible for the parents but not for the children. The parents looked at each other and did not stir. Each reached for and held one of the children. Their mother felt suddenly resigned, sure that her son and daughter would die and she and her husband would quickly follow. The house became buried to the eaves. Boulders sat on the roof. Thirteen automobiles were packed around the building, including five in the pool. A din of rocks kept banging against them. The stuck horn of a buried car was blaring. The family in the darkness in their fixed tableau watched one another by the light of a directional signal, endlessly blinking. The house had filled up in 6 minutes, and the mud stopped rising near the children's chins.

Source: Excerpted from McPhee, J. 1989. Los Angeles against the mountains. In *The Control of Nature.* New York: Farrar Straus & Giroux.

TABLE 10-2 MAJOR CALIFORNIA FLOODS

Date	Area Affected	Lives Lost	Damage and Cost
December 1861–January 1862	Entire state	Unknown	Much of inhabited parts of state submerged for many weeks
December 1937	Northern two thirds of state	Unknown	$15 million
March 1938	Coast south of San Luis Obispo, Mojave Desert	87	$79 million
December 1955	Northern two thirds of state	67	$166 million
December 1964	Northern half of state	24	Greatest known flood in northern California's recorded history; $239 million
January–February 1969	Southern, central coast, Mojave Desert	60	$400 million
January–February 1980	Central, southern coast	18	$350 million
January 1982	San Francisco	31	$75 million
February 1986	Northern half of state	14	$379 million
January and March 1995	Northern (January) central (March)	27	$3 billion; 42 (January), 57 (March) of 60 counties declared disaster areas; 10,000 residences damaged or destroyed
December 1996–January 1997	Sierra Nevada, Great Valley	8	$1.6 billion; 43 counties declared disaster areas; 16,000 residences damaged or destroyed

Sources: *California Floods and Droughts* 1991. U.S. Geological Survey Water Supply Paper 2375; and National Climatic Data Center, 1997, *www.ncdc.noaa.gov.*

WATER CONTROL AND TRANSFER

Most of California's 35 million inhabitants live in cities and suburbs that are far too arid to support them. The state's enormous agricultural industry, particularly in the San Joaquin, Imperial, and other southern valleys, grows crops that require far more water than what nature supplies. In addition, most of California's industries require large volumes of water for manufacturing, cleaning, and processing. It has been only been the efforts of humans to bring water to the people of California that have given this state the largest population and richest economy in the United States. Water development continues to be a major force that shapes the state's economy, politics, and history. Today, California relies on about 1,300 dams and reservoirs, with a storage capacity of about 43 million acre-feet, as well as on six major aqueduct systems and countless smaller structures to accomplish the task of water supply (Fig. 10-10). Unfortunately, basic lessons of the hydrologic cycle, stream equilibrium, and the variability of precipitation in California have often been overlooked during the development of California and the other states of the arid West, with the result that water crises are an almost annual event.

Fig. 10-10 California's manmade water network. (Source: California Department of Resources. 1987. *California Water: Looking to the Future,* Bulletin 160-87.)

Dams and reservoirs have been constructed for four different purposes, although many may serve more than one purpose simultaneously. In the Mississippi River basin, many dams have been built for flood control. Some of California's first water projects were built for the same reason. These consisted mainly of *levees* and bypass channels, built in the late 1800s on the American and Sacramento Rivers to protect Sacramento and surrounding areas. Dams built for flood control are designed to reduce peak flood discharges, retaining some of the flood runoff in the reservoir upstream of the dam. Levees are also commonly constructed along the banks of rivers, raising the height of the channel walls to contain high flows. In southern California, at the mouths of steep canyons flowing from the San Gabriel, San Bernardino, and other ranges, *debris basins* have been constructed to contain the huge volumes of sediment washed from canyons during floods.

A second purpose for constructing dams is for water storage and irrigation. In California, where peak runoff periods are directly out of phase with the main crop-growing season, it is essential to save runoff for the summer months. Runoff generated by the spring snowmelt can be caught in the reservoirs and later released. Many of California's streams have minimum flow increased or maintained by releases from reservoirs. Dams are also constructed to generate hydroelectric power. The energy released by the falling of the water through the dam's gates is converted to turbine-generated electricity. Finally, reservoirs can store water that is intended for transfer out of the drainage basin to another stream or to one of the aqueduct systems. Many of California's reservoirs provide recreation as an additional benefit, although they were not constructed for boating, fishing, or water skiing.

It is not surprising that the construction of dams has an impact on stream equilibrium. Because water flow ceases in reservoirs, streams drop their sediment loads there, with the predictable result that the reservoirs are filled by *deposition*, or siltation. Without dredging of the sediment or enlargement of the dam, many reservoirs would be filled within 20 to 30 years of construction. Downstream effects of dams are more variable. Reduction of flood flows might lead to downstream reaches becoming choked with sediment. On the other hand, removal of the sediment load behind the dam could cause the stream to become more erosive below the dam as it attempts to restore its load. Removal of part of a stream's discharge to another drainage basin has obvious effects on both streams as they attempt to adjust to the changes.

California's Aqueducts

The network of aqueducts that brings water to farms and people would not be possible without the system of water laws governing California. Water rights in California are established under two fundamentally different systems. *Riparian rights,* which governed California under Mexican rule, stipulate that those owning property adjacent to a stream or other body of water own the right to use the water in that stream. In the arid West, the lack of flowing streams made riparian law unworkable, particularly when the gold rush brought a great demand to divert water from streams to wash the gold-bearing gravels from the hill slopes. The miners developed a system of *appropriative rights*, which allow water to be appropriated and used in a beneficial manner at any site. Under this system, the holder of the first appropriation right in a stream system is entitled to use all of the water they have historically used before the second appropriator receives any. Those holding the rights receive their water allotment in the order of priority, and in time of drought, those with lower priority might receive no water. Both types of rights are recognized today in California, and a complex body of laws regulates water use in the state.

What has made California's water diversion system possible is that appropriative rights can be sold or transferred without losing their priority. This possibility, together with the legality of moving water away from the streams, has enabled cities, agricultural districts, water companies, and government agencies to move surface water to the farms and cities needing water. Recent legal decisions have required that minimum flows be maintained in some streams to preserve important habitats and populations, but historically the use of California's water by fish, birds, and other wildlife did not have legal protection. The Wild and Scenic Rivers Act of 1972 also currently prohibits additional construction of major dams and diversions on the Smith, Klamath, Trinity, and Eel Rivers in the North Coast region.

The first of the aqueducts built in California delivered its first water from the Owens Valley to the city of Los Angeles, 388 kilometers away (Fig. 10-10). The Los Angeles Aqueduct, opened in 1913, supplies about 80 percent of the Los Angeles water supply through a system of conduits, tunnels, and natural river channels. Because of the elevation of the Owens Valley, about 1,220 meters above sea level, the water flows largely under the influence of gravity.

HISTORY OF THE LOS ANGELES AQUEDUCT

The early pueblo of Los Angeles relied on the Los Angeles River for its water needs. At most times a very small stream nourished in part by groundwater from the San Fernando Valley, the Los Angeles River experienced occasional huge floods that overran the river's banks. As the city grew, the river's flow dropped because of pumping from groundwater wells, and city engineers began to look for additional water supplies. Fred Eaton and his successor William Mulholland, early superintendents of the Los Angeles City Water Company (later the Los Angeles Department of Water and Power), turned their sights eastward in a search for abundant and inexpensive water for their growing city. By 1900, the population of Los Angeles reached 100,000, and within 4 years, it doubled. The search for water grew more intense as Mulholland launched a scare-tactics campaign in the local newspapers to gather support.

In the 1860s, more than 400 kilometers away, along the eastern front of the Sierra Nevada, settlers began to irrigate the Owens Valley with water from the Owens River. By the late 1800s, the flourishing farmlands supplied the silver mines of Tonopah, Nevada to the east. Although the Owens River lies in the rain shadow of the Sierra Nevada, it collects abundant runoff from tributaries that flow eastward from the snowy High Sierra. The Owens River flowed into Owens Lake, a remnant of a formerly more extensive chain of lakes along the Sierra Nevada. Owens Lake lies at an elevation of about 1,200 meters above sea level, a fact that tempted Eaton to consider tapping the water of the Owens Valley for Los Angeles. The elevation difference meant that water could be delivered by gravity without any expensive pumping. At first, Mulholland was skeptical about the project, claiming that evaporation losses from the desert reservoirs and canals would be enormous. However, in 1905, the first exploratory visit to Owens Valley began one of the most colorful episodes in the West's water wars.

The U.S. government had been deeply involved in the irrigation of the American West since passage of the Reclamation Act. In the Owens Valley, the Reclamation Service had been developing the Owens Valley Reclamation Project, which would aid the farmers in irrigating the valley. In an arrangement of questionable morality, the Los Angeles Power and Water Company hired the Reclamation Service's senior engineer as a consultant to investigate acquisition of Owens Valley water for the city of Los Angeles. Convinced of the necessity of acquiring the Owens Valley water, Eaton and Mulholland began buying up land with senior priority water rights, often paying far more than the assessed value of the land. By the time that Owens Valley farmers realized that Los Angeles was conducting an all-out land grab, much of the water was controlled by Los Angeles. In

Continued.

HISTORY OF THE LOS ANGELES AQUEDUCT—cont'd

1907, after extensive lobbying by Mulholland, the Federal Reclamation Project was killed, and the Owens Valley water rights were effectively placed in the hands of Los Angeles.

The first water was delivered to Los Angeles by the Los Angeles Aqueduct in 1913 after 6 years of construction. The water-delivery agreements stipulated that the water could only be used by the city of Los Angeles, and that surplus water would remain in the Owens Valley. At the time of construction, the San Fernando Valley lay outside the city and was therefore ineligible to receive Owens Valley water. The actual water delivered by the aqueduct was far in excess of what was needed by the city, but the extent of the surplus was kept quiet. Mulholland had placed the end of the aqueduct in the San Fernando Valley, and, together with a group of wealthy investors, he had bought substantial land in the seemingly arid wasteland. The arrival of Owens Valley water made the land worth millions overnight, and the annexation of the San Fernando Valley to the City of Los Angeles completed the land and water grab.

The Owens Valley began to revert to its natural desert state, and the farms and orchards dried up. Irrigation ditches operated by small companies continued to tap water from the Owens River, but the City of Los Angeles fought for every drop. Things came to a head during the drought years of the 1920s, and frustrated ranchers blew up parts of the aqueduct system on several occasions. On November 16, 1924, a group of ranchers closed the Alabama Gates that send water from the lower Owens River into the aqueduct. This action temporarily reestablished the natural flow from the river into Owens Lake. The sabotages were unsuccessful, however, and by the mid-1930s, the city of Los Angeles owned 95 percent of the farmland in the Owens Valley.

In the 1970s the Los Angeles–eastern California water wars heated up again, when Los Angeles' diversions of water further to the north caused the level of Mono Lake to drop to new low levels. Biologists feared that the drop in the lake levels would destroy the safety of the breeding grounds of the Mono gull. The gull builds its nests on two islands in Mono Lake, inaccessible to predators unless the lake level becomes low enough for land bridges to form. A heated legal battle ensued, accompanied by a blast of environmental publicity, with the result that Los Angeles agreed to maintain sufficient flow into Mono Lake to protect the lake's habitats.

The Colorado River and the Salton Sea

A second major source of water in southern California is the Colorado River, which supplies water to 12 million people in seven states and parts of Mexico. Completed in 1940, the Colorado River Aqueduct carries water 400 kilometers to Los Angeles and, since 1947, via the San Diego Aqueducts to San Diego. Hoover Dam, completed in 1935, also supplies a large portion of southern California's electricity. Downstream on the Colorado River, the All-American Canal and the Coachella Canal irrigate the Yuma, Palo Verde, Imperial, and Coachella Valleys. California's total allotment from the Colorado River is currently 4.4 million acre-feet per year, of which 87 percent goes to irrigate the desert valleys, and 11 percent to the Metropolitan Water District, which serves 11 million people in San Diego and in communities adjoining Los Angeles.

Many water wars have developed over use of the Colorado River. In past years, southern Californians have used more than their legally allotted share of Colorado River water, relying on surplus water beyond that allocated to other users, including Mexico. However, on January 1, 2003, the Department of the Interior cut off the surplus flows to the state. The battle to use the allocated amount of Colorado River water puts agriculture (the Imperial Irrigation District and the Coachella Valley Water District) on one side and rapidly growing cities (the Metropolitan Water District of

Southern California and the San Diego County Water Authority) on the other. The environmental health of the Salton Sea ecosystem is a third party in the controversy. On January 1, 2003, a fragile Quantification Settlement Agreement collapsed, sending all parties back to the bargaining table.

Salton Sea Formation

Beginning in the late 1800s, the Imperial Valley had been irrigated by a system of canals taking water from the Colorado River. In 1901, a cut was opened along the river to bypass the intakes to the irrigation canals, which were choked with sediment. Unfortunately, five floods hit the area in 1905, causing the entire Colorado River to be diverted into the Alamo and New Rivers. Within weeks, the water formed the Salton Sea in the low-lying basin of the Salton Sink. In 1907, the river was rediverted back into its channel, but the Salton Sea remains as a part of the Imperial Valley's landscape (Fig. 10-11). Despite enormous losses to evaporation, the sea is maintained by water diverted from the Colorado River.

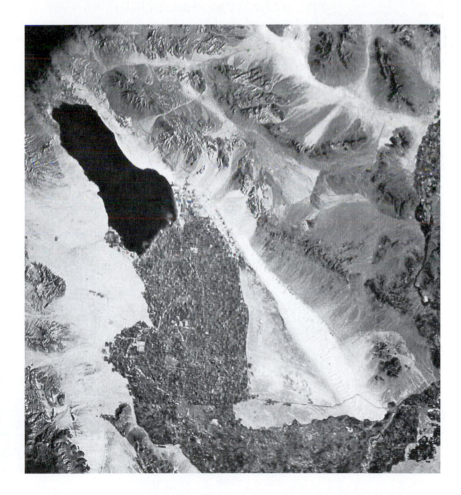

Fig. 10-11 An infrared satellite image taken by Landsat-I showing the Salton Sea and vicinity. Note the contrast between irrigated fields south of the Salton Sea and sand dunes east of the fields.

Other Aqueducts

In 1923, water behind O'Shaughnessy Dam flooded Hetch Hetchy Valley of the Tuolumne River, signaling the end of a long and bitter battle for its preservation. John Muir had stated, "Dam Hetch Hetchy? As well dam for water tanks the people's cathedrals and churches, for no holier temple has ever been consecrated to the heart of man." But San Francisco was thirsty for reliable water supplies, and in 1934, the first Sierra Nevada runoff flowed through the Hetch Hetchy system to San Francisco. Today the system has been expanded to supply San Francisco and 50 additional communities and water districts. Communities of the East Bay Municipal Utilities District receive water from the Mokelumne Aqueduct, which also brings Sierran water to the Bay area.

As groundwater supplies became scarce in the farmlands of the San Joaquin Valley, farmers began to examine the possibility of importing water from the north. Although the state approved construction of the Central Valley Project (CVP), which would bring water to the valley from a storage reservoir in the upper Sacramento watershed, California could not fund the project, and the Federal Bureau of Reclamation took it over. Shasta Dam, in the upper Sacramento Valley, was completed in 1951, and Friant Dam on the San Joaquin River was completed in 1954. The Delta-Mendota Canal, the Contra Costa Canal, Friant-Kern Canal, and the Madera Canal transported the stored water, and the San Luis Reservoir and Canal were added to the system in 1967. About 83 percent of the water supplied by the project is used for irrigation. Under the Federal Reclamation Act of 1903, the water supplied by the CVP can only be used by farmers holding 160 acres of land or less, but violations of the 160-acre rule by individuals and large corporations have been numerous. In 1992, a landmark bill, the Central Valley Project Improvement Act, was passed by Congress. The bill mandates more water for wetlands and maintenance of instream flows, allows water to be transported outside the project service area, and charges the farmers

Fig. 10-12 The California Aqueduct near Tracy, California. (Source: Sarna-Wojcicki, A. U.S. Geological Survey.)

more for their water. The bill also provides funds for restoration and enhancement of fish and wildlife habitats. Implementation of the bill could bring major changes to agriculture in the San Joaquin Valley.

California's most recent water project, the State Water Project, resulted in construction of the 444-mile-long California Aqueduct (Fig. 10-12). This system supplies water to the Napa, Livermore, and Santa Clara Valleys of northern California and to Kern County farmers in the San Joaquin Valley. The system also delivers water to southern California even though water must be pumped uphill over the Tehachapi Mountains and across the San Andreas fault.

GROUNDWATER

Precipitation that infiltrates into the ground can be stored in the spaces within sediments and rocks beneath the Earth's surface. This stored water is ***groundwater***, and it is an important part of California's water supply. At present, slightly less than half of the water used in California is groundwater, and in the drought years of 1987 through 1991, groundwater use rose to about 60 percent of California's total supply. Although no one knows the exact number of water wells in California, estimates are between 1 and 2 million.

As water seeps into the ground, some of it is held by the soil near the surface in the ***unsaturated zone*** (Fig. 10-13). This zone might be dry during periods when evaporation and transpiration are greater than infiltration. Beneath the zone of aeration, groundwater fills all available spaces in the rock or sediment in the ***zone of saturation***. The boundary between the unsaturated and saturated zones is the ***water table***, which marks the depth at which one encounters saturated materials while digging a hole or drilling a well. The water table can be at the Earth's surface in river valleys or at springs, or it can

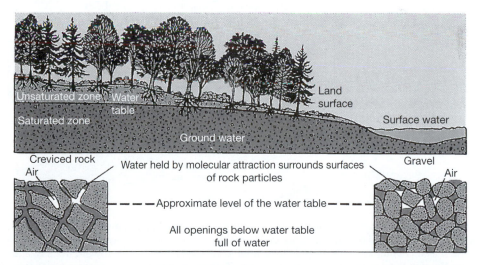

Fig. 10-13 Diagram illustrating the occurrence of groundwater. (Source: Waller, R. 1994. *Groundwater and the Rural Homeowner.* U.S. Geological Survey, General Interest publication 1994-380-615.)

be 100 meters or more below the surface beneath the floor of a desert valley. The position of the water table changes seasonally, rising as the groundwater is *recharged* with infiltrated precipitation, and falling during dry periods. During prolonged droughts in California, drops in the water table of tens of feet have been recorded.

The capacity of rocks and sediments as sources of groundwater is extremely variable. The *porosity* of a rock or sediment is a measure of the percentage of space in the material; porosity can be as low as 1 to 3 percent in granitic rocks or greater than 50 percent in some sediments. For groundwater (or petroleum or other fluids) to flow through rock or sediment, the pores must also be connected. *Permeability* of rocks or sediments is the ability of the material to transmit fluids. Clay and shale, which both have high porosity, are generally impermeable because the water is tightly held to the clay particles in very small pores. In contrast, fluids are easily transmitted in gravel, sand, and sandstone. Even crystalline rocks such as limestone or

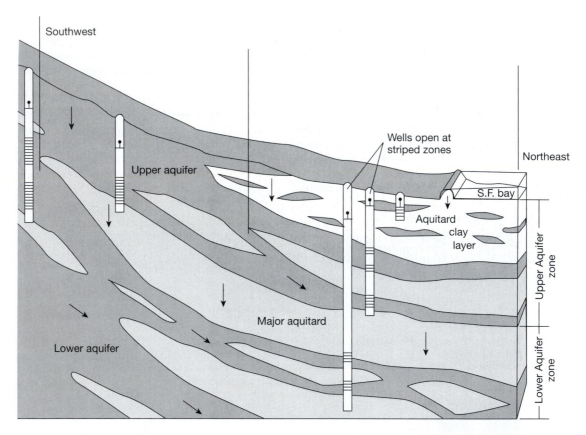

Fig. 10-14 Diagram showing aquifers and aquitards near San Francisco Bay in Santa Clara County. The aquifer layers are sandy alluvium, and the aquitards are clay layers deposited when the Bay was slightly more extensive than at present (see Chapter 12). Like the area topography, the layers slope gently toward the Bay, and the groundwater flows in the aquifers in the same direction. (Source: Adapted from Environmental Protection Agency diagram.)

granite can be permeable if fractures, weathering, or bedding create sufficient permeability (see Fig. 10-13). Permeable materials are called ***aquifers,*** whereas those that retard groundwater movement are termed ***aquitards*** or ***aquicludes*** (Fig. 10-14).

Groundwater flows very slowly from areas of high water pressure to low-pressure areas. Usually the direction of flow is very similar to the slope of the land surface, so that groundwater generally flows from higher elevations to low-lying areas. Groundwater can return to the land surface where it is discharged into streams or flows at the ground surface at springs, which form where the water table intersects the land surface. In California, the largest volume of groundwater is stored in unconsolidated, relatively young, sandy sediments. Commonly, these are alluvial deposits that underlie the valleys and basins of California. The aquifers used today for groundwater are relatively shallow, with water being pumped from depths of several meters to a few tens of meters. Some aquifers are ***confined*** between layers of less permeable sediment or rock. In this situation, the water might be stored under pressure. Humans extract groundwater by drilling a well into the saturated zone. In many areas of California, the groundwater was naturally confined under sufficient pressure that groundwater would flow to the surface without pumping, thus creating an ***artesian well*** (Fig. 10-15). When the water table drops, artesian conditions disappear, and mechanical pumps are required to lift the water to the surface. Of course, with continued drops in the water table, wells can dry up.

Fig. 10-15 A flowing artesian well in the northern Santa Clara Valley near Milpitas. Under normal conditions, groundwater is pumped into the pipe to the right of the well. Because of heavy rains from 1995 through 1997, the water table rose to the point that artesian conditions allowed water to flow under its own pressure from below the pipe. (Source: Calhoun, J.B. Santa Clara Valley Water District.)

Fig. 10-16 Percolation ponds along Los Gatos Creek, San Jose. Surplus runoff is released into Los Gatos Creek from Lexington Reservoir upstream, and then allowed to infiltrate into the groundwater system from these ponds. The groundwater is later pumped from the saturated zone as needed for municipal use. (Source: Santa Clara Valley Water District.)

It is important to recognize that the groundwater is an integral part of the hydrologic cycle. The level of the water table is maintained by recharge, and it will drop if water is pumped from aquifers at a rate exceeding the recharge. Changes to another part of the hydrologic cycle often have major impacts on the groundwater system. For example, if a significant part of a drainage basin is urbanized, the amount of recharge of the groundwater could be significantly reduced because asphalt and concrete do not allow infiltration to occur (Fig. 10-16). Because water that formerly seeped into the ground might now run off, streamflows could increase.

GROUNDWATER PUMPING, SUBSIDENCE, AND SALTWATER INTRUSION

In most large aquifers in California, water is stored in the pore spaces between grains of sand or gravel, and the aquifers are commonly confined by clay-rich layers. In the deeper aquifers, the pressure of the water helps to keep the grains separated. When pumping from an aquifer exceeds the rate of recharge, the water table drops and the water pressure decreases. The pore spaces collapse, the grains shift together, and compaction occurs (Fig. 10-17). Because pore spaces in the aquifer are lost, the capacity of the aquifer to hold water is reduced and will not recover, even with reduced pumping and a series of wet years.

The land surface above a compacted aquifer could also collapse, especially if young, unlithified sediments underlie the area. The process of land-surface sinking is known as **subsidence**. Evidence of subsidence includes ground cracking and damage to roadways, aqueducts, and structures. Subsidence caused by excessive groundwater pumping is a common occurrence in areas of California where groundwater is pumped for agricultural and municipal wells. The following are a few examples:

- In the China Lake Basin in the Mojave Desert, a series of long cracks developed during 1990, damaging the runways at the China Lake Naval Air station. At first,

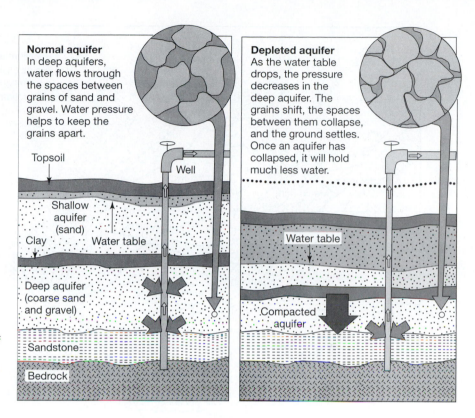

Normal aquifer
In deep aquifers, water flows through the spaces between grains of sand and gravel. Water pressure helps to keep the grains apart.

Topsoil

Well

Shallow aquifer (sand)

Clay   Water table

Deep aquifer (coarse sand and gravel)

Sandstone

Bedrock

Depleted aquifer
As the water table drops, the pressure decreases in the deep aquifer. The grains shift, the spaces between them collapse, and the ground settles. Once an aquifer has collapsed, it will hold much less water.

Water table

Compacted aquifer

Fig. 10-17 The effects of compaction on an aquifer. (Source: *San José Mercury News,* April 4, 1991.)

geologists were uncertain about the origin of the cracks and included faulting in their list of possible causes. Investigations of the geology and groundwater use in the basin revealed that the cracks are the expression of subsidence in the alluvial aquifer being pumped to supply the base and surrounding community.

- The San Joaquin Valley and the Delta area have a long history of subsidence due to groundwater pumping. On the western edge of the San Joaquin Valley near Mendota, an area of several square miles under the California Aqueduct and Interstate 5 experienced 25 centimeters of subsidence during 1991. Some areas of the San Joaquin Valley have sunk as much as 9 meters since the 1920s. In the Delta, farmlands now lie well below sea level because of subsidence, caused by the disintegration of organic layers within the sediments as well as by agricultural development. Only a system of dikes and levees stands between these farmlands and disastrous flooding.

- In the central third of the Santa Clara Valley, almost 2.5 meters of subsidence occurred between 1934 and 1967 (Fig. 10-18). As a result, areas near sea level around the margins of San Francisco Bay are more frequently flooded by storms and high tides. Some parts of the city of Santa Clara have subsided below sea level, and only levees along the shore of San Francisco Bay prevent flooding by bay water.

In aquifers along the coast, fresh groundwater normally flows to the sea to a point where it sits in contact with salt water. Because it is lighter than seawater,

Fig. 10-18 Subsidence in the Santa Clara Valley between 1934 and 1967. Subsidence values are given in feet. (Source: Poland, J. U.S. Geological Survey, Professional Paper 497-F.)

which contains more dissolved load, the fresh groundwater overlies the salt water (Fig. 10-19). Pumping fresh water at a rate greater than the recharge rate causes the water table to drop and the pressure in the aquifer to drop. Salty groundwater is then drawn inland and can replace the fresh water in wells near the coast. This process is known as *saltwater intrusion*, and it occurs in coastal aquifers along with subsidence. Well water in parts of the Santa Clara Valley, the Salinas Valley, and the Long Beach areas has been made unusable by saltwater intrusion. In parts of the Los Angeles Basin, groundwater levels fell to 30 meters below sea level in the 1950s. Although pumping fresh water back into the aquifers has halted the spread of saltwater intrusion in some areas, it has not managed to push back the intruding seawater.

In most large groundwater basins in California, pumping rates are beginning to be restricted to minimize the problems of overpumping, or *overdraft,* of the aquifers. With this management, water managers hope to attain a *safe yield* of groundwater extraction to minimize the problems of water table lowering, subsidence, and saltwater intrusion.

Fig. 10-19 Saltwater intrusion in coastal aquifers. (Source: Adapted from Coch, N., and Ludman, A. 1991. *Physical Geology.* New York: Macmillan.)

WATER QUALITY

Both surface water and groundwater naturally contain a variety of dissolved and solid materials. Dissolved substances can be carried in rainfall, or they might be derived from the chemical weathering of rocks within the drainage basin. The vegetation and animals of the drainage basin contribute a variety of organic compounds. Solid particles can be part of a stream's sediment load, but are not normally present in slow-moving groundwater. The solid particles in an aquifer generally act as a filter that removes some of the substances from the water.

The materials contained in water give it a taste and appearance unique to that water source. Many dissolved substances, such as iron, are essential for life. However, water pollution occurs when undesirable or toxic substances are present in the water, or when abnormally high levels of naturally occurring substances are introduced. Water can be naturally unsuitable for drinking due to the presence of volcanic gases, dissolved metals, or high amounts of salts, or it could become contaminated by human activities. Because of public concern about the safety of drinking water, the U.S. Environmental Protection Agency (EPA) has established safe levels of contaminants for public drinking water, known as *maximum contaminant levels (MCLs)*. These standards specify the maximum concentration allowed for particular substances, and the list is continually being revised as new contaminants are identified and as studies of the risk of exposures are conducted (Table 10-3).

In the United States we have come to take for granted the availability of unlimited safe drinking water. Potentially deadly problems of water contamination by sewage and animal waste are rare in California, and the state's water is generally abundant, clean, and cheap. Municipal water is usually treated and filtered to ensure its safety. However, concern about the safety of drinking water supplies has grown in recent years as problems have come to the public's attention:

- Agriculture, industry, and communities are all sources of water contamination in California. Runoff from agricultural lands in California's valleys introduces harmful levels of nitrates from fertilizers, as well as pesticides and high concentrations of potentially toxic elements flushed out of soils by excessive irrigation.
- Thousands of leaking underground storage tanks are sources of groundwater pollution by gasoline, solvents, and other hazardous chemicals. In 1984, the California Water Resources Control Board identified 165,000 leaking underground storage tanks, about 80 percent containing gasoline or other petroleum compounds.
- Landfills, both municipal and industrial, are sources of a host of contaminants ranging from animal wastes to used paint, batteries, and highly toxic solvents. In all, roughly 2 million tons of hazardous waste were produced by factories,

TABLE 10-3 SELECTED WATER CONSTITUENTS AND DRINKING WATER STANDARDS

Constituent	Federal MCL	Constituent	Federal MCL
Total dissolved solids (TDS)	500 ppm[1]	Coliform, per 100 ml	1–4
Asbestos	7.1 ppb[2]	1,1Trichlorethane (TCA)[4]	200 ppb
Cadmium	10 ppb (5)[3]	Trichloroethylene (TCE)[4]	5 ppb
Lead	50 ppb (5)[3]	Trihalomethanes (THM)[5]	100 ppb
Mercury	2 ppb	Enarin[6]	0.2 ppb
Selenium	10 ppb	Heptachlor[6]	0.4 ppb (.01)[3]
Benzene	5 ppb (1)[3]	Nitrate (ion)[7]	45
PCBs	0.5 ppb	Perchlorate[8]	1–32 ppb[9]

[1] Parts per million; [2] parts per billion; [3] number in parentheses is the lower California standard; [4] solvents; [5] by-product of chlorination; [6] pesticide; [7] fertilizer, animal waste; [8] oxidizing agent; [9] standard not set as of 2003.

Source: California's Drinking Water.

businesses, and government operations in California in 1989. In 1983, California had 1,800 surface storage impoundments for industrial wastes.

Current regulations require proper disposal or recycling of hazardous compounds, but enforcement has not yet been perfected. Current regulations also require liners for landfills and leak-proof containers for underground tanks, but a legacy of thousands of sites remains to be cleaned up in California.

Groundwater contamination has turned up in wells in all parts of California. In 1986, the Department of Health Services tested 538 wells and found that 18 percent showed some level of contamination. Cleanup of the sites with large concentrations of highly toxic chemicals is presently being managed by the EPA's Superfund program. These sites include the notorious Stringfellow Acid Pits in Riverside County, where 35 million gallons of chemical wastes were dumped into 22 ponds during the 1950s and 1960s. In southern California, other sites are contaminated by oil refinery waste. Santa Clara County has the dubious distinction of having the largest number of Superfund sites (28 in 1992) in the country, many of them involving leaking storage tanks for solvents used to manufacture silicon chips for the computer industry (Fig. 10-20; see Chapter 2). In the Sacramento Valley, Superfund sites

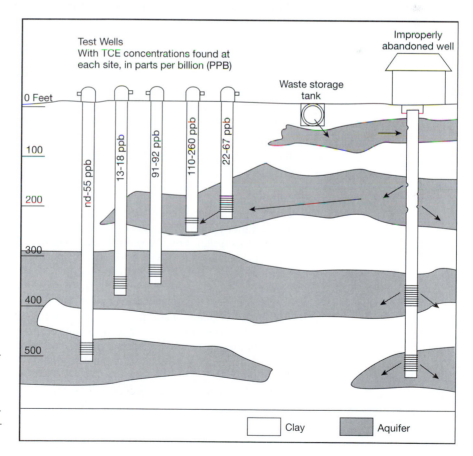

Fig. 10-20 Groundwater contamination by trichloroethylene (TCE) from a waste storage tank in Santa Clara County. (Source: *San José Mercury News.*)

include the abandoned Iron Mountain copper mine near Redding, an area near Oroville contaminated by wood-treatment chemicals, and a plume of solvents near Sacramento.

Although groundwater cleanup is progressing at numerous sites in California, it will take decades to complete the task. Once a contaminant site is identified, the extent of contamination must be determined by careful study of the aquifers. The direction and rate of groundwater flow must be precisely mapped, a difficult task that requires drilling and sampling to obtain detailed knowledge of the geologic picture in three dimensions. It is also important to estimate the rate of spreading of the contaminant plume to protect the groundwater in the vicinity of the spill. Cleanup might be accomplished by pumping out contaminated water and treating it, a costly process that can take several years of continuous pumping.

PERCHLORATE

Ammonium perchlorate is the most commonly used of a group of strongly oxidizing industrial chemicals that easily dissolve in water to produce the perchlorate (ClO_4^-) ion. Perchlorate is very mobile in water and can persist in the aqueous environment for decades. Because it interferes with the uptake of iodine by the thyroid gland, perchlorate contamination of drinking water is of serious concern to public health officials. As with many contaminants, perchlorate poses a particular hazard to developing fetuses and children. At the present time, no federal or state MCL for perchlorate exists. Until a formal level is adopted, the California Department of Health Services is using the MCL of 40 parts per billion as an advisory action level.

According to the EPA, ammonium perchlorate is the main ingredient in solid propellant for rockets, missiles, fireworks, matches, and highway flares. Perchlorate salts are used in manufacturing a variety of other materials, including airbag inflators and nuclear reactors, and in electroplating, rubber manufacturing, and aluminum refining.

Perchlorate has been recognized as a groundwater contaminant in California since 1997, although large amounts of perchlorate have been disposed of since the 1950s. In the Los Angeles region, groundwater within at least 38 drinking water systems has been contaminated with perchlorate, much of it originating from rocket testing facilities.

Perchlorate contamination has also been discovered in groundwater in southern Santa Clara County in the town of San Martin. The source of the perchlorate is a former manufacturing plant where highway flares and skeet were manufactured from 1956 until 1998. The corporation began testing of the groundwater in 2000, and when they confirmed the contamination, the Regional Water Quality Control Board led an investigation of areas outside the plant's borders. As of 2003, perchlorate had been found in at least 235 private wells (Fig. 10-21), causing residents to turn to other sources for their drinking water.

Fig. 10-21 Map of perchlorate contamination in groundwater wells, San Martin, Santa Clara County. (Source: Santa Clara Valley Water District.)

WATER FOR THE FUTURE

In a year with average runoff, California's current water supplies are adequate to meet the needs of the population. However, the drought of 1987 through 1992 stretched the ability of the system to supply water for much of the state and raised the issue of additional water for the future. As it has been during California's past, the issue of water development is controversial. Many feel that increased conservation, limited expansion of industry, and limited population growth are the steps necessary to preserved the ecosystems of California's rivers and wetlands. Others believe that continued development of water projects is a fundamental aspect of guaranteeing California's economic growth. As water supplies become scarce, communities are investigating the possibility of desalinization, conservation programs, and water purchases from other water districts. Many believe that agriculture will have to relinquish some of the 83 percent of California's water supply that it currently uses to facilitate the growth of cities and industry. The cost of the original Hetch Hetchy system was $100 million, and that of the Los Angeles Aqueduct approximately $200 million. As of 1988, the State Water Project had cost $3.7 billion. Even if other obstacles to further water development could be overcome, the cost of such projects is formidable.

SUMMARY

1. California receives on average only about 58 centimeters of precipitation annually, and most of it is brought by storms originating in the northern Pacific Ocean. The northwestern part of the state and the northwestern slope of the Sierra Nevada receive the most precipitation, whereas southeastern California and the southern San Joaquin Valley receive the least.
2. California has a Mediterranean climate, with warm, dry summers, and cool, wet winters. The dry season coincides with the time of greatest water demand by agricultural users, who consume about 85 percent of California's water. The largest water users are farms in the Central Valley, the Imperial Valley, and other valleys in central and southern California. California's largest municipal water users are also located in areas with relatively low precipitation.
3. Because of the variability in precipitation patterns, both droughts and floods are normal occurrences in California.
4. Precipitation that falls on the land surface can evaporate, be transpired by vegetation, infiltrate into the groundwater system, or become runoff. Runoff is the water that flows in California's rivers and streams. Streams carry dissolved ions in solution and a solid load, consisting of suspended sediment and bed load.
5. Stream channels develop over time to carry the average runoff and sediment supplied to an area. The channels are integrated into a network, with each stream carrying runoff from its drainage basin. All runoff from a given drainage basin flows past its lowest point, or base level.
6. Flooding occurs when runoff exceeds the capacity of a stream channel to carry it. Floodplains form by the deposition of fine-grained sediment on low surfaces adjacent to rivers.
7. California's 1,300 dams have been constructed for flood control, water storage for irrigation and municipal use, hydroelectric power, and water transfer. A complex

network of reservoirs and aqueducts transports water from areas of high runoff to areas of high water demand.

8. Groundwater is stored in permeable sediments and rocks, called aquifers, beneath the surface. At many localities in California, the rate of pumping of groundwater has exceeded the recharge rates, lowering the water table. In some areas, this has resulted in saltwater intrusion and land subsidence.

9. In general, California has had abundant, inexpensive, and safe water for drinking and other uses. However, there are many documented cases of water contamination from a variety of natural, agricultural, and industrial sources. Water is also a limiting factor faced by the state's growing population and economy.

REFERENCES

California Department of Water Resources. 1987. *California Water: Looking to the Future*, Bulletin 160-87, 122 pp.

CECH, T.V. 2003. *Principles of Water Resource*. New York: Wiley, 446 pp.

KAHRL, W.L., ed. 1978–1979. *The California Water Atlas*. Governor's Office of Planning and Research, in cooperation with the California Department of Water Resources, 118 pp.

——————————. 1982. *Water and Power: The Conflict Over Los Angeles' Water Supply in the Owens Valley*: Berkeley: University of California Press, 583 pp.

MOUNT, J. 1994. *California Rivers*. Berkeley: University of California Press, 359 pp.

PURIN, G., and FONO, E. 1989. *Consumer's Guide to California Drinking Water*. Sacramento, CA: Golden Empire Health Planning Center, 81 pp.

REISNER, M. 1986. *Cadillac Desert: The American West and its Disappearing Water*. New York, Viking Penguin, 582 pp.

Web Sites and Media Resources

Cadillac Desert, public television series based on the book, produced by KTEH, San Jose.

National Oceanic and Atmospheric Administration (Weather and Climate Information, including links to National Weather Service): *http://www.noaa.gov*

U.S. Geological Survey, Water Resources Division: *http://water.usgs.gov*

U.S. Environmental Protection Agency: *http://epa.gov*

U.S. Bureau of Reclamation, links to Shasta Dam, Hoover Dam, and others: *http://www.usbr.gov*

State Water Resources Control Board, including links to water quality and water rights information: *http://www.swrcb.ca.gov/*

California Department of Water Resources: *http://wwwdwr.water.ca.gov/*

California Flood Information, California Environmental Resources Evaluation System: *http://ceres.ca.gov/topic/flood2.html*

California Department of Fish and Game: *http://www.dfg.ca.gov/*

Association of California Water Agencies: *http://www.acwanet.com/*

11

The Great Valley: Sediments and Soils

Fig. 11-1 Map showing the location of the Great Valley Province.

Great Valley Highlights

Major geographic feature:	Great Valley, also called Central Valley
Major rivers:	Sacramento River, San Joaquin River
Other landscape features:	Sutter Buttes, Tulare Lake
Predominant rock types:	Late Mesozoic marine sedimentary rocks of the Great Valley sequence, Cenozoic marine and terrestrial sedimentary rocks, recent floodplain alluvium, Coast Range ophiolite

Plate tectonic events:	Mesozoic sea between subduction zone and Sierran arc
Famous faults:	Coast Range fault
Other geologic events:	Sedimentation during glacial periods; flooding of wetlands
Resources:	Soil and fertile agricultural lands; water, sand, and gravel
State Parks of interest:	Colusa, Great Valley Grasslands, San Luis Reservoir, Turlock Lake
Other sites of interest:	Cosumnes River Preserve, Shasta Dam

Viewed from above, California's Great Valley (Fig. 11-1) is a long, flat-bottomed hollow, smoothed out between the rugged mountains of the Coast Ranges and the Sierra Nevada (Endpaper 3). From Red Bluff to Bakersfield it is a relatively feature-less plain, broken most noticeably by the Sutter Buttes near Marysville. The Great Valley, also known as the Central Valley, is about 650 kilometers long and averages about 80 kilometers in width. Its 15 million acres of farmlands constitute the richest agricultural region in the history of the world. The 1986 agricultural harvest, worth $14 billion, exceeded the value of all the gold ever mined from California. In 2000, the San Joaquin Valley's harvest alone reached $10 billion.

The Sacramento Valley forms the northern third of the Great Valley, and the San Joaquin Valley makes up the southern two thirds of the province. The south-flowing Sacramento River and the north-flowing San Joaquin River come together in the Delta area where they flow into San Francisco Bay. Both valleys are watered by the large rivers flowing west from the Sierra Nevada, with smaller tributaries flowing eastward into the Sacramento and San Joaquin Rivers from the Coast Ranges (End-paper 3 and Fig. 11-2). South of Fresno there is a region of low-lying bottomlands where rivers drain onto the floors of *intermittent lakes*, the largest of which is Tulare Lake. Gently rolling plains and hills as high as 200 meters from the valley floor along the borders with the surrounding mountainous provinces.

Before cultivation and the introduction of cattle, the Great Valley was a vast area of native grasslands, riparian forests, and extensive tule marshes. It was home to one of the largest Native American populations in North America, estimated by histori-ans at more than 100,000. Because of the mild climate and ample food and water, the valley also supported a wide variety of wildlife. In 1852, James H. Carson, a writer for Stockton's *San Joaquin Republican,* described in *Tulare Plains,* the wildlife of the Tulares (San Joaquin Valley). A century and a half later, it is difficult to recognize even the traces of this former world:[*]

> Every beast and bird of the chase and hunt are to be found in abundance on the Tulares. Horses, cattle, elk, antelope, black tail and red deer, grizzly and brown bear, black and grey wolves, coyotes, ocelets, California lions, wild cats, beaver, otter, minks, weasels, ferrets, hare, rabbits, grey and red foxes, grey and ground squirrels, kangaroo rats, badgers, skunks, musk rats, hedge hogs, and many species of small animals not here mentioned; swan, geese, brant, and over twenty different description of ducks also cover the plains and waters in countless myriads, from the first of October until the first of April, besides millions of grocus [sand hill crane], plover, snipe, and quail. The rivers are filled with fish of the largest and most delicious varieties, and the sportsman and epicurean can find on the Tulares every thing their hearts can desire.

[*]Source: Browing, P., ed. 1991. *Bright Gem of the Western Seas: California 1846–1852.* Lafayette, CA: Great West Books, p. 90.

Fig. 11-2 Map of the geomorphic features of the Great Valley. (Source: Adapted from Bailey, E.H., ed. *Geology of Northern California.* California Department of Conservation, Division of Mines and Geology, Bulletin 190, p. 240; geomorphic units after Davis and others [1959] and Olmstead and Davis [1961].)

FLOODPLAIN SEDIMENTS

Most of the surface of the Great Valley is covered by recent and Pleistocene alluvium (Endpaper 1). Sediments eroded from the Sierra Nevada, and to a lesser extent the Coast Ranges, are deposited on the floodplains and bottomlands as the mountain streams greatly decrease their velocity in the flat-bottomed valley. The broad, flat floodplains of the Central Valley are fertile agricultural land, periodically replenished with fine-grained alluvium by spring floods, and these lands are one of California's greatest natural resources.

As discussed in Chapter 8, periodic glaciation of the Sierra Nevada had drastic effects on the landscape. The impacts of global climate change were also strongly felt by Great Valley rivers. Huge quantities of sediment were piled up in moraines at the edges of alpine glaciers during times of glaciation. Sierra Nevada rivers, made even more powerful by the abundant glacial meltwater and the considerably wetter climate during these periods, carried much of this sediment downstream to the Sierran foothills and into the Central Valley. Geologists term this episodically deposited alluvium *glacial outwash*. At least four major pulses of deposition took place as sediment washed out from the Sierra Nevada during the glacial episodes of the past 2 million years.

In the central lower parts of the Great Valley, the younger alluvium buried older deposits, which geologists can identify in cores obtained from beneath the Valley floor (Fig. 11-3). Today the floodplains of the Sacramento and San Joaquin Rivers and their

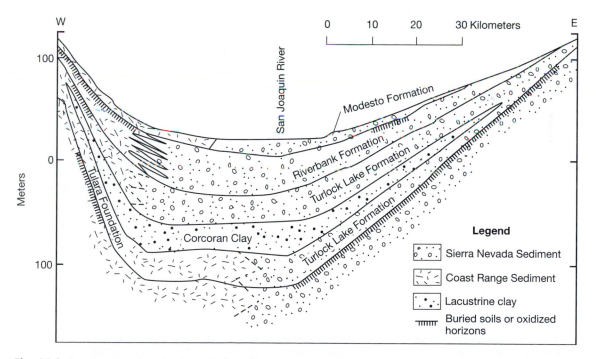

Fig. 11-3 Cross section showing buried alluvial layers identified in cores from the northern San Joaquin Valley. The Corcoran clay layer is discussed in Chapter 12. (Source: Adapted from Bartow, J.A. 1991. U.S. Geological Survey, Professional Paper 1501.)

tributaries receive sediment during times when the rivers overflow their channels, further burying the older deposits. However, along the Sierran foothills, the older alluvium can be seen at the surface. Because the Sierra Nevada continues to be uplifted relative to the Great Valley, rivers are constantly cutting downward (Fig. 11-4). As they cut or *incise* their channels, the rivers leave behind remnants of the older floodplain surfaces as **stream terraces**. Stream terraces can be recognized by their flat surfaces (Fig. 11-5; see also Fig. 10-7) and by the presence of alluvium beneath the surfaces. Along the edges of the main river valleys, several terrace levels are preserved at different elevations above the modern rivers. These terrace levels record a time sequence of river down-cutting. With increased elevation above the modern rivers, one finds progressively older valleys and deposits.

GREAT VALLEY SOILS

The soils of California's Great Valley are a direct result of the geological processes just discussed. Fertile soils are one of California's great resources, and they are particularly abundant on young alluvial surfaces. Soil forms as the result of both geological and biological processes that alter sediment and rock at the Earth's surface. In the Great Valley, most soils are formed on floodplain deposits. Following deposition of sediment on a floodplain, vegetation becomes established on the surface. As plant material decays because of the activity of microbes living in the soil, organic matter is added to the upper soil in the form of **humus**. The addition of humus to the original deposit increases the fertility of the soil, changes the

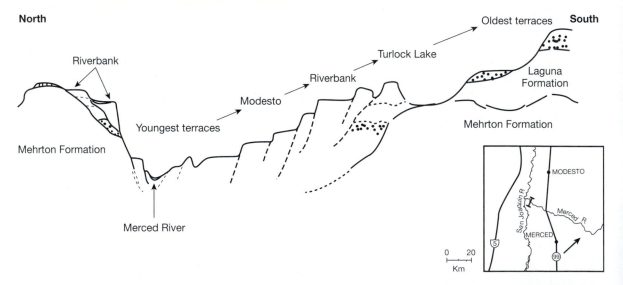

Fig. 11-4 A schematic cross section of river terraces above the Merced River near Snelling, Merced County, in the area shown on the inset map. As the Sierra Nevada has been uplifted and tilted, the Merced River has cut progressively downward to lower levels. The terraces and their deposits mark the former, now elevated, positions of the river floodplain and channel. The highest alluvium, deposited on the Laguna formation, is the oldest. (Source: Adapted from Marchand, D., and Allwardt, A. 1981. U.S. Geological Survey, Bulletin 1470.)

Fig. 11-5 A relatively flat younger terrace of the Merced River, near Dry Creek. In the background, an older terrace shows the rolling hills that are typical of older, more weathered surfaces. (Source: Marchand, D. U.S. Geological Survey.)

texture of the soil, and increases the potential for mineral weathering because of the formation of organic acids. The accumulation of organic matter in the soil is evident by the brown, gray, or black colors in the upper part of the *soil profile*. The layer or *soil horizon* characterized by organic matter accumulation is known as the *A horizon*. A horizons of soils can be thought of as the zones of maximum biologic activity.

At the same time that biological processes are altering the sediment, chemical and physical weathering processes act on the minerals. As water percolates downward, it becomes acidified by the organic acids in the A horizon. Beneath the A horizon, chemical weathering breaks down silicate minerals. Some easily dissolved elements might be moved completely through the profile if sufficient water moves through the deposit. Other, less soluble elements might simply be redeposited lower in the profile when the percolating water evaporates. Weathering of common silicate minerals like feldspar and mica produces clay minerals, which cause the soil to become stickier and better aggregated. Downward-percolating water also carries fine solid particles of clay, iron oxide, and salts lower in the profile. The soil beneath the dark A horizon becomes increasingly red as iron oxide (rust) builds up. The soil horizon characterized by the accumulation of one or more weathering products is known as the *B horizon*. B horizons can be thought of as the zones with the greatest evidence of weathering activity (Fig. 11-6).

Because the processes that form B horizons occur more slowly than the buildup of organic matter, young soils display A horizons but have not yet developed B horizons. In the Great Valley, soils on floodplain sediments deposited during the past several thousand years typically show A horizons overlying unchanged floodplain sediments. Soils on older floodplain sediments, found on higher terraces dating to the last glacial period or before, display progressively redder B horizons with progressively more accumulated clay (see Fig. 11-6). In some parts of the valley, these older soils also have developed *hardpans*, hard layers formed by the accumulation

Fig. 11-6 A close-up view of a soil profile of the San Joaquin soil formed on the Riverbank formation. The relative position of the Riverbank terraces is shown in Fig. 11-4. Note the quarter for scale in the upper right. The blocky aggregated zone immediately beneath the quarter is the B horizon, where sufficient clay has accumulated to form soil peds. At the base of the soil, accumulated silica in the hardpan area has turned the soil white. (Source: Marchand, D. U.S. Geological Survey.)

of silica, iron, or calcium carbonate. The hardpans pose problems for irrigation engineers in some areas. To plant orchards in some areas along the margin of the San Joaquin Valley, farmers have dynamited through the hardpans.

The oldest soils found in the Great Valley can be seen on ancient fans and terraces high above the modern rivers along the eastern side of the valley (see Fig. 11-4). Here one can see several meters of highly weathered, bright red soil that probably has taken more than a million years to form.

Soil profiles develop on other types of deposits, as well as on rock surfaces, in a similar manner to that just described. The type of soil that forms on a surface, the depth of the soil profile, and the rate at which it forms all depend on several soil-forming factors. These include the climate of the region, the organisms present (especially plants and microbial organisms), the topography of the surface, the nature of the rock or deposit, (called the ***parent material*** by soil scientists), and the time that has elapsed since the surface formed. The great fertility of California's alluvial soils in the Central Valley and other valleys throughout the state is the result of an ideal combination of all of these factors.

One of California's most agriculturally productive soils is the San Joaquin soil series, developed on older alluvium of the Great Valley. It is found in nearly a half-million acres of California and has been designated as California's state soil. A typical San Joaquin soil has a well-developed B horizon and a silica hardpan at about 1 meter below the surface (Fig. 11-6). The landscape formed on the San Joaquin soil and its alluvial parent material shows well-developed mounds and hollows (Figs. 11-7 and 11-8), known to geologists as ***mima mounds***. They are found on older, high terraces in the eastern Great Valley, as well as older marine terraces in California. The mounds and depressions are probably the result of weathering and the activities of colonial burrowing animals like pocket gophers.

Fig. 11-7 Mima mound topography on an older soil, south of Merced.

Fig. 11-8 A relatively well-developed B horizon of a soil profile on granitic rocks can be seen on the left side of this outcrop on Highway 120, western Sierra foothills. (Source: Sarna-Wojcicki, A. U.S. Geological Survey.)

The shallow depressions of the mima mound topography are home to a complex, fragile, and rare habitat known as ***vernal pools***. During the rainy seasons of winter and spring, the depressions between the mounds fill with direct rainfall and runoff from the surrounding mounds. The water forms ponds because it cannot percolate through the clay-rich B horizon and the hardpan layer of the soils (Fig. 11-9 and Color Plate 27). The pools then come alive with a complex ecosystem of plants and animals, including rare species like the vernal pool fairy shrimp and the frogs and birds that eat them. All of the life forms are adapted to the short-lived pond seasons; some remain dormant

Fig. 11-9 A vernal pool in the western Sierra Nevada foothills near Merced. The pool is formed in a hollow between mima mounds, which have formed in the China Hat soil developed on the highest terraces shown in Fig. 11-4. (Source: Durgarian-Ellsworth, A.)

SOILS, FOOD, AND THE ROCK CYCLE

Agricultural scientists have demonstrated that 16 chemical elements are essential for plant growth. Three of these *nutrients*—carbon, oxygen, and hydrogen—are obtained directly from the atmosphere by the leafy parts of plants. Nitrogen, sulfur, and phosphorous are made available to plant roots by the decay of organic matter. For the remainder of the essential nutrients, plants depend directly on mineral weathering. These elements are released from minerals into solution and then absorbed from the soil by roots. Many of the most essential elements—potassium, magnesium, and calcium—are common constituents of silicate minerals such as feldspar and mica, as well as other common minerals (see Chapter 2). The remaining essential micronutrients—iron, chlorine, manganese, zinc, boron, copper, and molybdenum—are also supplied by minerals.

Animals, including humans, that rely on plants for some or all of their food supply also require essential chemical nutrients for survival and growth. Small amounts of elements not required by plants, such as *selenium* and iodine, are necessary for the growth of many animal species. Studies have shown that soils deficient in these elements can lead to disease and death. On the other hand, large concentrations of some elements, including sodium, boron, selenium, and copper, could be toxic to animal species. Both deficiencies and high concentrations of certain elements in soils can be a natural result of the weathering of certain types of rocks and minerals.

From what we learned in Chapter 2 about the rock cycle, it is clear that the supply of nutrients to soils by organic decay, and especially by mineral weathering, is a slow process. In addition to direct solution from minerals, nutrients can be replenished by other geologic processes, including addition from wind-blown dust and deposition of fine-grained flood sediments. The natural plant community in a given area is well balanced with its nutrient supply. However, when humans cultivate the land, they often replace the natural plants with crops that require a greater nutrient supply. As soil nutrients are used more quickly than they are naturally replaced, soils become depleted in some nutrients, thus requiring farmers to use fertilizers, increasing the cost of farming. Most crops grown in the Central Valley today require the addition of nitrogen, phosphorous, and other nutrients.

during the dry season, others like leeches burrow into the mud, and many species complete their life cycle in a few short weeks or months. Vernal pools are extremely fragile and are destroyed by agricultural and residential development. Protection of vernal pool habitats has been provided as part of the construction of the University of California campus at Merced, and vernal pools have also been protected at Four Creeks Land Trust near Visalia and McKenzie Preserve, where the pools are formed in a table mountain basalt (see Chapter 8 for discussion of table mountains).

Continued intensive agriculture and irrigation have created several water-quality problems in the Great Valley and other agricultural areas of California. Partly as a result of fertilizers and irrigation runoff, high levels of nitrates, pesticides, and herbicides in streams pose a water-quality problem in some areas. However, not all water-quality problems are caused by the application of chemicals to farmlands.

In arid regions like the San Joaquin Valley, less water percolates through the soil profile every year than in more humid regions. Over time, relatively soluble minerals—calcium carbonate, sulfates, and other *salts*—accumulate in the lower part of the soil profile at the depth where percolating water evaporates. The buildup occurs because rainfall is not sufficient to flush these dissolved materials through the soil. In the western San Joaquin Valley, the bedrock source for the alluvial parent materials is the marine sediments of the Coast Ranges. As a result, the natural salinity of the soils is high, and soils in this area are particularly high in selenium-bearing minerals.

Irrigation of arid soils causes salts to be dissolved and carried to the lower horizons at a higher rate than would occur under natural conditions. In many areas, including the western San Joaquin Valley, problems occur because of the presence of clay layers in the alluvial parent materials. The clay layers form a barrier to percolating water, causing salt-rich irrigation water to build up near the surface. It is estimated that approximately 10 percent of the soils in the Great Valley have been damaged by the buildup of salts (Fig. 11-10). Salinization might inhibit plant growth, damage leaves, or even kill salt-sensitive crops.

Salinity problem areas

Great Valley

Fig. 11-10 Map showing the extent of salinization in the Great Valley. (Source: Johnson, S., and others. 1993. *The Great Central Valley.* Berkeley: University of California Press, p. 185.)

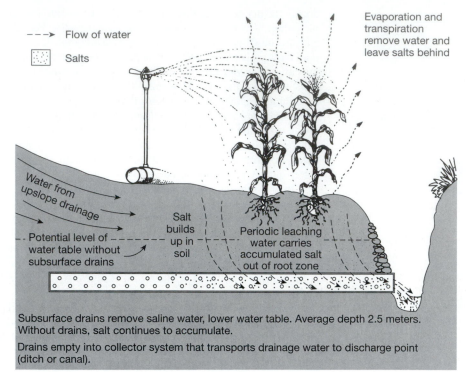

Evaporation and transpiration remove water and leave salts behind

- - - → Flow of water

Salts

Water from upslope drainage

Potential level of water table without subsurface drains

Salt builds up in soil

Periodic leaching water carries accumulated salt out of root zone

Subsurface drains remove saline water, lower water table. Average depth 2.5 meters. Without drains, salt continues to accumulate.

Drains empty into collector system that transports drainage water to discharge point (ditch or canal).

Fig. 11-11 The design of subsurface drains built to flush salts from soils in the western San Joaquin Valley. (Source: U.S. Bureau of Reclamation. 1984. *Drainage and Salt Disposal.* Sacramento, CA: Central Valley Project, Information Bulletin 1.)

One solution to the problem of salinization is installation of subsurface drains beneath fields to carry off salt-laden irrigation water (Fig. 11-11). Farmers used drains to remedy salt buildup at least 7,000 years ago in ancient Mesopotamia (present-day Iraq). In the western San Joaquin Valley, a 135-kilometer-long concrete canal known as the San Luis Drain was built in the early 1970s to carry agricultural drainage water from the fields north to Kesterson Reservoir near Los Banos. In the early and mid-1980s, wildlife biologists began to find alarming numbers of dead and deformed chicks of water birds at Kesterson National Wildlife Refuge, near the reservoir and surrounding ponds. Studies revealed that water in the reservoir and nearby ponds contained selenium concentrations as high as 400 parts per billion (ppb), far above the drinking water standard of 10 ppb. Remediation of the Kesterson site included filling of the selenium-contaminated ponds. The long-term solution to the problem of salinization of soils in the San Joaquin Valley has not been found, but it is of great importance to the economy of California.

SUTTER BUTTES

Rising dramatically from the flat floor of the Sacramento Valley, Sutter Buttes are the major topographic feature of the otherwise nearly flat Great Valley (Fig. 11-12). They were originally named the Marysville Buttes for the nearest city, about 12 kilometers

Fig. 11-12 Sutter Buttes, viewed toward the east. (Source: Photo by author.)

Fig. 11-13 A cross-sectional sketch of Sutter Buttes, looking toward the east from near Williams. (Source: Williams, H. 1929. *Geology of the Marysville Buttes, California.* Berkeley: University of California Publications, Department of Geological Sciences Bulletin, Vol. 18, pp. 103-220.)

to the southeast. Sutter Buttes are about 17 kilometers in diameter, and the highest peak is about 700 meters above the valley floor.

The geologists who first studied Sutter Buttes in detail, Professors Williams and Curtis of University of California, Berkeley, drew an analogy between the Buttes and a medieval castle. At the center of Sutter Buttes is a group of towering domes called the Castellated Core because of their resemblance to a castle (Fig. 11-13). Surrounding the core is a low circular area underlain by older, more easily eroded sedimentary rocks, appropriately called the Moat. The smooth slopes leading up to the Moat on all sides are formed by pyroclastic flows and alluvial deposits, the Rampart leading to the "castle."

Igneous activity at Sutter Buttes began during late Cenozoic time with the intrusion of silicic magma into the existing late Mesozoic and Tertiary sedimentary rocks. The first shallow intrusions did not reach the surface, but they caused the sedimentary rocks to be uplifted and tilted. Volcanic eruptions of intermediate and silicic lava followed, beginning about 1.6 million years ago. The first volcanic eruptions formed rhyolite domes along a circular system of faults and fractures that underlie the Moat (Fig. 11-14). The final volcanic activity formed the rugged core of Sutter Buttes, a group of andesite domes. At the center of the Core is a thick section of sediments that filled a deep crater lake during the eruption of the pyroclastic material. All volcanic activity apparently ceased at Sutter Buttes by about 1.3 million years ago.

The eruptions that produced the domes also generated pyroclastic materials that were deposited around the flanks of Sutter Buttes. As would be expected, the older pyroclastic materials found on the Rampart are more silicic, coinciding with the formation of the rhyolite domes. Younger pyroclastic materials are andesitic.

Fig. 11-14 Geologic map of Sutter Buttes (Source: Williams, H., and Curtis, G.H. 1977. *The Sutter Buttes of California*. Berkeley: University of California Press, Publications in Geological Sciences.)

THE GREAT VALLEY SEQUENCE

Beneath the young Cenozoic alluvium of the Great Valley lies an incredible thickness of older sediments and sedimentary rocks. The geology beneath the surface can be pieced together from the many wells that have been drilled to explore for oil and natural gas. Geologists are also able to examine the older rocks directly where they are exposed along the margins of the valley and around the margins of Sutter Buttes. The three-dimensional picture that emerges shows an elongated, asymmetrical *sedimentary basin* with approximately the same borders as the modern Great Valley. The basin is deeper on the western side of the valley, where it is filled by as much as 20,000 meters of sedimentary rock (Fig. 11-15). Along the western margin of the basin, the sedimentary formations have been steeply upturned so that many of the older units are exposed (Fig. 11-16). The rocks on the eastern flank of the valley dip inward more gently toward the axis of the basin.

The Mesozoic (Jurassic and Cretaceous) and earliest Cenozoic rocks of the Great Valley, known as the Great Valley sequence, consist of sandstone, shale, and conglomerate. Marine fossils (see Fig. 3-8) and characteristic sedimentary features indicate that

Fig. 11-15 A, A simplified cross section across the San Joaquin Valley near Bakersfield, showing the formations found in the upper part of the sedimentary basin. (Source: Adapted from Christiansen, R., and Yeats, R. *Decade of North American Geology,* Vol. G-3, p. 331.)

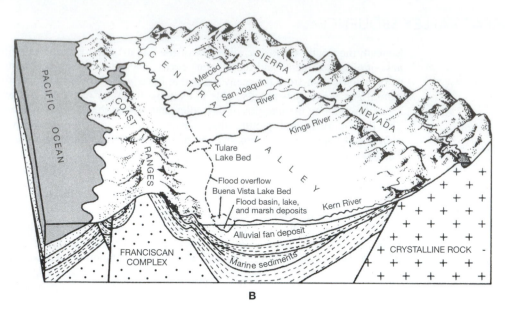

B

Fig. 11-15 B, A perspective drawing of the San Joaquin Valley region. (Source: modified from Page, R. 1986. U.S. Geological Survey, Professional Paper 1401-C.)

Fig. 11-16 Steeply dipping beds of the Great Valley Sequence along the western edge of the Sacramento Valley along Cache Creek at Monticello Dam. Note dog at far right for scale. (Source: Photo by author.)

they accumulated in an ocean basin that lay west of the Mesozoic North American margin. On the eastern side of the valley, the rocks are characteristic of deposits formed in deltaic and shallow ocean waters. In contrast, those found on the western side of the valley were apparently deposited deeper on the ocean floor by turbidity currents (which are discussed in more detail in Chapter 12). Using the nature of the sedimentary rocks to interpret the environments where they were originally deposited, geologists conclude that the shoreline of the ancient ocean basin lay along the western edge of the Sierra Nevada (Fig. 11-17).

The sedimentary basin where the Great Valley sequence accumulated was a *forearc basin* sandwiched between the Sierran arc (see Chapter 8) and the Mesozoic subduction zone. Today similar modern basins can be found along the margins of the Pacific Plate between volcanic arcs and ocean trenches; one example is the Sunda-Java forearc basin of Indonesia. Marine sediments of the Great Valley Sequence continued to be deposited in the forearc basin in early Cenozoic time, after the magmatic activity in the Sierra Nevada had ceased.

By examination of the minerals present in the Great Valley sequence, geologists can detect the sources of the sedimentary formations. Sandstone in the older, lower formations is rich in volcanic rock fragments eroded from the volcanoes of the Jurassic and early Cretaceous Sierran arc. The volcanic fragments are proof that the chain

Fig. 11-17 Map showing the approximate position of the Great Valley Sea during late Cretaceous time. The smaller arrows show the direction of sediment transport, as indicated by the sedimentary features in the rocks. The Great Valley Sea lay between the subduction zone trench and the Mesozoic Sierran arc. (Source: Adapted from Ingersoll, R. 1978. Paleogeography and paleotectonics of the Late Mesozoic forearc basin of northern and central California. In *Mesozoic Paleogeography of the Western U.S.* Los Angeles: Pacific Section, Society of Paleontologists and Mineralogists.)

of volcanoes, long since eroded from the Sierra Nevada, existed during Jurassic and Cretaceous time. Younger sandstones of the Great Valley sequence are rich in feldspar and quartz eroded from the granitic plutons. In the northern Great Valley, the Klamath Mountains were also a source of sediment.

CENOZOIC SEDIMENTS: THE END OF THE GREAT VALLEY SEA

During Jurassic and most of Cretaceous time, the Great Valley Sea persisted because the basin continued to subside as sediments filled it in. However, by the end of Mesozoic time, the northern part of the Great Valley basin began to fill up as uplift and deposition outpaced the subsidence of the basin. Sediments typical of increasingly shallow marine environments reflect this change. By about 24 million years ago, the sediments deposited in the Sacramento Valley were mostly of terrestrial origin. At that time, a basaltic lava erupted from the northern Sierra Nevada and flowed west and then south for at least 260 kilometers to the vicinity of Orland Buttes. Today, a volcanic flow called the Lovejoy Basalt can be seen near Oroville, where it forms Table Mountain, similar to the Table Mountain discussed in Chapter 8. The lava buried alluvium across today's Sacramento Valley, providing evidence that by about 24 million years ago, the basin had emerged from the sea.

Younger pyroclastic flows and ash erupted from the Cascades in Pliocene time are widespread in the Sacramento Valley, particularly at the northern end. About 3.5 to 4 million years ago, volcanic debris flows and pyroclastic lavas flowed into the northeastern side of the Sacramento Valley from sources in the Cascades; these deposits make up the Tuscan Formation (Fig. 11-18). During the same time, on the western side of the Sacramento Valley, alluvial deposits eroded from the Coast Ranges were deposited as the Tehama Formation. Both units contain distinctive, widespread ash layers erupted from the vicinity of the present-day Lassen Peak. One of these is the 3.4-million-year-old Nomlaki tuff (see

Fig. 11-18 Pyroclastic flows of the Tuscan Formation. (Source: Harwood, D. U.S. Geological Survey.)

Chapter 5). Streams continue to deposit alluvium along floodplains of the Sacramento Valley to this day.

In the south, the deep marine basin persisted much longer, as indicated by the continued deposition of marine shale and sandstone during early and middle Cenozoic time. In the longest-lived part of the Great Valley Sea, the San Joaquin basin, marine sediments from the Sierra Nevada and the newly formed Coast Ranges accumulated until late Pliocene time. Three million years ago, the San Joaquin basin was a large embayment in the coastline (Fig. 11-19). Today, the San Joaquin basin still persists as a lowland, and it is not mere coincidence that this is the site of the Great Valley's lake basins.

About 2 million years ago, the San Joaquin Valley emerged above sea level, and terrestrial sediments began to cover the marine sediments that had already been deposited. Between about 1 million and 600,000 years ago, after the San Joaquin Valley was closed off from the Pacific Ocean, a large lake periodically occupied much of the Great Valley. Geologists infer the existence of this lake from an extensive layer of clay, about 10 to 15 meters thick, but as much as 50 meters thick in some places, found at about the same depth beneath the San Joaquin and southern Sacramento Valleys. The clay contains Pleistocene fossils typical of shallow lakes. Known as the Corcoran clay, it is an important piece of evidence for the history of the Sacramento–San Joaquin river system, a topic discussed in more detail in Chapter 12.

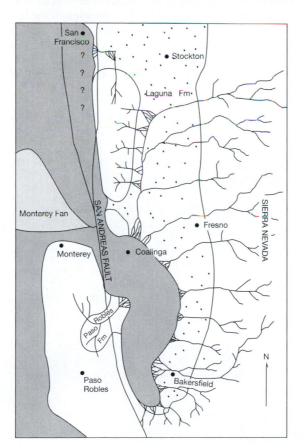

Fig. 11-19 The Sacramento–San Joaquin Delta and the marine embayment in the San Joaquin Valley about 5 million years ago. (Source: Modified from Bartow, J.A. 1991. U.S. Geological Survey, Professional Paper 1501.)

BENEATH THE GREAT VALLEY SEQUENCE

In the eastern half of the Central Valley, where the sediments are thinner, the sediments of the Great Valley sequence were deposited on the igneous and metamorphic rocks of the western Sierra Nevada. At Sutter Buttes, the Sierran rocks are covered by about 2,100 meters of sedimentary rocks of the Great Valley sequence and younger rocks. Further north, along the northern end of the Sacramento Valley, the Great Valley rocks were deposited on the rocks of the Klamath Mountains. Most geologists interpret the relations between the rocks of these areas as further evidence that the Great Valley forearc basin lay west of the North American continent when the Great Valley sequence was deposited.

Along the western margin of the Great Valley, where the rocks have been tilted steeply and exposed, geologists can observe the rocks that lay beneath the Great Valley basin when sediments first began to accumulate. The best exposure of the base of the Great Valley sequence is along the banks of the South Fork of Elder Creek, a small tributary to the Sacramento River, about 10 kilometers northwest of the town of Paskenta. Because the rocks are steeply tilted, the different units and the contacts between them can be seen by walking along the creek. At this locality, mudstone of the Great Valley sequence was deposited on a sedimentary **breccia** containing angular clasts of mafic and ultramafic rocks. The contact is an unconformity, indicating that a period of erosion separated the formation of the breccia and the initial deposition of mud in the Great Valley basin (see Chapter 3). The rubble of ophiolite debris in turn overlies layered ultramafic rocks of the Coast Range ophiolite (see Chapter 9). Near Mt. Diablo in the Coast Ranges east of San Francisco Bay, the sheeted dike section of the Coast Range ophiolite is about 2.5 kilometers thick and is overlain by 1.5 kilometers of pillow basalt.

Igneous rocks that intrude the Coast Range ophiolite in the western Sacramento Valley crystallized about 155 million years ago, and the oldest rocks of the Great

Fig. 11-20 Steeply dipping basalt dikes cutting through gabbro of the Point Sal portion of the Coast Range ophiolite near Santa Barbara. (Source: Bailey, E. U.S. Geological Survey.)

Valley sequence contain *Buchia* fossils that are about 152 million years old (see Fig. 3-8). The Coast Range ophiolite is therefore bracketed between these two ages, and it is about the same age as the Josephine ophiolite in the Klamath Province.

Another excellent exposure of the Coast Range ophiolite, termed the Point Sal ophiolite remnant, can be found along the California coast in northwestern Santa Barbara County (Figs. 11-20 and 11-21). The ophiolite at Point Sal contains many of the rocks typical of an ophiolite suite (see Chapter 9), including submarine basalt flows, sheeted dikes, gabbro, and serpentinized ultramafic rocks. The Point Sal ophiolite remnant has been displaced from the main Coast Range ophiolite by Cenozoic faulting (discussed in Chapter 16).

At the western edge of the Coast Range ophiolite, where it borders the rocks of the Coast Ranges, is a major fault system known as the Coast Range fault (Fig. 11-22). This Mesozoic fault system separates the rocks of the Great Valley sequence and its basement ophiolite from those of the subduction complex known as the Franciscan complex (see Chapter 12). Many geologists believe that the Coast Range fault marks the line along which oceanic rocks were thrust underneath the western edge of the fore-arc basin (Fig. 11-22). However, the Coast Range fault shows evidence of multiple types and ages of movement, and its complete history is far from understood.

Fig. 11-21 Geologic map of western California showing the Coast Range ophiolite. (Source: Adapted from Hopson and others. 1981. In Ernst, W.G., ed., *The Geotectonic Development of California*. Englewood Cliffs, NJ: Prentice Hall, Fig. 14-1, p. 428.)

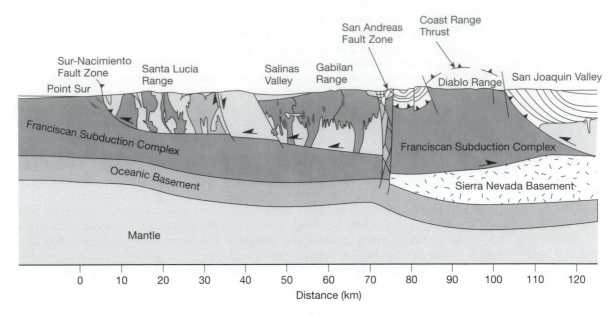

Fig. 11-22 Cross section from the central Coast Ranges near Big Sur to the western San Joaquin Valley, showing the Franciscan complex (discussed in Chapter 12), the Coast Range ophiolite (labeled Oceanic Basement), and the Coast Range fault. The interpretations of materials and relationships below the surface layers are controversial and are based on seismic surveys. (Source: Ross, D. 1979, and Coleman, R. 2000. *Revised Cross Section of the Southern Coast Ranges and San Joaquin Valley*. Geological Society of America, Map and Chart Series, Vol. 28H.)

SUMMARY

1. During Jurassic time, about 155 to 152 million years ago, a layered section of oceanic lithosphere formed as a result of magma generation in the mantle, forming the Coast Range ophiolite.

2. Beginning about 152 million years ago, and continuing through the entire Cretaceous period into early Cenozoic time, marine sediments accumulated in a fore-arc basin between the subduction zone west of North America and the Sierra Nevada volcanic arc (discussed in Chapter 8). These sediments make up the Great Valley sequence, which is up to 20,000 meters thick beneath the Great Valley. Rocks of the Great Valley sequence have been exposed by uplift and tilting along the western edge of the Great Valley.

3. During Cenozoic time, the Great Valley sedimentary basin became filled with sediment, and the Great Valley emerged from the sea, beginning about 24 million years ago in the Sacramento Valley. The Great Valley Sea persisted in the San Joaquin Valley until 2 million years ago.

4. About 1.6 to 1.3 million years ago, volcanism created Sutter Buttes.

5. During the past 1.6 million years, the Great Valley has experienced at least four episodes of deposition of glacial outwash. The outwash can be seen in river terraces along the eastern side of the Sacramento and San Joaquin Valleys. About 1 million to 600,000 years ago, a large lake was present in much of the Great Valley, forming the Corcoran clay.

6. Soils formed on the modern and ancient alluvium of the Great Valley show a progressive degree of profile development. Soils and landscapes formed on the older alluvial terraces are home to vernal pools.

7. Soils and floodplain surfaces of the Great Valley are California's richest agricultural resource, but they are increasingly threatened by residential development.

8. Today the lowlands of the Great Valley are subject to periodic flooding by the Sacramento and San Joaquin Rivers and their tributaries.

REFERENCES

ANDERSON, W. 1983. *The Sutter Buttes: A Naturalist's View.* Chico, CA: The Natural Selection Publishers, 326 pp.

BLAKE, M.C., JR., JAYKO, A.S., JONES, D.L., AND ROGERS, B.W. 1987. Unconformity between Coast Range ophiolite and part of the lower Great Valley sequence, South Fork of Elder Creek, Tehama County, California. In *Geological Society of America Centennial Field Guide.* Vol. 1, pp. 279–282.

DICKINSON, W.R. 1981. Plate tectonics and the continental margin of California. In Ernst, W.G., ed. *The Geotectonic Development of California.* Englewood Cliffs, NJ: Prentice Hall, 706 p.

HAUSBACK, B.P., AND NILSEN, T.H. 1989. *Sutter Buttes Field Trip Guide.* Washington, DC: American Geophysical Union, pp. 101–111.

JOHNSON, S., HASLAM, G., AND DAWSON, R. 1993. *The Great Central Valley: California's Heartland.* Berkeley: University of California Press, 253 pp.

PAGE, R.W.. 1986. *Geology of the Fresh Ground-Water Basin of the Central Valley, California, with Texture Maps and Sections.* U.S. Geological Survey, Professional Paper 1401-C.

Salinity in California (Special Issue). 1984. *California Agriculture* 38(10).

SINGER, M.J., AND MUNNS, D.N. 1987. *Soils: An Introduction.* New York: Macmillan, 492 pp.

Web Sites and Media Resources

California Heartland, Public Television Series on California Agriculture: *http://www.californiaheartland.org*

Cosumnes River Preserve: *http://www.cosumnes.org*

U.S. Department of Agriculture, Natural Resources Conservation Service, California Office: *http://www.ca.nrcs.usda.gov*

The Fresno Bee, 1999, Rescuing the San Joaquin: *http://www.fresnobee.com/man/projects/savesjr/*

Black Gold: The Oil Experience, Kern County Museum, Bakersfield: *http://www.kcmuseum.org/blackgold*

Buena Vista Museum of Natural History: *http://sharktoothhill.com/california.html*

San Joaquin Valley Geological Societies and links to geology sites: *http://www.sjgs.com/*

University of California, Agricultural Issues Center: *http://aic.ucdavis.edu*

U.S. Bureau of Reclamation, Shasta Dam: *http://www.usbr.gov/dataweb/dams/ca10186.htm*

CHAPTER

12

The Coast Ranges:
Mountains of Complexity

Fig. 12-1 Map showing the location of the Coast Ranges Province.

Coast Ranges Highlights

Major geographic feature:	Coast Range Mountains
Named mountain ranges (north to south):	Yolla Bolly, King, Diablo, Santa Cruz, Gabilan, Santa Lucia, Sierra Madre, San Rafael, Temblor
Other major landscape elements:	Klamath River, Eel River, Russian River, Salinas River, Humboldt Bay, San Francisco Bay

Selected high points:	Solomon Peak, Trinity County: 2,312 m (7,583 ft.)
	Mt. Tamalpais, Marin County: 784 m (2,571 ft.)
	Mt. Diablo: 1,173 m (3,849 ft.)
	Cone Peak, Monterey County: 1,572 m (5,155 ft.)
	Big Pine Mountain, Santa Barbara County: 2083 m (6,828 ft.)
Predominant rock types:	Mesozoic Franciscan complex (marine sedimentary, volcanic and other); Cenozoic marine sedimentary
Plate tectonic events:	Mesozoic subduction and accretion, Cenozoic transform motion
Other geologic events:	Sea level changes, formation of estuaries
Famous faults:	San Andreas, Rogers Creek, Hayward, Calaveras, San Gregorio–Hosgri, Sur-Nacimiento
Resources:	Water, sand, and gravel; mercury (1800s), natural gas
National Parks:	Redwood, Pt. Reyes National Seashore, Golden Gate National Recreation Area
State Parks of interest:	Jedediah Smith, Prairie Creek, Humboldt Redwoods, Big Sur, Guadalupe Dunes
Other parks of interest:	Almaden Quicksilver County Park, San José

California's Coast Ranges (Fig. 11-1) do not reach the heights of the Sierra Nevada, but in places they rise steeply from the Pacific Ocean to elevations of almost a mile within a mile of the coast. To visitors from the plains of the Midwest or the rolling mountains of the eastern United States, these are rugged and impressive mountains. The Coast Range Province extends about 1,000 kilometers from the Transverse Ranges to the Oregon border and about 130 kilometers from the Pacific Ocean to the Great Valley in the south and the Klamath Mountains in the north.

The northwest-oriented grain of the landscape (Color Plate 17) is a striking characteristic of the Coast Range Province. The Coast Ranges are a series of long, northwest-trending ranges separated by parallel river valleys. Prominent ranges include the King Range in the north, the Diablo Range east of San Francisco Bay, the Santa Cruz Mountains west of the Bay, and the Santa Lucia Range and Gabilan Range, which flank the Salinas Valley. California's large coastal rivers flow in the valleys between the ranges; these include the Eel, Russian, and Salinas Rivers.

Like many of California's landscape features, the orientation of rivers and ranges is controlled by tectonics. The boundaries between ranges and valleys are generally defined by faults that separate more resistant rocks from weaker ones. In some cases, the faults are active members of the San Andreas system; these faults are discussed in Chapter 14. In other areas, the faults are older and mark the junction between two accreted blocks of rock. A landscape where the orientations of different rock types and faults control the drainage pattern is called ***structurally controlled topography*** (see Color Plate 17 for an excellent example).

PLATE TECTONICS AND COAST RANGE GEOLOGY

Beginning about 140 million years ago, in the mid-Jurassic period, a new plate boundary formed along the western edge of the North American Plate. The accretion of the oceanic terranes in the Sierra Nevada and the Klamath Mountains had been accomplished (see Chapters 8 and 9), and the subduction zone had moved west of the modern Great Valley. From this time until about 28 million years ago, oceanic crust was accreted to North America along a subduction zone that ran the length of western North America (Fig. 12-2A). Fragments of the oceanic plates that collided with the North American Plate along this subduction zone are preserved today: These are the Mesozoic and early Cenozoic rocks of the Coast Ranges. The exact movement of the ancient plates is not completely understood, but at least some of the terranes might have traveled north from near the equator. The distant oceanic plate that is best understood by geologists was an eastward-moving ancient plate named the Farallon Plate.

Fig. 12-2 Changing plate tectonics in California during the past 30 million years. **A,** At about 30 million years ago, the Farallon Plate was being subducted beneath the North American Plate along the length of California. When the Pacific Plate came into direct contact with North America, the San Andreas transform boundary was created. **B and C,** As more of the Pacific Plate made direct contact with the North American Plate, the San Andreas transform lengthened and the Mendocino (M) and Rivera (R) Triple Junctions moved north and south. **D,** Today, the Juan de Fuca Plate, the northern remnant of the Farallon Plate, is north of Cape Mendocino, the present location of the Mendocino Triple Junction. (Source: U.S. Geological Survey, This Dynamic Earth: *http://www.pubs.usgs.gov/publications/text.*)

As the Farallon Plate was destroyed along the subduction zone, the Pacific Plate came into direct contact with the North American Plate (Fig. 12-2B–D). This contact began about 28 million years ago, but it did not occur at the same time along the entire length of western North America. When the meeting did occur, the plate boundary along the western edge of the North American Plate changed from a subduction zone to a transform boundary. As more of the Farallon Plate came into contact with North America, the transform boundary grew. Today, the transform boundary runs from Cape Mendocino to the Gulf of California—the San Andreas fault (Fig. 12-2D). The consistent northwest orientations of the faults, the different rock units, and the topography demonstrate that the orientation of the plate boundary remained fairly constant from Mesozoic time until the present.

As we saw in Chapter 1, the type of active plate boundary controls the geologic processes that take place along it. The change from subduction and accretion to transform motion is recorded in the rocks of the Coast Ranges. In fact, the brief plate tectonics history given in the preceding paragraphs was put together only after decades of studying the rocks and the faults in the Coast Ranges.

Even subtle changes in the movement of the plates are reflected in the geology of the Coast Ranges. During the past 28 million years, the Pacific Plate has shifted its course slightly, and at times its rate of movement has also changed. One of these shifts produced the mountains of the Coast Ranges about 3.5 to 5 million years ago. At that time, the Pacific Plate began to move obliquely past the North American Plate, rather than sliding past it. The change caused some transpression between the plates (see Chapter 1), and compression squeezed up the mountains. Like the Sierra Nevada, the Coast Range mountains are geologically young; in fact, they are still being uplifted today. During the 1989 Loma Prieta earthquake, for example, the Santa Cruz Mountains west of the San Andreas fault grew approximately 1.2 meters. As we will see in the following pages, the geology of the Coast Ranges reveals much about the changing plate boundary along western North America.

HOW OLD IS A MOUNTAIN? THE STORY OF MT. DIABLO

Mt. Diablo is a 1,173-meter high peak in Contra Costa County where a variety of rocks can be found (Fig. 12-3). At the center of the mountain is Mesozoic serpentine, flanked by a variety of younger Mesozoic and Cenozoic sedimentary rocks. Because the landscape cuts across all of these rocks, the mountain itself must be younger than the rocks. Once again, the principles of superposition and cross-cutting discussed in Chapter 3 can be applied.

But how much younger than the rocks is the mountain? South of Mt. Diablo is a bed of volcanic rock that represents an ancient ash flow, similar to those forming on the flanks of Cascade volcanoes today. By analyzing its chemistry, geologists have traced it to similar rocks erupted from the Sonoma volcanic field to the north (Fig.12-4). The ash flow, called the Lawlor tuff, is about 4.8 million years old. Applying the principles of gravity, it is clear that the ash could not have flowed over Mt. Diablo. In fact, it could not have flowed across the Carquinez Straits of San Francisco Bay either. The logical conclusion is that neither obstacle was there when the volcano erupted. The uplift of Mt. Diablo, and the folding and faulting of the Lawlor tuff and the other rocks, must have less than 4.8 million years ago. On the sides of Mt. Diablo, the alluvial layers containing the Lawlor tuff are tilted away from the mountain. The layers were flat when deposited, so their tilting must have resulted from the growth of Mt. Diablo, again confirming that the mountain rose after these young layers were deposited.

Fig. 12-3 Aerial view of Mt. Diablo from north of the Carquinez Straits of San Francisco Bay, looking south. (Source: Sarna-Wojcicki, A. U.S. Geological Survey.)

Fig. 12-4 Cross section from north to south across Mt. Diablo, showing the 4.8-million-year-old Lawlor Tuff (indicated by arrows and in black) as it occurs beneath the modern landscape. When the ash flow erupted, it must have flowed downhill from its source in the Sonoma volcanic field. At that time, neither Suisun Bay (part of San Francisco Bay) nor Mt. Diablo was present. (Source: Adapted from Sarna-Wojcicki, A. 1972. *Friends of the Pleistocene Field Trip Guidebook.*)

The Franciscan Assemblage: A Geologic Puzzle

Geologic maps of California show that the Franciscan Assemblage of Mesozoic age forms the heart of the Coast Range Province. This unit forms two northwest-trending belts, one extending almost the length of the province along the eastern edge of the San Andreas fault. The second belt of Franciscan rocks is found along the western edge of the province and appears on land only in the southern Santa Lucia Range (Fig. 12-5).

Fig. 12-5 The distribution of the Franciscan complex, the Great Valley Sequence, and rocks of the Salinian Block. (Source: Irwin, W.P. 1990. Geology and plate-tectonic development of the San Andreas Fault. In Wallace, R., ed. U.S. Geological Survey, Professional Paper 1515.)

The bewilderment of early geologists who encountered the rocks of the California Coast Ranges was probably not much different than that of a student who completes a general course in geology, and then attempts to explain the rocks seen in a typical road cut through the Coast Ranges. By the early 1900s, the "textbook" principles of geology (outlined in Chapters 2 and 3) had developed to explain the origin of many different rock types. However, the rules of classical geology were not sufficient for geologists to explain the origin of the rocks they saw in the Coast Ranges. Without the theory of plate tectonics, these rocks are a baffling mixture of different rock types, jumbled together with enormous complexity. The solution of one geologist, Andrew Lawson, who first mapped these rocks in 1895 and prepared a geologic map of San Francisco in 1914, was to make unity out of the chaos and name the rocks the Franciscan Formation.

Until the early 1970s however, the origin of Franciscan rocks remained a mystery. It is interesting to view the Franciscan problem by examining the different rock types and reviewing their origin. From a perspective enlightened by our knowledge of plate tectonics, it now is clear that the Franciscan rocks represent pieces of oceanic lithosphere that have been accreted to North America by subduction and collision. Although plate tectonics explains much about the Franciscan, geologists working today in the Coast Ranges are still developing new ideas.

Most abundant in the Franciscan Assemblage are sandstone and shale (or mudstone). In addition to quartz and feldspar, the sandstone contains fragments of shale, chert, and other rocks, giving it a gray, brown, or "dirty" color. The color gives rise to its name, *greywacke sandstone*. The shale or mudstone is also typically dark gray and is commonly interlayered with sandstone beds. Geologists now know that the sequences of sandstone and shale accumulated as marine sediments in relatively deep water by a process that was completely unknown during Lawson's time.

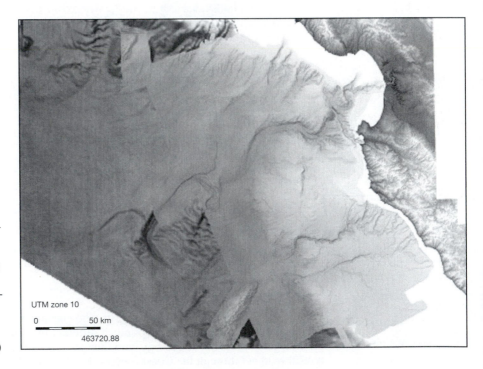

Fig. 12-6 The relief (shaded bathymetry) of the continental margin near Monterey submarine Canyon, offshore central California. (Source: Image created by Gardner, J.V. U.S. Geological Survey. Illumination azimuth 315°, elevation 45°. Horizontal reference UTM zone 10. Data sources, NOAA Geodas CD-ROM and MBARI Digital Data Series No. 1 CD-ROM. Land topography USGS 30-min. DEMs.)

UTM zone 10

0   50 km

463720.88

MONTEREY CANYON AND FAN: A MODERN ANALOG FOR FRANCISCAN SANDSTONE AND SHALE

Just a kilometer off the coast near Monterey is the head of Monterey Canyon, a winding gorge that reaches depths of 3,650 meters. The canyon, deeper in places than Arizona's Grand Canyon, is well known to divers, fishermen, and oceanographers as a haven for diverse and abundant marine life. Monterey Canyon extends from the continental shelf onto the floor of the deep ocean, and during the geologically recent past it has carried enough sediment to form the giant Monterey submarine fan at its lower end. The fan is a submarine delta similar in shape and origin to a large river delta (Fig. 12-6).

The processes that carry sediment from the beach and shelf into the canyon and out onto the deep ocean floor differ from sediment transport in a river. Sediments are carried to the canyon head from nearby rivers by nearshore ocean currents flowing on the shallow continental shelf. The sediments accumulate on the gentle slopes near the outer edge of the continental shelf until they become unstable. Then a mass of water-laden sediment breaks loose in a *turbidity current*, a dense, fast-moving mixture that travels into deeper water like a submarine landslide (Fig. 12-7). Turbidity currents can reach velocities of more than 90 kilometers per hour and can carry sediment as far as 1,000 kilometers down the continental slope. Along the shelf near Monterey Canyon, turbidity currents can be triggered by earthquakes on nearby faults as well as by the gradual buildup of unstable sediment.

As a turbidity current travels from the submarine canyon onto the flat surface of the ocean floor, its velocity decreases, and particles settle out onto the sea floor. Heavier, large particles settle more quickly and form the base of the turbidity current's deposit. The finest particles settle last, perhaps days after the first particles. The deposit that results from a single turbidity current is thus coarse-grained at the base, gradually becoming finer grained toward the top. This type of bedding, *graded bedding*, is diagnostic of turbidity currents and thus identifies a deposit that is formed in an offshore marine environment. Each subsequent turbidity current produces a similar deposit with gradually smaller particles toward the top, known as a *fining-upward sequence* (Fig. 12-8). Turbidity currents can rip up portions of the underlying mud layers, incorporating chunks of mud known as *rip-up clasts*. As the submarine channel shifts laterally over time, repeated turbidity currents build up a submarine fan with a shape similar to that of a river-mouth delta. The Monterey submarine fan contains more than 100 times the volume of sediment that would be required to fill Monterey Canyon, indicating that it has taken as long as a few million years for the fan to form and that the sediment was derived from a very large drainage area.

Over geologic time, a rhythmic sequence of alternating sand and mud, representing the deposits of hundreds or thousands of turbidity currents, can be lithified and preserved as sandstone and shale. These sequences, common to many formations in the Coast Range Province, are known as *turbidites* (Fig. 12-9). Their diagnostic characteristics, such as graded bedding and rip-up clasts, have been preserved to tell geologists that turbidity currents were active in the geologic past. The greywacke sandstone and shale today seen as part of the Franciscan Assemblage were deposited in environments similar to those found today in the Monterey submarine canyon and fan.

A third type of sedimentary rock found in the Franciscan Assemblage is chert, a dense, fine-grained rock composed almost entirely of silica (SiO_2). Because chert is a very hard rock, it resists erosion. In places like the Marin Headlands on the northern side of the Golden Gate in the Golden Gate National Recreation Area, chert holds up higher ridges and hills in the Franciscan landscape (Fig. 12-10 and Color Plate 18). Chert originates as siliceous *ooze*, a very fine-grained rain of radiolarian skeletons

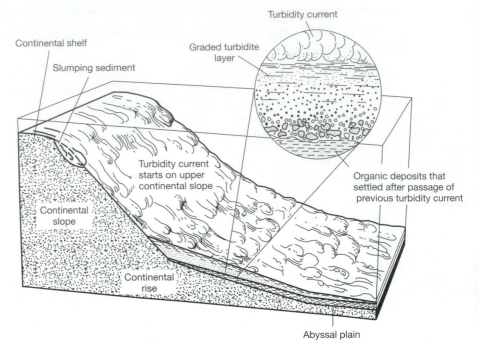

Fig. 12-7 Diagram showing turbidity currents at the edge of the continental shelf. (Source: Coch, N., and Ludman, A. 1991. *Physical Geology.* New York: Macmillan.)

Fig. 12-8 Formation of graded beds by repeated turbidity currents. (Source: Coch, N., and Ludman, A. 1991. *Physical Geology.* New York: Macmillan.)

Fig. 12-9 Outcrop of nearly vertical turbidites, showing the characteristic rhythmic alternation of sandstone and shale beds. The graded beds are now almost vertical because of recent uplift and tilting. Cretaceous Pigeon Point Formation, San Mateo County coast. (Source: Photo by author.)

Fig. 12-10 View of the Golden Gate Bridge, looking toward the north at the rugged Marin Headlands. Resistant Franciscan chert and greenstone make up much of the Marin Headlands. The southern bridge abutment, seen at the right edge of the span, is anchored in a less resistant block of serpentinite. (Source: Sarna-Wojcicki, A. U.S. Geological Survey.)

onto the deepest ocean floors. Although it makes up only about 1 percent of all Franciscan rocks, the chert has provided some of the most important clues about the origin of the Franciscan. The clues lie in the remains of radiolaria whose skeletons make up the chert (Fig. 12-11; see also Fig. 3-9). The types of radiolaria present in a chert sample provide evidence about the water depth, temperature, and chemistry of the ancient Mesozoic ocean that was their home. Two important and surprising conclusions have resulted from analyzing Franciscan cherts during the past two decades: The Franciscan Assemblage consists of a number of separate blocks of oceanic plate, and some of the blocks might have originally formed in oceanic water near the equator.

A last sedimentary rock type found in the Franciscan in very minor amounts is *limestone*, which occurs as isolated outcrops. It is most abundant on the San Francisco Peninsula, where it occurs together with volcanic rocks indicative of an ancient submarine volcanic plateau. Limestone is also found near Laytonville in Mendocino County, where it appears to represent a deep-sea ooze composed of foraminifera. Limestone in the Franciscan has been of great commercial interest because it is a relatively rare rock in much of California. Limestone is necessary for making cement, and many of the limestone blocks within the Franciscan are rock quarries. A good illustration of the value and scarcity of these limestone blocks is the fact that Permanente Quarry, located on the San Francisco Peninsula, was first opened to supply the concrete to build Shasta Dam, 500 kilometers to the north. Like the chert, the limestone appears to have originally formed at latitudes far south of present-day California.

Volcanic rocks make up about 10 percent of the total area where Franciscan rocks occur. Most of the volcanic rocks are basalt that has been slightly metamorphosed by interactions with seawater. These changes produced green metamorphic minerals, resulting in a characteristic Franciscan volcanic rock logically named *greenstone*. At some localities, the greenstone shows well-developed rounded forms

Fig. 12-11 Resistant chert beds at Baker Beach, Golden Gate National Recreation Area. (Source: Sarna-Wojcicki, A. U.S. Geological Survey.) A photomicrograph showing the microscopic radiolaria that make up much of the chert is shown as Fig. 3-9.

called *pillows* (Fig. 12-12; see also Chapter 2). The origin of these features would have been a mystery to early geologists, but in recent decades, geologists have explored active submarine volcanoes, like the mid-Atlantic ridge and the Hawaiian islands, using submersible diving vessels. They have observed pillow lavas being formed as hot lava enters the ocean and cools into rounded blobs. Studies of the chemistry of modern lavas and comparison with Franciscan greenstones confirm that they formed in various submarine volcanic environments, including ancient midocean ridges, sea mounts, and oceanic islands. Like chert, greenstone is a relatively hard rock, and both have been extensively quarried for road base.

Metamorphic rocks of the Franciscan Assemblage are perhaps the most unusual component of the mixture. These are most abundant in the northeastern Coast Range, but can also be found on the Tiburon Peninsula in Marin County, near Cazadero in Sonoma County, and in the eastern Diablo Range. These rocks contain rare silicate minerals, including the blue mineral *glaucophane*, which gives a blue color to the schist in which it occurs. The rocks are known as *blueschist*, and early geologists were very puzzled by their origin. The rare metamorphic mineral *jadeite* is also found in Franciscan metamorphic rocks, and small amounts of gem-quality jadeite (one of the two jade minerals) have been found on central California beaches.

During the 1970s, geologists conducted laboratory experiments to determine the pressure and temperature conditions that would produce the blueschist minerals. Rocks were heated and cooled under different pressures and temperatures to determine which groups of silicate minerals were stable under different conditions. These experiments indicated that conditions of high pressure but relatively low temperature were required for blueschist formation. Plate tectonics has given us an ideal environment for these unusual conditions: a subduction zone where an oceanic plate is being quickly forced underneath another plate.

Fig. 12-12 Pillow lavas in Franciscan greenstone, Marin County near Nicasio Reservoir. A rock hammer provides scale. (Source: Photo by author.)

A final important component of the Franciscan mixture is serpentinite, a soft green rock composed mainly of the serpentine group of minerals. As discussed in Chapter 9, serpentinite originates in the Earth's mantle as a more brittle, ultramafic rock known as peridotite. Alteration of peridotite resulted in the formation of plastic, lower density serpentinite. The weak serpentinite of the Coast Ranges is notorious for creating landslide problems, and in outcrops it commonly shows an extreme degree of shearing and flow (Color Plate 4). Serpentinite is also noted for the rare plant species that grow only on it because of its unusual chemistry. These plant communities are known as serpentine barrens, and examples can be viewed at the San Francisco Presidio in the Golden Gate National Recreation Area, as well as other places in California where serpentinite occurs. *Eclogite* is another rare metamorphic rock found associated with Franciscan rocks. Eclogite contains garnet and probably originates in the mantle. The ultramafic rocks of the Franciscan complex probably represent disrupted pieces of the Coast Range ophiolite, which was discussed in Chapters 9 and 11.

If you review the various types of rocks just described, you can well imagine the confusion of early geologists when they attempted to explain how all of these diverse rocks came to be found together. Once the plate tectonics theory came to light, however, the task became less difficult. We now know that submarine volcanic rocks, deep-sea chert, marine limestone, and sediments deposited by turbidity currents could all form on the dynamic surface of an oceanic plate. By observation of modern subduction zones, where oceanic lithosphere can be driven beneath a continent, geologists came to realize that the Franciscan rocks could represent fragments of such a collision. Pieces of the subducting oceanic crust, together with metamorphic rocks formed in the subduction zone and fragments from the upper mantle, could be scraped off against the margin of the continent and not be completely consumed in the subduction zone. As early as the 1970s, geologists recognized that the Franciscan Assemblage represented the remains of an ancient oceanic plate, which they named the Farallon Plate, accreted to North America during Mesozoic time (Fig. 12-13). The accretion process was the same one that produced the metamorphic belts of the Sierra Nevada foothills during earlier Mesozoic time (see Chapter 8).

Fig. 12-13 A simplified cross section showing the Mesozoic subduction and accretion that formed the Franciscan complex in the area of today's Coast Ranges. (Source: Blake, M.C. U.S. Geological Survey.)

Continuing detailed studies of the Franciscan have created an even more complex picture of its formation. It appears that Franciscan rocks represent not one, but several, accreted terranes. These separate Franciscan blocks represent the remnants of several oceanic plates that met their end against the edge of North America. So far, at least nine different blocks or terranes have been identified (Fig. 12-14). Studies of radiolaria contained in the chert, paleomagnetic analysis of the limestone, and detailed studies of the mineral content of the greywacke sandstone are some of the tools that geologists are using to complete this picture.

The area where Franciscan rocks are seen can be broadly divided into three belts (Fig. 12-15). In the Coastal Belt, a northwest-trending zone between Eureka and the Russian River, Franciscan rocks are mainly greywacke and shale sequences that are not as disrupted as other parts of the Franciscan. The Central Belt, including the San Francisco Bay Area, is found east of the Coastal Belt, and this part of the Franciscan Assemblage is a jumble of rocks. Isolated small blocks of exotic greenstone, blueschist, eclogite, chert, or greywacke "float" in a matrix of highly sheared mudstone. Geologists term this mixture *mélange*, adopting a cooking term used to describe a soup that contains floating chunks of ingredients. In the Franciscan mélange, the soup base or *matrix* is sheared mudstone, and the floating chunks are the more resistant blocks (Fig. 12-16 and Color Plate 28). Geologists now realize that different terranes are present even within the mélange (see the box below). For example, at the northern end of the Golden Gate Bridge, sequences of greenstone, chert, and sandstone represent deposition on a Mesozoic sea floor over a span of 100 million years.

The Eastern Belt of the Franciscan extends east to the Coast Range Fault, the major fault boundary with rocks of the Great Valley Sequence (see Chapter 11). Here the rocks are typically blueschist and related metamorphic rocks. Faulted slices of these rocks are also found in the San Francisco Bay area on the Tiburon Peninsula and west of Los Banos. The Eastern Belt and the mélange matrix of the central Belt are older than the blocks within the mélange, and the Coastal Belt contains the youngest Franciscan rocks, with the entire Franciscan spanning a period of about 150 million years of accretion from middle Mesozoic (late Jurassic) to early Tertiary (mainly Eocene) time.

WHY DOES SAN FRANCISCO HAVE HILLS? THE FRANCISCAN LANDSCAPE

The steep hills that pop up from the flat streets of San Francisco remain in the minds of visitors from all over the world. The city's landscape is a striking example of how geology shapes topography. The hills are more resistant blocks of Franciscan rock: For example, Telegraph Hill is resistant greywacke, Twin Peaks is a sequence of radiolarian chert and basalt, and in the Bay, Angel Island contains metamorphosed Franciscan greywacke and chert (Fig. 12-17). The flatter parts of the city and some areas submerged under San Francisco Bay are underlain by weaker Franciscan rocks, typically sheared shale of the mélange. The weaker Franciscan rocks have been worn down by erosion. Younger sediments, such as dune sand, alluvium, and bay mud, then covered these lower parts of the landscape.

Throughout the Coast Ranges where Franciscan rocks occur, particularly the mélange, the mountains have a characteristic appearance. Where the rocks are weak, large, deep landslides are common, and grasslands are the dominant vegetation. The landslides give the topography a lumpy and bumpy look. Resistant blocks commonly appear as isolated rocky knobs along ridges, or even poke up from the sides of hills. Geologists familiar with the Franciscan topography commonly refer to it as a "melted ice cream" landscape (Fig. 12-18).

Fig. 12-14 Franciscan terranes in the San Francisco Bay area. (Source: Wahrhaftig, C., and Sloan, D. 1989. *Geology of San Francisco and Vicinity.* American Geophysical Union, Field Trip Guidebook T105.)

Fig. 12-15 Belts of Franciscan rocks in the northern and central Coast Ranges (FZ = fault zone). (Source: McLaughlin, R., and others. 1988. Tectonics of formation, translation, and dispersal of the Coast Range ophiolite of California. *Tectonics* 7[5]: 1033–1056.)

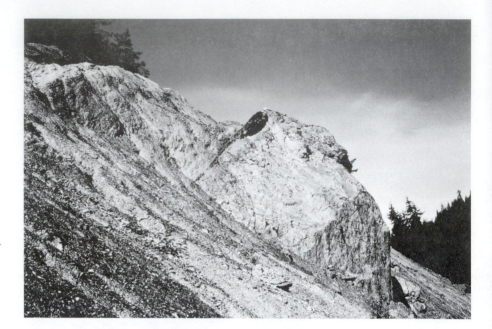

Fig. 12-16 Typical Franciscan mélange, Van Duzen River, Humboldt County. (Source: Blake, M.C. U.S. Geological Survey.)

Fig. 12-17 Angel Island, San Francisco Bay, with the city of Tiburon in the foreground and the Berkeley Hills on the far horizon. (Source: Photo by author.)

THE SALINIAN BLOCK

During the early years of work in the San Francisco area, geologists also recognized that Franciscan rocks are completely absent in an entire section of the Coast Ranges. This northwest-trending belt is bordered by the San Andreas fault on the east and the Sur-Nacimiento fault on the west. Geologists have long referred to it as the Salinian

Fig. 12-18 The Geysers steam field, Sonoma County, illustrating a typical Coast Range landscape.

Fig. 12-19 Granitic rocks of the Salinian Block form rounded outcrops at Point Reyes National Seashore, Marin County. (Source: Photo by author.)

Block, named for the city of Salinas. Much of the northern Salinian Block lies offshore, but parts of it can be seen on the Point Reyes Peninsula and at Point Arena (see Fig. 12-5 and Endpaper 1). The older rocks in the Salinian Block are fundamentally different from the Franciscan Assemblage, except in one way: Most of the older rocks in both blocks are of Mesozoic age.

The major component of the Salinian Block is granitic plutonic rocks (Fig. 12-19). These crystallized between about 110 and 78 million years ago, during late Mesozoic

QUICKSILVER: MERCURY MINING IN THE COAST RANGES

In 1845, Captain Andres Castillero, a Mexican Army officer who had studied at the College of Mines in Mexico City, picked up some curious red rock from the yard of Mission Santa Clara about 5 kilometers northeast of the present city of San Jose. After questioning the mission's padre, Castillero found that the Ohlone Indians of the area used the red rock to paint both their bodies and the mission buildings. The people called the red dirt, which we know as mercury ore, *mohetka*. Although other explorers had discovered the curious red mineral as early as 1824, they had not been able to identify it. Urged on by the Mexican government's offer of the equivalent of $100,000 for discovery of valuable minerals, Castillero found some mission Ohlones to guide him to the source of the red rocks, about 25 kilometers southwest of the mission. After performing tests on the local rocks, he realized that they contained neither gold nor silver, but he did notice their similarity to rocks he had seen at La Mancha, Spain, the site of the world's largest mercury mines. By heating the red mineral and then condensing the escaping vapor, Castillero extracted small silver globs of native mercury. Pure mercury is liquid at room temperature, and the silver globs were identical to those seen in today's thermometers. Its common name, *quicksilver*, describes its appearance very well. Castillero filed a claim to the mining site in 1847, becoming the first person to file for a legal claim in California.

Later in 1847 the mines came under the direction of Alexander Forbes, who named them Nuevo (New) Almaden after the famous Almaden quicksilver mine in Spain. By 1848, New Almaden was producing 45 to 70 kilograms of mercury daily, which was stored in flasks holding 34 kilograms each. The mercury was used mainly to extract gold and silver from crushed ore. The ore was washed with liquid mercury, which combined with the gold and silver. By vaporizing the mercury, miners could extract the pure gold or silver. Obviously, the gold rush in the Sierra Nevada created a boom in the quicksilver industry. Mercury from the New Almaden mines was taken in wagons to Alviso at the southern end of San Francisco Bay, where it was ferried by boat to San Francisco and on to the gold fields, Mexico, or South America.

At the height of production, the New Almaden mines produced over $2 million of quicksilver annually, and $15 million worth was mined between 1847 and 1864. In 1865, the New Almaden mines were home to 2,400 people from many countries. However, the demand for quicksilver fell as gold production dropped, and by 1912, the Quicksilver Mining Company had filed for bankruptcy. Brief flurries of demand for mercury to produce bombs

A   B

Fig. 12-20 A, The New Almaden mercury reduction works in the early 1880s. **B**, Abandoned works at New Almaden in 1991. (Sources: **A**, New Almaden Quicksilver County Museum; **B**, Ly, R. San José State University.)

Continued.

QUICKSILVER: MERCURY MINING IN THE COAST RANGES—cont'd

during the World Wars resulted in activity at New Almaden, but in 1975 the mines closed. Today some of the historic buildings and a small museum preserve remnants of the New Almaden mines, but much of the hillside area has returned to chaparral (Fig. 12-20). Most of the New Almaden area is now part of Santa Clara County's Almaden Quicksilver Park, and visitors can hike or ride horses in some of the former mine areas.

New Almaden is one of the three largest mining districts that produced mercury from the Coast Ranges. New Idria to the southeast was the largest producer of quicksilver, and the region around Clear Lake, known as the Mayacamas District, was a third major district. Mercury has been mined from the southernmost Coast Ranges near Santa Barbara north to Lake County. Most of the mercury is found as the red sulfide mineral cinnabar (chemical formula HgS_2; see Color Plate 5), although droplets of pure mercury are found in some samples. The cinnabar is most commonly found in a Franciscan host rock called **silica carbonate rock**. This unusual rock is associated with the serpentinite within the Franciscan and is thought to have formed by alteration of serpentinite by hot springs active along faults. The formation of the veins, masses, and disseminated bodies of cinnabar is thought to have occurred during mid-Tertiary (Miocene) and younger time. Applying the fundamental principles of cross-cutting relationships (see Chapter 3), the fact that significant mercury deposits occur in Miocene and younger rocks proves that the mineralization was younger. In fact, mercury

mined in Lake County is found in Pleistocene sediments. All together, the mercury mines of the Coast Ranges account for about 85 percent of the total U.S. mercury production.

Although it has a variety of uses, mercury has long been known to be highly toxic to humans and other organisms. Prolonged exposure to mercury vapors has a cumulative toxic effect on humans, one of the major effects being permanent brain damage. During the 1800s when tall beaver-skin hats were the height of fashion (picture Abraham Lincoln's hat), workers used mercury to tan the beaver hides, and the common occurrence of dementia caused by mercury poisoning gave rise to the expression "mad as a hatter." The levels of mercury present in the sediments of San Francisco Bay and in some of the streams of the Coast Ranges are naturally higher than the drinking water standards. A century of mercury mining increased the mercury levels, and many of the abandoned mines continue to pose a hazard.

The following excerpt from Brewer's *Up and Down California in 1860–1864* shows that the contamination problem at New Almaden was recognized even in 1861:

> An old furnace has been taken down, and the soil beneath for 25 feet down (no one knows how much deeper) is so saturated with the metallic quicksilver in the minutest state of division, that they are now digging it up and sluicing the dirt, and much quicksilver is obtained in that way. Thousands of pounds have already been taken out, and they are still at work.

Source: Farquar, F., ed. 1966. *Up and Down California in 1860–1864: The Journal of William H. Brewer.* Berkeley: University of California Press, 583 pp.

(Cretaceous) time. The mineral composition, chemistry, and age of the granitic rocks in the Salinian Block are very similar to those of the central Sierran Batholith. Like the Sierran Batholith, the Salinian Block also contains remnants of older rocks intruded by the magmas (see Chapter 8). Some of these are roof pendants within the granitic rocks. These are seen today as schist, gneiss, marble, and other high-grade metamorphic rocks formed from a variety of sedimentary rocks. They are abundant in the Gabilan and Santa Lucia Ranges near Big Sur. The metamorphic rocks are difficult to date, but geologists estimate that the original sediments were deposited in Paleozoic time. However, in the southeastern part of the Salinian Block, near Mt. Pinos, metamorphic rocks are of Precambrian age.

The similarities between rocks of the Salinian Block and the Sierra Nevada have convinced geologists that the granitic rocks of the Salinian Block are indeed pieces of the Sierran batholith that have been separated and transported north by the San Andreas fault system (see Chapter 14). The granitic and metamorphic rocks represent part of the Mesozoic batholith intruded along the entire length of the North American continent as oceanic crust was being subducted to the west. The Franciscan rocks, pieces of Mesozoic oceanic crust, and the Salinian Block, pieces of continental crust, were brought together in the Coast Ranges by the San Andreas fault system long after they formed.

UNDER THE COAST RANGES: BASEMENT GEOLOGY

The older dense, crystalline rocks in any area are commonly referred to as the *basement rocks* because they form the foundation for the geologic package of rocks and sediment. In the Coast Ranges, faulting brought together two very different *basement complexes* after they formed. The Franciscan basement consists largely of accreted ocean-floor and subduction zone rocks. The vastly different granitic-metamorphic basement of the Salinian Block represents continental crust intruded by granitic plutons. It is important to remember that all of the basement rock in the Coast Ranges is relatively young, predominantly Jurassic and Cretaceous in age. Geologists and geophysicists are currently debating the relationships between the basement complexes of the Coast Ranges and the basement rocks underlying the Great Valley and the Sierra Nevada (see Chapter 11).

CENOZOIC SEDIMENTARY BASINS

Cenozoic sedimentary rocks can be found covering the basement rocks on both sides of the San Andreas fault. Most of the rocks were deposited in marine basins, and some of the basins remain offshore along the continental shelf today (Fig. 12-21). The formation of these basins, some of which received more than 6,000 meters of sediment, was controlled by motion along the transform boundary. Some geologists believe that basins formed by extension during a period when the Pacific Plate was moving in a more northwesterly direction relative to North America. As we will see in Chapter 16, the sedimentary rocks in California's late Cenozoic basins are important producers of petroleum.

Some of the older marine basins, which were being filled with sediment during Eocene time about 25 million years ago, were later separated by as much as 315 kilometers of right-lateral movement on the faults of the San Andreas system (see Chapter 14). The marine sedimentary rocks seen today in the central and northern Coast Ranges west of the San Andreas fault were adjacent to the San Joaquin Basin when the sediments were deposited in Eocene time. Between about 22 and 11 million years ago, sediment rich in silica and organic matter, including sediments rich in diatoms, accumulated in deep marine basins. These sediments were subsequently uplifted and are seen today as the Monterey Formation, a distinctive and widespread sedimentary unit discussed in Chapter 16 (Fig. 12-22).

Because of recent faulting, compression, and uplift near the Mendocino Triple Junction, part of the Eel River Basin in northwestern California now lies on land. The sedimentary rocks that were deposited in the basin have been highly folded and

Fig. 12-21 Late Cenozoic sedimentary basins of the California Coast Ranges and adjacent offshore areas. (Source: Christiansen, R., and Yeats, R. 1992. In Geological Society of America. *Decade of North American Geology,* Vol. G-3.)

Within the figure:

0   80 kilometers

Legend

⬭ Major late Cenezoic sedimentary basin

—— Major fault

SAF = San Andreas fault

SGH = San Gregorio–Hosgri fault

MF = Mendocino fault

PACIFIC OCEAN

MF   Eureka

Basins
1 Sacramento } Great
2 San Joaquin } Valley
3 Livermore
4 San Benito
5 Cuyama (Caliente)
6 Huasna-Pismo
7 Santa Maria
8 Salinas
9 Santa Cruz (La Honda)
10 Outer Santa Maria
11 Sur
12 Outer Santa Cruz
13 Bodega
14 Point Arena
15 Eel River

San Francisco

SAF

SGH

Santa Barbara

Coastline

Fig. 12-22 Vertical chert beds in the Miocene Monterey Formation, Berkeley Hills, Alameda County. (Source: Photo by author.)

Fig. 12-23 Dipping beds of the Rio Dell Formation, south of the Eel River, Centerville Beach, Humboldt County. (Source: Photo by author.)

faulted. Along the beach south of the mouth of the Eel River, the rocks have been tilted almost vertically (Fig. 12-23). The vertical tipping of the originally horizontal layers makes a walk along the beach from south to north a tour through the basin's history. One can "see" the Eel River basin fill with sediment as younger sedimentary formations represent increasingly shallow-water deposits (Fig. 12-24). The youngest unit is nonmarine, signaling the end of marine conditions in this part of the basin.

Marine sedimentary rocks of Cenozoic age in other parts of the Coast Ranges reveal histories similar to that of the Eel River Basin. In the Berkeley Hills, for example, sedimentary units contain large amounts of rock fragments eroded from Sierra Nevada volcanoes. One of these formations is the Neroly sandstone, made up of sand-sized pieces of volcanic rock eroded from Miocene volcanoes that once covered the northern Sierra. In the Santa Cruz Mountains, submarine volcanic rocks, known as the Mindego Formation, are part of the basin-filling sequence. The San Joaquin Basin was a very deep marine basin between 35 and 15 million years ago, and then filled with material washed in from the rising Coast Range and Sierra Nevada. A shallow marine embayment persisted in the San Joaquin until 2.5 to 3 million years ago (see Chapter 11).

CENOZOIC VOLCANIC ACTIVITY IN THE COAST RANGES

On September 8, 1769, the expedition commanded by Don Gaspar de Portola camped along a creek near a bay in the southern Coast Ranges. At the entrance to the bay stood a large, rounded hill, or ***morro***. The Spanish word for this round hill of rock gives its name to both Morro Rock and the town of Morro Bay (Fig. 12-25). Morro Rock is actually the westernmost in a chain of 14 morros that form a northwest-trending line between San Luis Obispo and Morro Bay. The hills are the remains of plugs, domes, and volcanic necks of a chain of volcanoes, and the rocks have been dated between 20 and 27 million years old. Volcanic rocks with similar

Fig. 12-24 A simplified geologic map and stratigraphic column showing formations of the Eel River basin, Humboldt County. The approximate positions of Fig. 12-23 and Color Plate 8 are shown in the cross section. (Source: McLaughlin, R., and others. 1982. *Friends of the Pleistocene Field Trip Guidebook.*)

Fig. 12-25 Photograph of Morro Bay and Morro Rock, San Luis Obispo County. (Source: Photo by author.)

chemistry and ages are also found in the Santa Cruz Mountains north of Santa Cruz and at Point Arena, where they are known as the Iverson Basalt.

At Pinnacles National Monument at the southern end of the Gabilan Range southeast of Hollister, visitors can see a spectacular array of volcanic towers and caves (Fig. 12-26 and Color Plate 10). The rocks at the Pinnacles are mostly pyroclastic

A

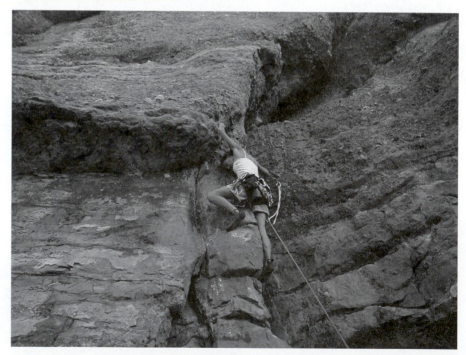

B

Fig. 12-26 **A,** Volcanic rocks at Pinnacles National Monument. **B,** Close-up of pyroclastic layers at Pinnacles. (Sources: **A,** Photo by author; **B,** Jordan, A., Friends of Pinnacles.)

flows and banded flows of rhyolite. Because erosion was concentrated along the original cooling cracks in the lava, the interiors were less eroded and left standing as towers (Fig. 12-26A). As discussed in Chapter 14, the rocks at the Pinnacles, which are about 23 million years old, appear to match well with the Neenach volcanics 315 kilometers to the southeast near Lancaster, on the North American side of the San Andreas fault in the Mojave Desert.

Younger Cenozoic volcanic rocks are also found in the Coast Ranges (Fig. 12-27). East of the San Andreas fault, rocks from a series of volcanic fields get systematically younger from south to north along the Coast Ranges. Southernmost of the group are the 9- to 15-million-year-old Quien Sabe volcanics in the central Diablo Range, followed by the 12- to 8-million-year-old basalt and rhyolite in the Berkeley Hills.

Volcanic rocks in the Sonoma volcanic field are between 8 and 3 million years old, and at the northern end of the series are the Clear Lake volcanics, which are between 2 million and 10,000 years old. In both the Sonoma and Clear Lake volcanics,

Fig. 12-27 Map showing locations and ages of Cenozoic volcanic rocks in the Coast Ranges. (Source: Wahrhaftig, C., and Sloan, D. 1989. *Geology of San Francisco and Vicinity.* American Geophysical Union, Field Trip Guidebook T105.)

early eruptions produced basalt, whereas later volcanism produced andesite and rhyolite. The Clear Lake area was an important source of obsidian for early Native Americans in the region (see Fig. 5-8).

Most geologists agree that the northward movement of volcanic activity over time is related to the northward movement of the Mendocino Triple Junction. However, the volcanic activity is not as clearly related to position of the subduction zone west of the North American Plate as are the presently active Cascade volcanoes. The chemistry of the volcanic rocks is more diverse than the subduction-zone volcanoes, and the volcanic activity lags behind the passing of the triple junction. It is possible that a period of extension and volcanism followed the movement of the triple junction past each of these areas.

GEYSERS GEOTHERMAL FIELD

Beneath the central part of the Clear Lake volcanic field, a large magma chamber appears to remain somewhat active. The area, which is south of Clear Lake along Geyser Creek, is known as The Geysers (see Fig. 12-18). Here magma close to the surface results in the production of large amounts of steam and water heated by the *geothermal* energy of the magma. The steam is produced when groundwater percolates downward and is heated by magma. The heated groundwater travels toward the surface along faults and fractures in the overlying rock, which is mostly Franciscan greywacke. Numerous hot springs and boiling springs are active in the area, and in places the hot springs have produced hydrothermal mercury deposits within the volcanic field. At The Geysers area, relatively permeable greywacke is capped by impermeable rock that traps the geothermal energy in a steam reservoir located several thousand feet below the surface. Since 1960, the geothermal energy in The Geysers area has been tapped for electric power. Wells have been drilled into the steam, and the steam is then converted by turbines into electricity.

THE HISTORY OF SAN FRANCISCO BAY

Today San Francisco Bay is the outlet for the entire drainage of the Sacramento and San Joaquin Rivers. About 40 percent of California's runoff passes through the Golden Gate to the Pacific Ocean (Fig. 12-28; see also Chapter 10 and Endpaper 3), and most of this fresh water enters the Bay from the Sacramento–San Joaquin Delta. The southern portion of the Bay, a northwest-trending extension of the Santa Clara Valley, receives fresh water from streams that flow from the Santa Cruz Mountains and the Diablo Range. San Francisco Bay is California's largest *estuary*, which is a body of water, partly isolated from the open ocean, where both fresh water and marine water circulate. Cold, saline marine water flows into San Francisco Bay from the Pacific Ocean. Because salt water is denser than fresh water, it sinks beneath the fresh water flowing into the Bay from the Delta. The movement of the marine water is controlled by the tides, so that currents beneath the Golden Gate Bridge reverse twice daily. During major winter storms, large volumes of fresh water flow through the Sacramento–San Joaquin Delta and circulate into the South Bay. The floodwaters are important there because they flush contaminants from an area where circulation is normally poor.

Fig. 12-28 San Francisco Bay (*left, expanded view*) and the tributarties to the Sacramento–San Joaquin Delta (*right*). (Source: Conomos, T.J. 1979. *San Francisco Bay: The Urbanized Estuary.* Pacific Division, American Association for the Advancement of Science.)

Because San Francisco Bay is such a vital part of northern California's economy and its landscape, it is important to investigate its geologic past. Three forces have combined to shape San Francisco Bay and the surrounding landscape: tectonics, recent sea level changes, and human activities since about 1850.

Tectonics, River Systems, and San Francisco Bay

When the present Sierra Nevada mountains were uplifted, approximately 5 to 9 million years ago, a system of rivers carried runoff from the highest parts of the range westward to the Great Valley, just as today's rivers do (Chapter 10 and Endpaper 3). However, the exact configuration of this original river system is not known. Because the faults of the San Andreas system are so active, today's topography in the Coast Ranges is quite different from what the landscape would have looked like 3 million years ago. It is likely that the Carquinez Straits and the Golden Gate did not exist at that time. In addition, before the recent tilting in the southern San Joaquin Valley, the rivers there would have flowed further south than at present.

During late Tertiary time, the streams of the Great Valley flowed south to a marine embayment in the San Joaquin Valley. The embayment, the remains of the

Fig. 12-29 Map showing the Kettleman Hills and Monterey Fan today.

Great Valley Sea (see Chapter 11), was connected to the open Pacific Ocean by a *strait* or narrow inlet. The strait crossed the San Andreas through a low area of the Coast Ranges, which had created a barrier by the time they were uplifted 3 to 5 million years ago. The rivers must have flowed through a low pass or gap in the range.

Geologists searching for this former position of the Sacramento–San Joaquin outlet have found clues in the sedimentary formations along the western margin of the San Joaquin Valley near Coalinga (Fig. 12-29). In the Kettleman Hills, sediments of Pliocene age—the San Joaquin, Etchegoin, and Tulare Formations—contain minerals and rock fragments derived from sources in the southern Cascades, Sierra Nevada, and the northern Sacramento Valley. For example, pumice and fine ash, transported by water, are concentrated in a thick bed in the San Joaquin Formation. Chemical "fingerprinting" of the ash showed that it was the Nomlaki tuff, produced by a very large eruption 3.5 million years ago, from a source near today's Lassen

Peak. The sediments in the Kettleman Hills also show the sedimentary characteristics of delta deposits.

The sediments were deposited in shallow marine water that became shallower over time, so that the Tulare Formation is mostly marshy coastal plain deposits. Remembering that a marine embayment persisted in the San Joaquin Basin until Pliocene time (see Chapter 11), one can envision a river emptying into the embayment and forming a delta (see Fig. 11-19). The age of the deposits in the Kettleman Hills is fixed between about 5 and 2 million years before present by fossils and volcanic ashes interbedded with the sediments.

One of the most fascinating ideas about this ancient river delta is that it was continuous with the Monterey submarine fan at the time that it was forming. The Monterey Canyon and fan are very large features to have formed from the relatively small Salinas River (see the box on page 311). Geologists have observed that moving the San Andreas fault "backwards" in time at the rate it presently moves would place the fan and the deposits of the Kettleman Hills close to each other. Although this theory has not yet been proven, recent coring in the younger sediments of the Monterey fan might reveal that it is indeed composed of Sierra Nevada materials.

Geologists speculate that it was movement on the San Andreas fault that caused the Kettleman Hills outlet to be blocked, forming Lake Clyde (named in honor of the late geologist Clyde Wahrhaftig). Right lateral movement and associated compression on the fault progressively distorted and constricted the strait connecting the embayment with the open ocean. The geography of the central Coast Ranges and Great Valley must have been vastly different than it is today.

Sediments found in the beach cliffs south of San Francisco provide evidence that the Sacramento–San Joaquin river system had established an outlet to the Pacific through San Francisco Bay by about 600,000 years ago. Marine sediments found in the Merced Formation show an increase in the supply of sediment and a distinctive change from minerals and rocks that originated in the Coast Ranges to materials that must have come from the Sierra Nevada. The Sierra Nevada-type sediments are found below the Rockland ash (Fig. 12-30), which was erupted from the Mt. Lassen area 600,000 years ago, as discussed in Chapter 5. The Rockland ash was also discovered in *estuarine* sediments from 120 meters beneath the modern Bay. The ash and the sediments were collected in cores drilled during construction of the bridges that span the Bay. The location and thickness of the sediments in the San Francisco region suggest that the original outlet was probably a few kilometers south of the Golden Gate. Geologists believe that Lake Clyde might have drained because it overflowed during an episode of glacial melting.

Recent Sea Level Changes

The global climate fluctuations that resulted in episodic glaciation of the Sierra Nevada during the last million years were reviewed in Chapter 8. It is not surprising to find that these worldwide fluctuations had a dramatic effect on San Francisco Bay. During glacial periods, more of the Earth's water is stored in the ice sheets and alpine glaciers, and worldwide sea level is low. When the glaciers melt, more water is stored in the oceans, and sea level rises. Worldwide sea level was approximately

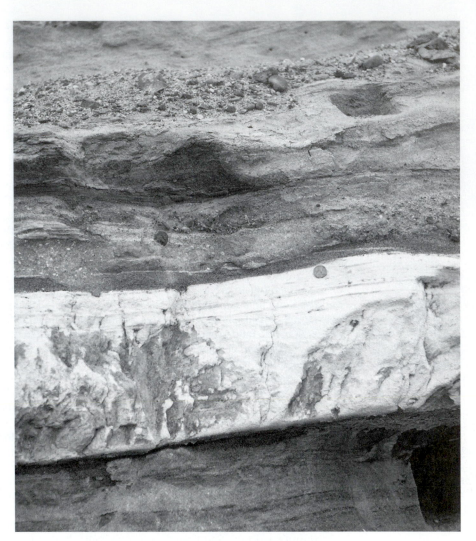

Fig. 12-30 Close-up view of the Rockland ash within sand and pebble layers of the Merced Formation. The penny provides scale. Beach cliffs near Daly City, San Mateo County. (Source: Sarna-Wojcicki, A. U.S. Geological Survey.)

120 meters lower than at present during the last major glacial episode 18,000 to 22,000 years ago. The 120-meter lowering of the sea level in the San Francisco area placed the shoreline just west of the Farallon Islands, about 32 kilometers west of today's beaches (Fig. 12-32). At that time, the combined Sacramento and San Joaquin Rivers, as well as the smaller streams flowing from the Santa Clara Valley, flowed past San Francisco, depositing their sediment load on what is now the continental shelf. During times of low sea level, rivers flowed through areas that are now San Francisco Bay.

Sediments deposited in the river valleys occupying San Francisco Bay during times of low sea level were river alluvium and, in some areas, dune sand. Alluvial deposits are typically sandy and similar in composition to the sediments carried by the streams that flow into the Bay today. Sand dunes are common in western San Francisco and Alameda. The sand that makes up the dunes was probably blown in

THE CORCORAN CLAY: THE SAN ANDREAS DEFEATS CALIFORNIA'S MIGHTY RIVERS

If the San Joaquin River once flowed to the Pacific in the vicinity of the Kettleman Hills, how and why did the drainage pattern change? The evidence is incomplete, but sediments in cores drilled throughout the San Joaquin and southern Sacramento Valleys provide some clues. A distinctive clay layer, named for the town of Corcoran, documents the presence of an ancient large lake in the San Joaquin and the southern Sacramento Valleys. Fossils present in the clay indicate that the lake was not necessarily very deep, but it covered a very large area (Fig. 12-31). The 758,000-year-old Bishop Ash is found within the older lake deposits, and an ash layer erupted from Yellowstone about 665,000 years ago is found near the top of the clay. At the time the Corcoran clay was deposited, the Sierran streams had no direct outlet to the Pacific Ocean.

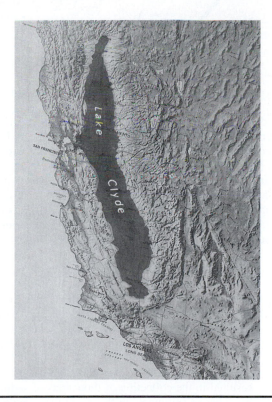

Fig. 12-31 The possible maximum extent of Lake Clyde about 600,000 to 700,000 years ago. (Source: Sarna-Wojcicki, A. U.S. Geological Survey.)

Fig. 12-32 The shoreline near San Francisco about 20,000 years ago, during the Tioga glaciation. (Source: Conomos, T.J. 1979. *San Francisco Bay: The Urbanized Estuary.* Pacific Division, American Association for the Advancement of Science.)

from the area west of San Francisco during times when the continental shelf was exposed.

The present estuary that Bay Area residents take for granted formed less than 10,000 years ago, as the Earth's climate warmed and the melting of glaciers caused worldwide sea level to rise. By 10,000 years ago, the rising marine water began to flood the Bay, and by about 4,000 years ago, the Bay occupied its historic limit. The inland Delta had formed at the position of the higher sea level. Sediments deposited in the estuary are typically dark-colored clay and silt, and they contain foraminifera and other marine microfossils that could not survive in fresh water. The sediments are commonly called ***Bay mud***. Sandier sediments deposited by channels in the estuary are commonly interfingered with the estuarine deposits.

After the Sacramento–San Joaquin outlet was established in its current location, San Francisco Bay came and went with climatically controlled sea level rises and falls. During glacial episodes, sea level was low, and river valleys occupied the Bay (Fig. 12-33). During warm periods, sea level rose and created an estuary. The oldest known estuary existed in San Francisco Bay 600,000 years ago, based on the presence of the Rockland ash in estuarine deposits identified beneath the Bay.

Drilling and coring beneath the Bay in preparation for construction of several of the bridges resulted in the collection of thousands of sediment samples. By careful analysis of the types of sediments in these samples, geologists were able to reconstruct the history of the Bay and confirm the alternating pattern of estuary and alluvial valleys (Fig. 12-34).

Fig. 12-33 Climatic controls on the sedimentation in San Francisco Bay.

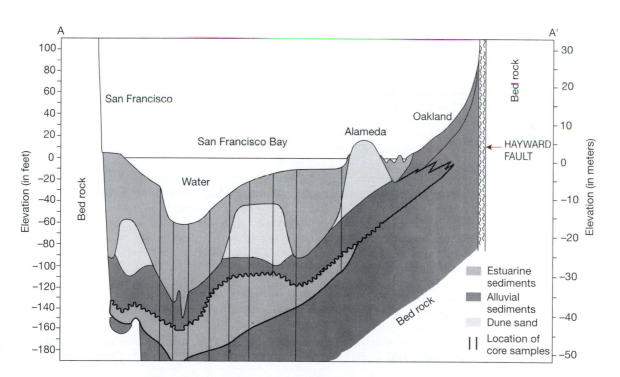

Fig. 12-34 Cross section showing the sediments beneath San Francisco Bay between San Francisco and Alameda. The vertical lines show the locations of cores. (Source: Atwater, B. 1977. U.S. Geological Survey, Professional Paper 1014.)

IMPLICATIONS FOR GROUNDWATER CONTAMINATION AND CLEANUP

Bay mud, sometimes containing shells, forms the upper layers of sediment along much of the margin of the southern part of San Francisco Bay. The heaviness and stickiness of the clay-rich mud, a great frustration to many gardeners in the area, make the Bay Mud an effective aquiclude. In contrast, the alluvial deposits in the area, found in small channels and uphill from the areas where Bay Mud is found, are permeable and much looser material because of their coarser grain size.

Beneath the surface of the Bay margins, alternating layers of alluvium and Bay mud record the sea level changes discussed previously. The alluvial layers, which slope from the hills toward the Bay, are excellent aquifers, and they met local demands for municipal and irrigation water until the 1950s. The aquifers are effectively confined by the intervening Bay mud, and artesian wells were abundant in parts of the Santa Clara Valley (see Chapter 10). Unfortunately, wells in the upper aquifer and even some in the lower aquifer have become contaminated by leaking storage tanks in Santa Clara County.

A clear understanding of the geologic history of the Bay sediments is essential to understanding the flow of groundwater in the Santa Clara Valley and to cleaning up contamination of the aquifers. By recognizing that the confining Bay mud layer is present only in areas that were covered by the older estuaries, it is easy to see how contaminants could leak into the lower aquifer. Channels of permeable sand are present even in the modern estuary; channels within the older Bay mud also enable contaminants to spread. The resulting three-dimensional picture of aquitards and aquifers is very complex, and a thorough understanding of the recent geologic history of the Bay is required to ensure the future protection of groundwater resources.

Human Activities

Since about 1850, human activities have made enormous modifications to the San Francisco Bay region and to the estuary itself. The first drastic impact was the effect of hydraulic mining in the Sierra Foothills (see Chapter 8). In 1917, pioneering geologist Grove Karl Gilbert calculated that more than a billion cubic yards (0.9 billion cubic meters) of this sediment had made its way to the upper part of San Francisco Bay, where it filled in the estuary by several feet.

Various activities along the margins of the Bay have reduced the wetlands habitats in the Bay by approximately 85 percent. Beginning in the 1800s, farmers in the Delta constructed dikes and levees to drain the marshy wetlands for agriculture. These low-lying areas (Fig. 12-35) are vulnerable to flooding during major storms, and concern is mounting about the seismic stability of the old levees and dikes. Wetlands have also been diminished by salt evaporating ponds and urban development on filled lands (Fig. 12-36).

Human activities have also affected the Bay's water. Because of water diversions, the flow of fresh water into the Bay is only about 60 percent of its prediversion volume. The combined effect of drought and the diversions has been increased salinity in some areas and decreased circulation in the South Bay. Biologists and geologists first became concerned about the ecological health of San Francisco Bay as early as the turn of the century, when the threat to world-famous

Fig. 12-35 Areas below sea level in the San Joaquin–Sacramento Delta. (Source: California Department of Water Resources. 1995. *Sacramento–San Joaquin Delta Atlas.*)

commercial oyster beds was first realized. In the 1960s a grassroots effort succeeded in limiting construction activities in wetlands, and in recent years, some wetland areas have been restored. Recent legislation requires maintenance of minimum flow to the Delta, and the discharge of municipal and industrial waste is now strictly controlled.

Fig. 12-36 Aerial view of San Francisco Bay near Foster City, San Mateo County. Wetlands were filled in to construct the city during the 1960s. The sloughs flowing into the Bay can still be seen. The western end of the San Mateo Bridge appears in the background. (Source: Photo by author.)

SUMMARY

1. Between about 190 and 66 million years ago, during the Jurassic and Cretaceous periods, sedimentary and volcanic rocks formed on the Farallon Plate west of present-day California. The rocks included pillow basalt erupted onto the ocean floor at oceanic spreading ridges, radiolarian chert formed on the floor of the deep ocean basin, greywacke sandstone and mudstone deposited by turbidity currents on the continental slope, and minor amounts of limestone formed at tropical coral islands.

2. The Farallon Plate migrated northward and eastward to a subduction-zone boundary with the North American Plate. By 100 million years ago, the subduction zone was in the approximate position of today's Coast Ranges. At the same time, a large chain of andesitic volcanoes existed in the position of today's Sierra Nevada (see Chapter 8). Parts of the Farallon Plate were accreted to the North American Plate during subduction, creating the subduction-zone complex known as the Franciscan Complex. In addition to oceanic sedimentary and volcanic rocks, the Franciscan Complex contains belts of blueschist metamorphosed during subduction and ultramafic rocks originally formed in the mantle.

3. Today, Franciscan rocks are found throughout the Coast Ranges. Some terranes within the Franciscan Complex are relatively coherent blocks of ocean-floor rocks, whereas others are a complex tectonic mélange.

4. About 22 to 24 million years ago, the Farallon Plate was completely subducted at one point along the western edge of North America. When the Pacific Plate came into direct contact with the North American Plate for the first time, the San Andreas fault was created. Transform motion began between the Pacific and North American Plates.

5. As more of the Farallon Plate disappeared along the subduction boundary, the San Andreas fault lengthened, and the Mendocino Triple Junction migrated north to its present position. Today's Gorda and Juan de Fuca Plates are remnants of the Farallon Plate. Cenozoic volcanic rocks in the central Coast Ranges

are evidence that the Mendocino Triple Junction was further south and subduction was active during Cenozoic time.

6. During the same time that the Farallon Plate was being subducted and accreted to North America, the Salinian Block formed southeast of its present location as part of the Mesozoic Sierran batholith. Cenozoic motion on the San Andreas fault system has moved the Salinian Block to its present position west of the San Andreas fault in the Coast Ranges.

7. During Cenozoic time, sedimentary rocks accumulated in basins within the Salinian Block and parts of the Franciscan Assemblage of the Coast Ranges. Today these rocks, mostly of marine origin, are found throughout the Coast Ranges and in offshore sedimentary basins.

8. About 3 to 5 million years ago, the present mountain ranges of the Coast Ranges were created. The northwest–southeast orientation of the present topography of the Coast Ranges Province is the result of structural control of the topography.

9. The present outlet for the Sacramento and San Joaquin River systems, through the Golden Gate, formed about 600,000 years ago, as dated by the Rockland ash in sediments of the Merced Formation. Before this, the drainage had been dammed, and a large lake covered much of today's Great Valley. Prior to the formation of the lake, the Sacramento–San Joaquin drainage outlet was further south.

10. San Francisco Bay and smaller estuaries at the mouth of coastal rivers formed less than 10,000 years ago as a result of glacial melting and sea level rise. During worldwide glacial periods, sea level was approximately 130 meters lower, and San Francisco Bay was a system of river valleys. The most recent glacial period, termed the Tioga glaciation in California, occurred about 20,000 years ago. Older estuarine sediments are evidence of the previous warm period about 125,000 years ago, and alluvial deposits about 160,000 years old provide a record of low sea level during the Tahoe glaciation.

11. Since 1850, the tidal wetlands of San Francisco Bay have shrunk as a result of human activities. These include hydraulic mining in the Sierra Nevada, agricultural reclamation, residential and commercial development, and salt mining.

REFERENCES

AALTO, K.R., AND HARPER, G.D. 1989. *Geologic Evolution of the Northernmost Coast Ranges and Western Klamath Mountains, California.* Washington, DC: American Geophysical Union, Field Trip Guidebook T308, 82 pp.

BAILEY, E.H., IRWIN, W.P., AND JONES, D.L. 1964. *Franciscan and Related Rocks and Their Significance in the Geology of Western California.* California Division of Mines and Geology Bulletin, Vol. 183, 177 pp.

BURCHFIEL, B.C., LIPMAN, P.W., AND ZOBACK, M.L., ed. 1992. The Cordilleran Orogen: Conterminous U.S. In *The Geology of North America,* Vol. G-3. Boulder, CO: Geological Society of America.

CONOMOS, T.J., ed. 1979. *San Francisco Bay: The Urbanized Estuary.* Pacific Division, American Association for the Advancement of Science, 493 pp.

KARL, H., AND OTHERS. 2000. *Beyond the Golden Gate: Oceanography, Geology, Biology, and Environmental Issues in the Gulf of the Farallones.* U.S. Geological Survey, Circular 1198, 84 pp.

SCHNEIDER, J. 1992. *Quicksilver: The Complete History of Santa Clara County's New Almaden Mine.* San Jose, CA: Zella Schneider, 178 pp.

STOFFER, P.W., AND GORDON, L.C., 2001, *Geology and Natural History of the San Francisco Bay Area.* U.S. Geological Survey, Bulletin 2188, pp. 61–86.

WAHRHAFTIG, C. 1984. *A Streetcar to Subduction and Other Plate Tectonic Trips by Public Transport in San Francisco.* Washington, DC: American Geophysical Union, 76 pp.

WAHRHAFTIG, C., AND SLOAN, D. 1989. *Geology of San Francisco and Vicinity.* American Geophysical Union, Field Trip Guidebook T105, 69 pp.

Web Sites and Media Resources

U.S. Geological Survey, San Francisco Bay Region Geology Home Page: *http://sfgeo.wr.usgs.gov*

U.S. Geological Survey, Rocks and Minerals of the San Francisco Bay Region: *http://geopubs.wr.usgs.gov/bulletin/b2195*

U.S. Geological Survey, San Francisco Bay: *http://sfbay.wr.usgs.gov*

National Park Service, Golden Gate National Recreation Area: *http://nps.gov/goga*

National Park Service, Redwood National Park: *http://nps.gov/redw*

National Park Service, Point Reyes National Seashore: *http://nps.gov/pore*

Association of Bay Area Governments: *http://abag/ca.gov*

California Department of Water Resources, Bay-Delta Office: *http://baydeltaoffice.water.ca.gov/*

Monterey Bay Aquarium Research Institute: *http://www.mbari.org/*

Almaden Quicksilver County Park: *http://www.geocities.com/almadenqs/*

U.S. Department of Energy, The Geysers geothermal area: *http://www.eere.energy.gov/geothermal/geysers*

13

Earthquakes, Faults, and Seismic Safety

Fig. 13-1 Map showing the epicenters and magnitudes of major California earthquakes, 1906–1999. (Source: Adapted and updated from *Geotimes*, 1993.)

Northern
California
4-17-1991
7.1

Petrolia
4-25-1992
7.1

Mendocino fault
9-1-1994
7.1

San Francisco
4-18-1906
7.8

Loma Prieta
10-17-1989
6.9

Calaveraas Fault

San Andreas Fault

Big Bear
6-28-1992
6.4

Hector Mine
10-16-1999
7.1

Tehachapi
7-20-1952
7.5

Garlock Fault

Landers
6-28-1992
7.3

Northridge
1-16-1994
6.7

San Fernando
2-9-1971
6.6

Joshua Tree
4-23-1992
6.1

Long Beach
3-10-1933
6.4

TABLE 13-1 "BIG ONES": MAJOR CALIFORNIA EARTHQUAKES

Date	Name	Magnitude	Fault	Motion	Deaths	Damage
12/08/1812	Wrightwood	7.5 (est.)	San Andreas (?)	Right lateral (?)	40	Unknown
12/21/1812	Santa Barbara	7.1	Unknown	Unknown	1	Unknown
1/09/1857	Fort Tejon	8.0	San Andreas	Right lateral	2	Unknown
10/21/1868	Hayward	7.0	Hayward	Right lateral	Unknown	Unknown
3/26/1872	Lone Pine	7.6	Owens Valley	Normal, right lateral	57	$250,000
2/23/1892	Laguna Salada	7.0	Laguna Salada	Right lateral	Unknown	Unknown
4/18/1906	San Francisco	7.8	San Andreas	Right lateral	3,000	$524 million
11/04/1927	Lompoc	7.3	Hosgri (?)	Unknown	0	Unknown
3/10/1933	Long Beach	6.9	Newport-Inglewood	Right lateral	120	$50 million
5/18/1940	Imperial Valley	6.9	Imperial	Right lateral	8	$6 million
7/21/1952	Kern County	7.5	White Wolf	Reverse, left lateral	12	More than $50 million
2/09/1971	San Fernando	6.6	San Fernando	Thrust	65	$500 million
5/2/1983	Coalinga	6.7	Nunez	Blind thrust	0	$30 million
10/01/1987	Whittier Narrows	5.9	Puente Hills	Blind thrust	8	$358 million
10/17/1989	Loma Prieta	6.9	San Andreas	Right lateral	63	$6-10 billion
4/25/1992	Petrolia	7.1	Offshore	Thrust	0	$6.1 million
6/28/1992	Landers	7.3	Johnson Valley, others	Right lateral	1	Unknown
1/17/1994	Northridge	6.7	Northridge	Blind thrust	57	$20-40 billion
10/16/1999	Hector Mine	7.1	Lavic Lake	Right lateral	0	Unknown

Note: Figure 13-1 shows the location of earthquake epicenters.

THE NORTHRIDGE EARTHQUAKE

It was 4:30 on the morning of Martin Luther King Day, January 17, 1994, when, for the first time since 1933, an earthquake struck directly beneath an urbanized part of California. When the shaking was over, the Northridge earthquake had killed 33 people (the final death toll was 57), injured more than 7,000, left more than 20,000 homeless, and caused between $20 and $40 billion in damage (Fig.13-2). The Santa Susana Mountains and the northern San Fernando Valley gained as much as 70 centimeters in elevation. The Northridge earthquake was followed by 3,000 aftershocks during the next 3 weeks, and the personal and financial aftershocks continued for several years. According to seismologists, the Northridge earthquake was unusual for only one reason: Thankfully, it hit at 4:30 in the morning. Lucky timing probably saved thousands of lives that would have been lost in collapsed parking structures, malls, and freeways if it had struck during the day or evening.

Fig. 13-2 Collapsed parking garage, California State University, Northridge Campus, as a result of the 1994 Northridge earthquake. (Source: Rymer, M. U.S. Geological Survey.)

Most Californians will experience at least one significant earthquake during their lifetimes, and even a small tremor can be an unsettling experience (Table 13-1). For most people, a general understanding of earthquakes alleviates their fears and helps them make more effective earthquake preparations. A constantly improving understanding of earthquakes has also made California's civic agencies among the world's best prepared for a large earthquake.

CAUSES OF EARTHQUAKES

Earthquakes occur when rocks in a section of the Earth's crust are stressed beyond their breaking point by tectonic forces. As discussed in Chapter 1, California's rocks are subjected to stress because of the continual movements of the Pacific and North American Plates. Because the rocks that make up the Earth's plates are brittle, they do not move continuously with the asthenosphere beneath them. As plate motion progresses, the blocks of rock above are strained, or deformed, accumulating **strain energy**. The strain energy builds up until the strength of the rocks is overcome, and the rocks break at their weakest points along fault planes. The rocks give way, then spring back into a new position (Fig. 13-3). Some of the stored strain energy is released in the form of frictional heat as the rocks grind against each other, and some is released as vibrations or **seismic waves**. This explanation for earthquakes was first developed by H.F. Reid, who studied the patterns of displacements along the San Andreas fault before and after the 1906 earthquake.

The behavior of rocks before and during an earthquake is frequently likened to the actions of a bow. As an archer pulls on the bowstring of a bow, strain energy is being stored. When the string is released, the arrow is propelled forward by the release of stored energy, and the bow and string spring back into position. The distance

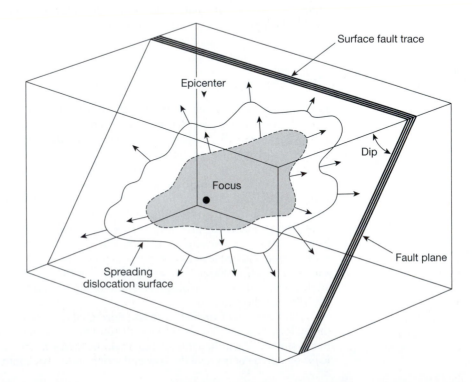

Fig. 13-3   Block diagrams showing rock behavior during an earthquake. **A,** Starting position of no strain and no displacement. **B,** Strain accumulates and the layers are bent without breaking as the rocks are subjected to stress. **C,** Rupture occurs on the fault, and the rocks are displaced. (Source: Keller, E., and Pinter, N. 1996, *Active Tectonics.* Upper Saddle River, NJ: Prentice Hall.)

that the arrow flies is proportional to the amount of strain stored in the bow, just as the size of an earthquake reflects the amount of strain energy accumulated in the rocks.

An earthquake is initiated when fault rupture begins, and the point of original rupture is known as the earthquake's *focus* (Fig. 13-4). Earthquakes do not originate right at the Earth's surface, but 90 percent of all earthquakes originate at depths of less than 100 kilometers. Shallow-focus earthquakes have focal depths

Fig. 13-4   Focus and epicenter of an earthquake. (Source: Bolt, B. 1993. *Earthquakes,* 3rd ed. New York: W.H. Freeman.)

between a few kilometers and 70 kilometers. All of California's large historic earthquakes fall into this category, and most earthquakes along the San Andreas fault system originate in the upper 10 kilometers of the crust. The 1989 Loma Prieta earthquake, with a focal depth of 18 kilometers, was unusually deep for the central San Andreas fault. The 1994 Northridge earthquake originated at a depth of 19 kilometers.

Seismologists classify intermediate-focus earthquakes as those originating at depths between about 70 and 300 kilometers, and deep-focus earthquakes at depths between 300 and 700 kilometers. Below these depths, temperatures and pressures are so high that rocks are plastic, rather than brittle. However, deep earthquakes are generated at subduction zones. Even along subduction zones, earthquakes with relatively shallow focal depths cause the greatest destruction. The 1964 Alaska earthquake, the strongest recorded in the United States, originated at a depth of about 30 kilometers.

EARTHQUAKES AND FAULTS

A casual look at almost any road cut, quarry, or sea cliff in California will reveal a *fault*—an abrupt line or zone breaking through the rocks or sediments (Fig. 13-5). The rocks or sediments on either side of a fault do not match up, indicating that there has been movement along the fault of the materials on opposite sides. Geologists recognize faults by the evidence of displacement or offset across them (see also Fig. 3-2). Sudden movement along a fault can take place during earthquakes, or the movement might be gradual.

Many faults seen at the Earth's surface have not moved in millions, or even hundreds of millions of years. These ***inactive faults***, generated by forces that are no longer operating, are unlikely to be the source of earthquakes today. Other faults are

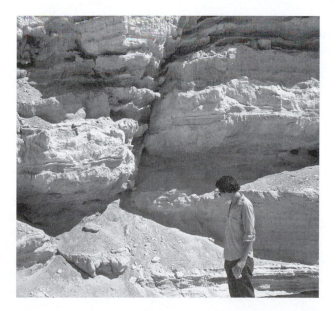

Fig. 13-5 A normal fault cuts through sediments and volcanic ash in the Owens Valley near Bishop. (Source: Photo by author.)

considered to be *active faults*; that is, capable of generating earthquakes in today's tectonic environment. Active faults can be recognized in several ways:

- Earthquakes occur on them today.
- Earthquakes have occurred on them in historic time.
- They displace young features or materials.

Using the principle of cross-cutting relations (see Chapter 3), it is clear that fault movement must be more recent than the materials it displaces (see Figs. 3-2 and 7-5). Repeated slippage weakens the rocks along a fault zone, making them the likely site of repeated earthquakes. For this reason, recognition of active faults is an important step in evaluating earthquake hazards in a region.

Faults can be meters up to hundreds of kilometers long, and the offset or movement along them can be centimeters up to hundreds of kilometers. During a single earthquake, only a part of a fault might slip; the slipped surface is known as the *rupture surface* for that earthquake. As would be expected, large-magnitude earthquakes result from ruptures along a great area of a fault. Many faults have ruptured to the Earth's surface, allowing geologists to examine their direction and age of movement. Other troubling active faults lie concealed beneath the surface; these *blind faults* pose particular seismic hazards to California's urban areas (see Table 13-1 and Chapter 16).

The boundary between the Pacific and North American Plates is far from a simple geometric line. In previous chapters, we have seen that California is made up of a complex assortment of rocks, many of which are separated by faults. It is not surprising, then, that California's active faults are also complex (Fig. 13-6 and Endpaper 2). Along the San Andreas system, many faults exhibit right-lateral motion. In some areas, this motion is accompanied by vertical movement as rocks are pushed together by compressional forces. In contrast, normal faults in the Basin and Range Province result from tensional forces (see Chapter 7). A more detailed view of the San Andreas fault system is presented in the following chapter.

EARTHQUAKE WAVES

During an earthquake, seismic waves travel outward from the point of original rupture. They arrive at the Earth's surface first at the *epicenter*, the point closest to the earthquake's focus. All earthquakes generate four types of seismic waves: two types of *body waves* that propagate through solid rock, and two types of *surface waves* that travel at or near the ground surface. The *P wave* (or primary wave) travels the fastest of all earthquake waves. Its motion is an alternating push and pull (compression and dilation) like that of a sound wave (Fig. 13-7A). When P waves emerge at the Earth's surface, some of them might be transmitted into the atmosphere as sound waves. These can sometimes be heard by animals or humans at the beginning of an earthquake as a rumbling or a sonic boom. The second, slower type of body wave is the *S wave* (or secondary wave). It propagates through rock by a sideways motion at right angles to the direction of travel, in a fashion similar to the shaking of a jump rope from one holder to another (Fig. 13-7B). When it arrives at the surface, the S wave shakes the ground surface with both vertical and side-to-side motion. S waves are responsible for much of the damaging ground shaking during earthquakes. The surface waves generated by earthquakes are called *Love waves*

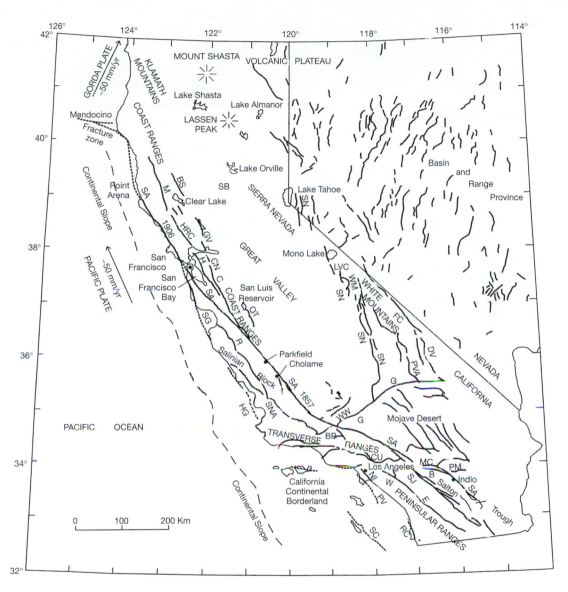

Fig. 13-6 A map of active faults in California. Compare with Endpaper 2 for activity level and location of California counties B = Banning; BP = Big Pine; BS = Bartlett Springs; BZ = Brawley seismic zone; C = Calaveras; CN = Concord; CU = Cucamonga; DV = Death Valley; E = Elsinore; FC = Furnace Creek; G = Garlock; GV = Green Valley; H = Hayward; HG = Hosgri; HRC = Healdsburg-Rodgers Creek; IM = Imperial; LVC = Long Valley Caldera; M = Maacama; MC = Mission Creek; NI = Newport-Inglewood; OT = Ortigalita; PM = Pinto Mountain; PV = Palos Verdes; PVA = Panamint Valley; R = Rinconada; RC = Rose Canyon; SA = San Andreas; SC = San Clemente Island; SG = San Gregorio; SJ = San Jacinto; SN = Sierra Nevada; SNA = Sur-Nacimiento; W = Whittier; Wm = White Mountains; WW = White Wolf. An interactive map of recent seismicity showing California faults and topography can be found at *http://quake.usgs.gov/recenteqs/latestfault.htm*. The arrows and numbers indicate direction and rate of Pacific and Gorda Plate motion relative to the North American Plate. (Source: Adapted from Wallace, R. 1990. U.S. Geological Survey, Professional Paper 1515.)

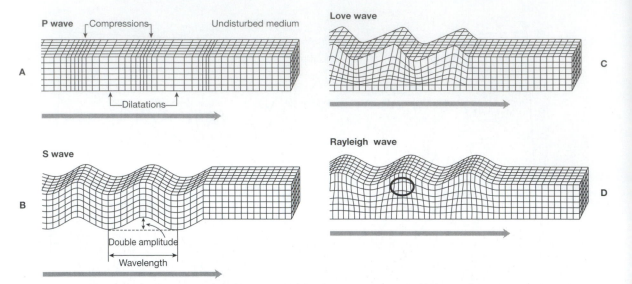

Fig. 13-7 Diagram illustrating the motions of four types of earthquake waves. (Source: Bolt, B. 1993. *Earthquakes,* 3rd ed. New York: W.H. Freeman.)

(Fig. 13-7C) and ***Rayleigh waves*** (Fig. 13-7D). Both are slower than the body waves, and the Rayleigh wave is the slower of the two. The Love waves move in a manner similar to S waves, except that their motion is only side-to-side. Like S waves, the motion of Love waves can be very damaging to structures. Rayleigh waves move across the Earth's surface like ocean waves, producing both vertical and horizontal motion.

Because the S wave is slower than the P wave, it arrives at the Earth's surface later than the P wave. As the distance away from the earthquake's focus increases, the S wave lags farther and farther behind the P wave. If an observer is 100 kilometers from the earthquake's epicenter, the S wave might arrive 10 to 15 seconds after the P wave. Seismologists use the time difference between the first arrival of P waves and the first arrival of S waves to determine the distance between an earthquake's focus and a ***seismograph station*** (Fig. 13-8). Knowing this distance for three or more stations, it is possible to locate the earthquake's point of origin. Because the type of rock through which they travel affects the speed of body waves, the exact location of an earthquake's epicenter might not be pinpointed until the data from tens of seismograph stations are processed, sometimes days after the quake.

An important characteristic of both types of body waves is that they are reflected or refracted (bent) when they encounter boundaries between different rock layers, including the boundary at the ground surface. The reflected waves amplify ground shaking at the surface. The patterns of ground shaking near the epicenter of an earthquake are very complex, as waves travel outward from different areas of the rupture zone, reflect and refract at geologic boundaries, and are amplified by particular conditions like water-saturated soils.

Fig. 13-8 Array of seismograph recorders at the U.S. Geological Survey, Menlo Park. The drums display the signals from seismographs at locations throughout northern California. (Source: Photo by author.)

Several federal, state, and local agencies have established networks of instruments designed to measure earthquakes. A network of seismograph stations records large earthquakes worldwide; data from these stations are transmitted to the National Earthquake Information Center in Golden, Colorado (*http://neic.usgs.gov*). The U.S. Geological Survey and California Institute of Technology in Pasadena are responsible for the seismograph network in southern California, and the U.S. Geological Survey in Menlo Park and the University of California, Berkeley are responsible for northern California stations. Arrays of highly sensitive, portable *seismometers* are positioned in several localities where seismologists are studying the behavior of specific faults or areas. Other instruments measure the frequencies of shaking in structures. Strong-motion seismographs are designed to record *ground acceleration* during an earthquake. More than 1,000 strong-motion sensors are in place throughout the state.

EARTHQUAKE SIZE

Seismographs record earthquakes by capturing the relative motion between a suspended pendulum and the vibrating earth. Incoming seismic waves are recorded as electrical signals on tape or as lines on a paper drum (Fig. 13-8). Seismographs can record horizontal or vertical motions, and they are capable of detecting small earthquakes that take place at great distances from the station.

Professor Charles Richter of the California Institute of Technology developed the first system for measuring earthquake size using the wiggly lines of seismograph records (Fig. 13-9). The ***Richter magnitude*** of an earthquake uses the amplitude of waves recorded by a seismograph as an indicator of the energy released during a quake. Because the Richter scale is related to strain energy release, the magnitude is a reflection of the earthquake itself. The Richter scale was developed specifically for southern California, based on the behavior of seismic waves in that region. Since its

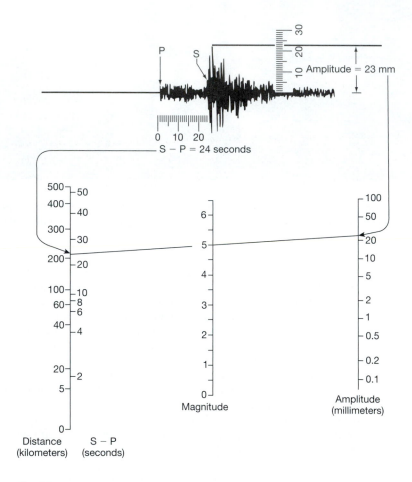

Fig. 13-9 An example of the calculation of Richter Magnitude (M_L) from a seismogram. To calculate the magnitude, follow these steps:

1. Use the time difference between the arrival of the S and P waves (P – S=24 sec) to measure the distance to the focus.
2. Measure the height of the maximum wave motion on the seismogram (23 mm).
3. Place a straight edge between the correct points on the distance (left) and amplitude (right) scales to obtain the magnitude (M_L = 5.0). (Source: Bolt, B. 1993. *Earthquakes,* 3rd ed. New York: W.H. Freeman.)

TABLE 13-2 COMPARISON OF RICHTER AND MOMENT MAGNITUDES OF EARTHQUAKES

Earthquake	Modified Magnitude (M_L)	Moment Magnitude (M_w)
Alaska, 1964	8.3	9.2
San Francisco, 1906	8.3	7.8
Loma Prieta, 1989	7.1	6.9
San Fernando, 1971	6.4	6.6
Northridge, 1994	6.4	6.7

invention in 1935, the Richter scale has been used worldwide as a measure of earthquake size, but different methods or modifications of the scale have been adapted for measuring deep earthquakes and earthquakes affecting different regions. New techniques of analyzing seismograph records have led to several refinements of the magnitude scale. In California, Richter magnitudes are generally equivalent to the reported *local magnitude* (M_L).

Since the 1990s, California seismologists have used the ***moment magnitude*** (M_w), as a measure of earthquake size. The moment magnitude scale, formally defined in 1972, is based on the seismic moment of an earthquake. An earthquake can be thought of as a moment: a force couple created by equal but opposite forces acting across a fault, and separated by a distance that is proportional to the size of the earthquake. The moment magnitude takes into account both the area of the rupture surface and the amount of displacement during an earthquake along the fault. The moment magnitude assesses the release of energy along the entire rupture zone of the fault, which can be hundreds of kilometers during a great earthquake. The moment magnitudes differentiate between earthquakes of all sizes, whereas the Richter magnitude values saturate for earthquakes above about 7.5. The Richter and moment magnitudes are very similar for earthquakes up to about magnitude 7.2, but moment magnitudes are much more consistent for larger events (Table 13-2). Seismologists have calculated moment magnitudes for historic earthquakes (Table 13-1) as well as for new events. New values are frequently assigned to earthquakes following refined calculations, so it is important to use the same magnitude measurement when comparing large earthquakes.

Earthquake magnitudes are *logarithmic*, meaning that each whole-number increase on the scale is a tenfold increase in the amplitude of the seismic waves. Each whole-number increase in magnitude also signifies approximately a thirtyfold increase in the amount of energy released (Fig. 13-10). "Great" earthquakes of magnitude 8.0 release about 1,000 times more energy than earthquakes with magnitudes of 6.0! A great earthquake strikes the Earth about once every 5 years, whereas 27 quakes with moment magnitudes between 5.7 and 7.0 have struck California since 1979.

Earthquake Intensity and the Mercalli Scale

If you ask someone to rank the earthquakes they have experienced, chances are good that they would use a scale based on earthquake *intensity*. The intensity of an earthquake is determined from the types and patterns of damage and people's experiences.

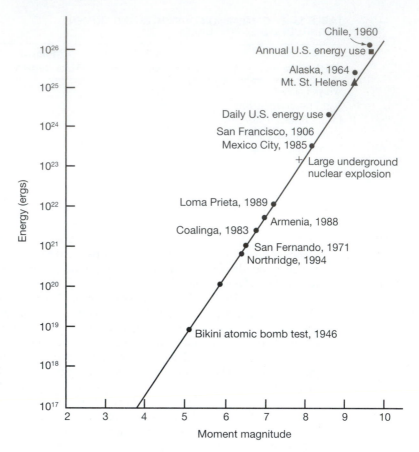

Fig. 13-10 Moment magnitudes and energy release of some moderate and large California earthquakes. (Source: Adapted from Bolt, B. 1993. *Earthquakes,* 3rd ed. New York: W.H. Freeman.)

Although intensity is linked to the magnitude of an earthquake, other factors are also important. For example, the intensity of shaking, as perceived by an observer or as evaluated from structural damage, decreases with distance from an earthquake's epicenter. In general, earthquake intensity decreases, or *attenuates,* away from the epicenter.

Two additional factors unrelated to the earthquake are equally important in determining earthquake intensity. The nature of the soil or bedrock in an area has a great effect on the severity of ground shaking. Areas underlain by unconsolidated sediments could sustain greater damages than bedrock areas much closer to an earthquake's epicenter, particularly if the sediments are water-laden. A final factor that influences intensity is a human factor: the type and density of structures near the epicenter. In many regions of the world, thousands of people are often killed when moderate earthquakes destroy their poorly constructed residences. In California, similar earthquakes only rattle nerves and knock things off shelves.

TABLE 13-3   MODIFIED MERCALLI INTENSITY SCALE (ABRIDGED)

Intensity	Effects
I	Not felt by people; detected by seismometers.
II	Felt only by a few persons at rest, or those who are on upper floors of buildings. Delicately suspended objects may swing.
III	Felt quite noticeably indoors, especially on upper floors of buildings; hanging objects swing. Standing cars may rock slightly; vibration like the passing of a light truck.
IV	During the day felt indoors by many, outdoors by a few. At night some awakened. Dishes, windows, doors rattle; walls make cracking sounds. Vibration like a heavy truck passing; standing cars rock noticeably; car alarms triggered.
V	Felt by nearly everyone; many awakened. Some dishes, windows, etc., broken; a few instances of cracked plaster; unstable objects overturned. Disturbance of trees, poles, pictures, and shutters sometimes noticed. Doors swing open or closed.
VI	Felt by all; many are frightened and run outdoors. Some heavy furniture moved; books fall off shelves; some fallen plaster or damaged chimneys. Windows, glassware, dishes broken.
VII	Difficult to stand. Damage negligible in buildings of good design and construction; slight to moderate in well-built ordinary structures; considerable in poorly built or badly designed structures. Some chimneys broken. Loose bricks and tiles fall. Noticed by persons driving cars.
VIII	Damage slight in specially designed structures; considerable in ordinary substantial buildings with partial collapse; great in poorly built structures. Houses moved off foundations if not bolted. Chimneys, water tanks, columns, monuments fall. Heavy furniture overturned. Sand and mud ejected in small amounts. Changes in well water. Difficulty steering cars.
IX	General panic. Damage considerable in specially designed structures; well-designed frame structures thrown out of plumb; some buildings collapse. Buildings shifted off foundations. Ground cracked. Underground pipes broken.
X	Some well-built wooden structures destroyed; most masonry and frame structures with foundations destroyed; ground badly cracked. Rails bent. Landslides from river banks and steep slopes. Shifted sand and mud. Water splashed (slopped) over banks.
XI	Few, if any, masonry structures remain standing. Bridges destroyed. Broad fissures in ground. Underground pipelines completely out of service. Earth slumps and land slips in soft ground. Rails bent greatly.
XII	Damage total. Waves seen on ground surfaces. Lines of sight and level distorted. Objects thrown upward into the air.

Source: Modified from U.S. Geological Survey, 1974. Earthquake Information Bulletin 6(5):29.

The *modified Mercalli scale* is used to categorize earthquake intensity (Table 13-3). Following an earthquake, seismologists use questionnaires to obtain information from inhabitants. An intensity value is assigned to each response, and the values are plotted on a map. The values are contoured, producing a generalized view of the damage pattern (Fig. 13-11). In recent years, the U.S. Geological Survey has produced intensity maps based on responses to an online questionnaire, "Did You Feel It?" The form can be completed by anyone who feels an earthquake (*http://pasadena.wr.usgs.gov/shake/ca/*), and the results are compiled into a Community Internet Intensity map (Fig. 13-12).

Fig. 13-11 Intensity map for the 1989 Loma Prieta earthquake (Source: Plafker, G., and Galloway, J. 1989. *Lessons Learned From the Loma Prieta Earthquake.* U.S. Geological Survey.)

DESTRUCTIVE EVENTS DURING AN EARTHQUAKE

If a fault ruptures to the ground surface during an earthquake, then the ground surface is broken (see Fig. 7-5). The ground slips in the same direction as the fault, displacing everything that lies across the surface trace. In a major earthquake, this ***ground rupture*** could involve several meters of vertical or horizontal separation (Fig. 13-13).

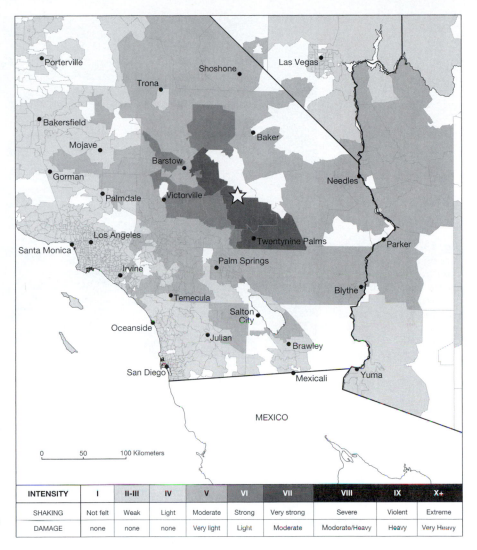

Fig. 13-12 Community Internet Intensity map for the October 16, 1999 Hector Mine earthquake. The maximum intensity was VII for this event, which was based on 25,215 responses. (Source: U.S. Geological Survey, Pasadena, *http://pasadena.wr.usgs.gov/shake/ca/.*)

INTENSITY	I	II-III	IV	V	VI	VII	VIII	IX	X+
SHAKING	Not felt	Weak	Light	Moderate	Strong	Very strong	Severe	Violent	Extreme
DAMAGE	none	none	none	Very light	Light	Moderate	Moderate/Heavy	Heavy	Very Heavy

Buildings and other structures lying directly on the ruptured surface clearly could not withstand such motion if the displacement is large. In open areas, surface rupture might be seen as a crumpled line often called a moletrack because it resembles the burrowed passage of a mole (Fig. 13-14). If differential vertical motion occurs, a linear cliff called a *fault scarp* results. As illustrated in Chapter 7, a fault scarp several meters high can be produced during a single, great earthquake. Surface rupture can also produce much more subtle zones of cracks, rumples, and horizontal displacements. Examples are discussed further in the following chapter. Ground rupture can cause great destruction to structures, but it is confined to the zones directly above the rupturing faults.

Fig. 13-13 Ground rupture on the Lavic Lake fault created by the 1999 Hector Mine earthquake, showing right lateral displacement. The dirt road is offset 3.7 to 4.8 meters (Source: Otis-Diehl, P. U.S. Marine Corps, 29 Palms.)

Fig. 13-14 The "mole-track" created by ground rupture in the 1906 earthquake, Olema, Marin County, near the epicenter. (Source: Gilbert, G.K. U.S. Geological Survey.)

Ground shaking generally causes the greatest damage during an earthquake. Even a moderate earthquake shakes the ground many kilometers away from the earthquake epicenter. In areas where unconsolidated sediments are saturated with water, ground shaking is more severe and lasts longer than the shaking in bedrock areas. Artificially filled land, created by filling in natural wetlands, might experience the most severe shaking of all. Ground shaking is responsible for collapsed and damaged buildings, elevated freeways, and other structures. Ground shaking can rupture

Fig. 13-15 Fire created by ruptured gas lines resulting from the 1994 Northridge earthquake, Balboa Boulevard, Granada Hills. (Source: Rymer, M. U. S. Geological Survey.)

gas pipelines and tanks, causing fires to start (Fig. 13-15). Furthermore, it might be difficult to put out fires if water mains are ruptured. About 80 to 90 percent of the destruction in San Francisco during the 1906 earthquake resulted from fire.

Liquefaction is a dangerous result of ground shaking in areas underlain by water-saturated fine sand. The wet sand becomes a slurry of sandy liquid as repetitive

Fig. 13-16 Liquefaction of sandy sediments along the Pajaro River floodplain near Watsonville during the 1989 Loma Prieta earthquake resulted in the collapse of the Highway 1 bridge. Note that the bridge pilings punched up through the pavement on the right side of the photo. (Source: Nakata, J. U.S. Geological Survey.)

A

Fig. 13-17 Landslides generated during the 1994 Northridge earthquake. **A**, Pacific Palisades, Santa Monica. **B**, Santa Susana Mountains. (Source: **A**, Jibson, R.; **B**, Harp, E. U.S. Geological Survey.)

B

ground shaking causes the sediment to lose strength. A walk along the wet sand at the edge of a beach can reproduce this effect. Repeated tapping on the wet sand turns it to a jelly-like mush. Pushing on the wet sand can also force water to the surface in adjacent areas. Both phenomena are possible during liquefaction. Water and sand can be forced to the ground surface, producing sand boils.

Liquefaction was widespread during the 1906 earthquake, which struck at the end of a wet winter, when soils and sediments were water-laden. During the 1989 Loma Prieta earthquake, liquefaction was common around the margins of San Francisco Bay and along the low-lying, sandy floodplains of coastal rivers (Fig. 13-16). The most notable liquefaction damage was in the Marina area of San Francisco, where apartment buildings are built in an area where sandy material had been used to fill in a lagoon. In an ironic example of history repeating itself, the liquefied fill brought with it to the surface some of the old rubble from 1906. Recognizing the serious threat posed by liquefaction, the California Geological Survey recently released maps of liquefaction for much of the San Francisco Bay Area (available at *http://www.abag.ca.gov/bayarea/eqmaps/liquefac/liquefac.html*).

Another potentially deadly side effect of ground shaking is the generation of earthquake-created ***landslides***. Many of California's steep hillsides are underlain by relatively weak rocks. Ground shaking can push these areas over their threshold of stability, resulting in multiple landslides during a single earthquake. For example, the 1994 Northridge earthquake triggered widespread landslides along the coast (Fig. 13-17A) and in the steep Santa Susana Mountains to the north (Fig 13-17B).

Fig. 13-18 Creation of a tsunami. In the deep ocean (rear of the diagram), tsunamis barely raise the sea surface, and the crests of successive waves can be up to 30 kilometers apart. When the waves reach shallower water, they are slowed down by interactions with the sea floor (see Chapter 15). The waves then pile up into a wall of water, creating a tsunami. (Source: *Earth Magazine*, July 1993.)

Tsunamis

Areas along California's coast are vulnerable to *tsunamis*. A tsunami is created when an earthquake ruptures the ocean floor, or as a result of huge submarine landslides or volcanic eruptions. Earthquakes with vertical displacement cause the water above the sea floor to be displaced. A series of long waves radiates outward from the point of rupture. The waves are long and low in the open ocean, but they can pile up along the shoreline to create a wall of water more than 30 meters high (Fig. 13-18). Because they can originate anywhere in the Pacific Ocean, tsunamis can arrive at the California coast without warning.

Following the 1964 Alaska earthquake, the harbor at Crescent City, California was destroyed by a 7-meter-high tsunami that killed 11 people and caused more than $7 million in damage to the city. Further south, tsunamis from the same earthquake

Fig. 13-19 Map showing the potential areas of tsunami flooding, liquefaction, and landsliding in Eureka, Humboldt County, following a large earthquake in the Cascadia subduction zone. (Source: National Oceanic and Atmospheric Administration. 1995. Pacific Marine Environmental Laboratory, Seattle, WA: *http://www.pmel.noaa.gov/ ~bernard/senatec.html*.)

damaged boats in San Francisco Bay, the Santa Cruz harbor, and as far south as Los Angeles and Long Beach. Other tsunamis generated by an Alaskan earthquake hit California in 1946.

Faults of the Cascadia subduction zone are also potential tsunami generators. In 1992, a small tsunami generated by a magnitude 7.1 earthquake near Cape Mendocino arrived in Eureka only 15 minutes after the quake. Following this event, the Federal Emergency Management Agency (FEMA) and the National Oceanographic and Atmospheric Agency sponsored preparation of tsunami inundation maps for Eureka and Crescent City (Fig. 13-19).

Along the San Andreas fault system, most of California's earthquakes involve primarily horizontal motion, so their tsunami potential is lower than that of subduction-zone faults. Nevertheless, the 1906 San Francisco earthquake, whose (my preference) epicenter was just southwest of San Francisco, generated a wave about 1 meter high in San Francisco Bay.

Scientists are increasingly concerned about the tsunami potential of faults in southern California. Several active faults lie off the coast of southern California, including the San Clemente fault system, the San Isidro fault zone, the offshore part of the Palos Verdes fault, and the Rose Canyon fault (see Fig. 13-6 and Chapter 16). In 1812, an earthquake centered in the Santa Barbara Channel apparently generated a tsunami that affected the Santa Barbara and Ventura areas (Table 13-1), and an offshore earthquake in southern California in 1927 created small tsunamis in the nearby coastal towns. Scientists have predicted that earthquakes with magnitudes of 7.0 or greater could generate tsunamis that would travel to the coast in less than 20 minutes.

Today, the Pacific Tsunami Warning Center records seismographic data from around the Pacific Ocean, giving warnings to coastal areas whenever a Pacific earthquake with tsunami potential strikes. However, many coastal communities are beginning to plan for emergency notification and evacuation in the event of locally generated tsunamis.

EARTHQUAKE PREDICTION AND EVALUATION OF SEISMIC HAZARDS

During the past 25 years, researchers have begun a number of programs designed to increase our understanding of California's earthquakes, with the ultimate goals of hazard reduction and earthquake prediction. The most comprehensive of these programs is the National Earthquake Hazards Reduction Program (NEHRP), established in 1977 to reduce the risk to life and property from earthquakes in the United States. Although the scientific community is far from the goal of reliable earthquake prediction, significant progress has been made. The stated goals of the NEHRP are to:

- Understand the earthquake source.
- Determine earthquake potential.
- Predict the effects of earthquakes.
- Disseminate research results.

The first step in evaluating earthquake hazards is to identify active faults. The California fault map (Endpaper 2) is a detailed compilation of all of California's known and suspected active faults that can be mapped at the surface. The fault map and its accompanying explanation summarize the historic and geologic history of movement along each fault. More detailed maps, published by the California Division of Mines and Geology, are available to cities, counties, and development agencies.

The best way of predicting the future behavior of California's faults is to better understand the patterns of their past behavior. The patterns cannot be seen by looking just at the historic earthquakes of the past century or two, because many of the most dangerous fault segments break less frequently. By remembering that strain energy accumulates between strain-releasing earthquakes (Fig. 13-20), we can easily conclude that the seismically quiet sections of active faults might pose the greatest danger.

To study suspected active faults, geologists might excavate trenches across them. By carefully mapping the sediments and soils in the excavated trench, geologists can measure the displacement of layers across the fault. They can also determine the youngest materials that are displaced by faulting, thus pinpointing the most recent ground-rupturing earthquake (Fig. 13-21, Fig. 3-2). In some trenches, geologists can find evidence of earthquakes that disturbed older sedimentary layers in the lower walls of the trench before the more recent layers were deposited. By dating the materials using radiocarbon dating or other methods, the timing of past earthquakes can then be determined with some accuracy. These detailed studies of *paleoseismology* provide critical information about the characteristic frequency and

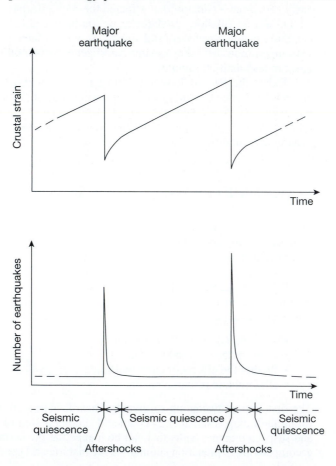

Fig. 13-20 The earthquake cycle. (Source: U.S. Geological Survey, National Earthquake Hazards Reduction Program.)

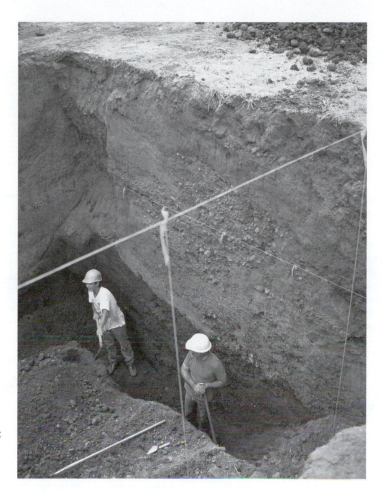

Fig. 13-21 Geologists have excavated and then peeled back one side of a trench to examine sediments offset by the San Andreas fault near Woodside, San Mateo County. (Source: Clahan, K.)

magnitude of earthquakes along different segments of California's active faults. Based on these studies, scientists estimate that there is a 62 percent probability of an earthquake with magnitude 6.7 or higher in the San Francisco Bay area within the next 30 years. Southern California has an estimated probability of about 85 percent within the next 30 years.

Seismologists from several agencies maintain a network of seismograph stations and other instruments designed to monitor California's active faults. Highly sensitive seismometers (portable seismographs) monitor the microearthquakes along some fault segments to detect possible *foreshocks* of a large earthquake. To date, foreshocks have not proved to be reliable predictors of major quakes, although the 1975 Haicheng, China earthquake was predicted by foreshocks that began 4 days before the 7.3 magnitude event.

Scientists are also examining other possible earthquake precursors, including changes in electrical resistivity, changes in the amount of dissolved radon gas in groundwater, changes in water-well levels, and bulging of the ground surface. As discussed in Chapter 1, very small changes in distances and elevations are being recorded by a network of surveyed points, called the Southern California Integrated

Fig. 13-22 Thirty-year earthquake probabilities for the San Andreas fault system for the period 1988 through 2018. Maps showing updated probabilities for California are shown in Endpaper 4. (Source: U.S. Geological Survey, National Earthquake Hazards Reduction Program.)

GPS network. Some of these phenomena might result from changes in the rocks in the area of imminent rupture, particularly an influx of water prior to the quake. Other possible precursors include radio-wave signals and anomalous animal behavior.

In 1989, the U.S. Geological Survey compiled a map showing the conditional probability of a major earthquake along various segments of the San Andreas fault system (Fig. 13-22). The map showed the probabilities during a 30-year period from 1988 to 2018, based on the known patterns of historic and prehistoric ruptures along various faults. An earlier map of this type indicated a high probability of rupture on the segment of the San Andreas fault that later produced the Loma Prieta earthquake. The 30-year earthquake probability forecasts were updated for southern California in 2001 (*http://www.scecdc.scec.org/PhaseII.html*) and for the San Francisco Bay area in 2003 (*http://quake.usgs.gov/research/seismology/ wg99/*).

PREDICTING GROUND SHAKING

Several characteristics of motion determine how much damage will result from ground shaking during an earthquake. These include the ***amplitude*** of ground motion, the duration of shaking, the frequency of the vibrations, and the peak acceleration of the ground. In general, large-amplitude waves cause greater damage to structures than low-amplitude waves. Duration of shaking is a critical factor, because as shaking continues, many structures lose their ability to withstand the motion. Engineers estimated that if the 15 seconds of shaking during the Loma Prieta earthquake had lasted a few additional seconds, the Embarcadero Freeway in San Francisco would have entirely collapsed. During large earthquakes like the 1964 Alaska event, ground shaking can last for 3 or more minutes.

The frequency of earthquake vibrations is one of the most important factors controlling damage because some materials and structures respond to vibrations by amplifying the motions at particular frequencies. If the frequency of some of the earthquake vibrations matches the natural *resonant frequency* of the soil beneath a structure or the structure itself, then the stress on the structure is significantly increased. Unconsolidated sediments or artificial fill with a high water content can greatly amplify ground shaking. Amplification is increased further if the natural frequency of the building matches that of the materials beneath it. For example, during the 1985 Mexico City earthquake, the greatest damage occurred in the areas where an ancient lake had deposited thick layers of clay. The clay layers amplified the motion

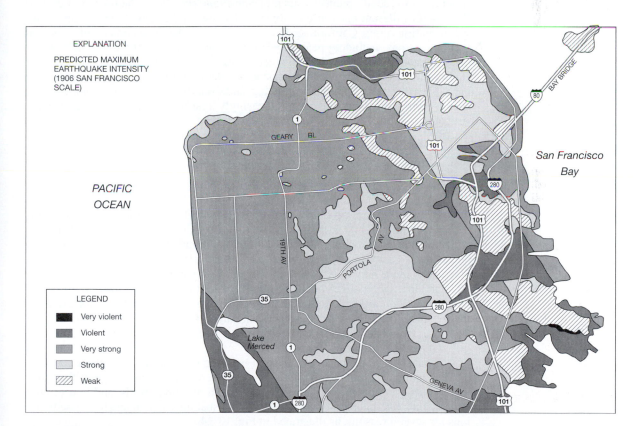

Fig. 13-23 Predicted severity of ground shaking in San Francisco based on the distribution of types of rock, sediment, and artificial fill. (Source: Association of Bay Area Governments.)

of waves with a 2-second period. However, not all structures within the ancient lake area suffered equally. Because waves with a 1- to 2-second period match the natural frequency of 10- to 14-story buildings, these buildings suffered far greater damage than either shorter or taller structures.

The peak acceleration of the ground during an earthquake is another critical factor that influences the destructiveness of earthquakes. A structure exposed to vertical accelerations greater than $1g$, or 100 percent of the force of Earth's gravity, would be airborne unless it was fastened to the ground. Until the 1971 San Fernando earthquake, seismologists had measured accelerations between 0.35 and $0.5g$ near the rupture zones of moderate-magnitude earthquakes. Until 1971, building engineers used the peak acceleration of $0.5g$ as the expected peak value. However, a peak horizontal acceleration of $1.25g$ was recorded during the 1971 San Fernando earthquake (discussed later). During the 1994 Northridge earthquake, surprisingly high horizontal accelerations of $1.8g$ and vertical accelerations of $1.2g$ were recorded at Tarzana, about 7 kilometers from the epicenter. The strong accelerations contributed to the extensive damage of modern structures during this earthquake.

Understanding the correlation between geologic materials and the amplification of earthquake waves is an important step in designing earthquake-resistant structures. A recently published map of San Francisco shows the predicted ground shaking for different parts of the city (Fig. 13-23). At the same time, continued instrumentation along California's faults is providing better estimates of expected ground motions, including peak acceleration, peak velocity, and duration of shaking. Endpaper 4 combines the most recent estimates of earthquake probabilities for California fault with estimates of ground acceleration to show the overall shaking hazard for the state. With this improving understanding of how the ground shakes during earthquakes, engineers are increasingly able to design earthquake-resistant structures.

BUILDING DESIGN

During an earthquake, structures respond to ground shaking in ways that reflect the overall building architecture and the characteristics of the materials used in construction. It is not possible to build structures that remain motionless during an earthquake, but it is possible to build structures that can withstand strong ground shaking. One of the more important factors in designing an earthquake-resistant structure is structural integrity. If all elements of a building are tied firmly together, the structure will shake with a unified motion. If different elements of a structure—for example, the foundation and the walls, the floors and supporting columns, two wings at right angles, or two ends of a bridge or freeway overpass—vibrate with different motion, then the building might shake apart.

Certain design elements can make a building particularly susceptible to earthquake damage. Examples are multistory houses with an open garage at the street level, split-level houses with large open areas on the lower walls, and houses constructed on stilts. In urban areas, strong ground shaking can damage high-rise buildings for several reasons, as illustrated in Fig. 13-24.

Fig. 13-24 Effects of ground shaking on high-rise buildings of various types. **A,** A building with wings at right angles experiences damage at the connection points because the wings respond differently. **B,** Building elements with varying heights have different resonant frequencies. **C,** Shaking is amplified at the top of a tall building. **D,** Adjacent buildings with varying heights and shaking frequencies experience pounding. **E,** Swaying is enhanced because of alignment relative to the direction of incoming waves. **F,** Flexible high-rise buildings with different architecture: Design 1 remains elastic, whereas the soft ground floor in Design 2 has no shear resistance. **G,** A pair of buildings on different soils: 3 on rock, which transmits higher frequency waves, and 4 on softer sediments, which are subject to wave resonance and require special bracing. (Source: Bolt, B. 1993. *Earthquakes*, 3rd ed. New York: W.H. Freeman, p. 251.)

Wood-frame buildings are relatively resistant to earthquake damage because wood is relatively light and flexible. However, even these buildings can be destroyed or badly damaged if they have inadequate load-bearing walls, improper shear-wall bracing, or weak foundations. Many wood-frame buildings suffer needless damage during earthquakes because they are not securely anchored to the building foundation. Many pre-1940 buildings were not originally connected to their foundations. Fortunately, it is relatively inexpensive to attach wood-frame buildings to their foundations using anchor bolts.

The most dangerous structures in California are unreinforced brick buildings built before 1940. Masonry is heavy, and the strength of brick walls depends on the mortar holding the individual bricks together. For the same reason, brick chimneys and fireplaces, slate or tile roofs, decorative stonework hanging from upper stories, and masonry veneers can pose deadly hazards during an earthquake. Modern masonry is strengthened with reinforcing rods, and other brickwork built within the past 20 years is generally well anchored.

SEISMIC SAFETY AND GOVERNMENT REGULATIONS

The Long Beach earthquake, centered about 6 kilometers offshore of Newport Beach on the Newport-Inglewood fault zone, struck on March 10, 1933. Although it was a moderate event with a moment magnitude of 6.9, the earthquake killed 120 people, injured hundreds, and caused about $50 million in damage (Fig. 13-25).

A

B

Fig. 13-25 Effects of the 1933 Long Beach earthquake: **A**, Damaged building. **B**, Refugee camp at Monterey Park. (These photographs were taken by the mother of Audrey Wright, who was 7 years old at the time of the earthquake. Audrey Wright was a student in the author's California Geology course in 2002.)

Water-saturated alluvial sediments underlying the Long Beach area were partially responsible for the destruction. However, it was poor construction that accounted for the "tragic and scandalous failures of school structures" in the Long Beach area (Gordon B. Oakeshott, California Division of Mines and Geology, as quoted in *California Geology,* March 1973, p. 57). Luckily, the schools were almost vacant at 5:54 P.M. when the earthquake struck: Officials estimated that the casualties would have been ten times greater if schools had been in session. Following the Long Beach earthquake, the California Legislature passed the ***Field Act***, which requires state approval and inspection of both plans and construction of school buildings. This was California's first attempt to ensure seismic safety through government regulation. The Field Act has been amended twice since its passage. A 1967 addition required inspection of older school buildings, with replacement or strengthening mandated by 1975, although many financially strapped school districts have not yet come into full compliance. A second amendment, also adopted in 1967, requires all proposed school sites to be inspected to avoid siting within a fault zone or in a landslide area.

Early on the morning of February 9, 1971, a moderate earthquake of magnitude 6.6 shook the densely populated San Fernando Valley of southern California. Although the shaking lasted only 15 seconds, the earthquake killed 65 people, destroyed two hospitals and a modern freeway overpass, and caused more than $500 million in damage. Had the shaking lasted 5 seconds longer, the tragedy would have been much worse. Overlooking the 80,000 residents of the San Fernando Valley stood the Van Norman Dam, built in 1915. During the earthquake, the upstream side of the dam surface broke away (Fig. 13-26). Partly because the reservoir was only slightly over half full, the dam held, and the residents below were spared from a disastrous flood.

If a single earthquake can be credited for California's current legislation regarding seismic safety, it would have to be this 1971 San Fernando earthquake.

Fig. 13-26 The Van Norman Dam, San Fernando Valley, following the 1971 earthquake. The geologists are standing immediately below cracks caused by the earthquake. (Source: U.S. Geological Survey.)

Because this earthquake and its effects were more accurately measured and thoroughly studied than any previous California earthquake, the San Fernando earthquake provided the basis for a number of legislative actions and code modifications designed to reduce earthquake hazards. The San Fernando earthquake and the 1983 Coalinga earthquake also helped to alert scientists to the threat posed by blind thrust faults beneath the Los Angeles basin and other parts of California (see Chapter 16).

In 1927, the International Conference of Building Officials established the *Uniform Building Code* (UBC) in an attempt to bring standardization to all aspects of building construction. The UBC is updated and revised every 3 years, and since 1975, California cities and counties have been required by the California Health and Safety Code to adopt the most recent edition of the UBC. Specific sections of the UBC are designed to increase the resistance of structures to ground motion, for example with design requirements for resisting lateral forces. Building officials and municipal departments follow UBC guidelines when permitting and building new construction.

Extensive studies of relatively modern structures that were partly collapsed or seriously damaged during the 1971 earthquake revealed some troubling problems. Modern concrete-frame structures had inadequate reinforcing or insufficient shear bracing, and the walls and roofs of "tilt-up" concrete buildings were insufficiently

Fig. 13-27 Alquist-Priolo earthquake fault zone along the Hayward fault in Fremont, Alameda County. (Source: California Geological Survey.)

tied together. Because modern buildings were damaged—despite being built to the standards in effect at the time—the UBC was extensively modified in 1973 and further strengthened in 1976. Future modifications are likely after studies of structural damage during the 1994 Northridge and 1995 Kobe, Japan earthquakes are completed. The surprising amount of damage to the welds in steel-frame buildings has been of particular concern.

The *Alquist-Priolo Earthquake Fault Zoning Act* (before 1994, called the Special Studies Zones Act) was passed into law in 1972. The Alquist-Priolo Act prohibits the construction of most human-occupied structures within 50 feet of an active fault. The faults zoned under the Alquist-Priolo Act are shown on official maps issued by the California Geological Survey (Fig. 13-27). These maps show the location of all faults with known surface displacement during the past 11,000 years, as well as faults considered to be potentially active with a high potential for ground rupture. Because the Alquist-Priolo Act prohibits most new construction across active faults, owners of pre-1972 buildings within the 50-foot setback zone find themselves unable to obtain permits for any new construction, including remodeling. Another important provision of the Alquist-Priolo Act is that property owners and their agents are required to disclose the existence of any active faults in the vicinity of a property being sold. Enforcement of the Alquist-Priolo Act and the granting of exemptions are the responsibility of local planning agencies.

The *Hospital Safety Act,* also passed following the San Fernando earthquake, called for the strengthening of construction standards for hospitals. The *Dam Safety Act* of 1973 required evaluations of the safety of existing dams in California. Part of the evaluation included preparation of maps showing the potential areas that would be flooded if a dam did fail. The Dam Safety Act also requires preparation of evacuation plans in the event of dam failure. The statewide dam inspections that followed passage of this act resulted in a number of existing dams being strengthened for increased seismic safety. An amendment to the California Government Code, also passed following the 1971 earthquake, requires every county and city to adopt a Seismic Safety Element as part of its general plan.

Despite the development of seismic safety elements in the state's laws and building codes, older unreinforced brick or masonry (URM) buildings still pose one of our greatest hazards. For example, 46 URM buildings were demolished in the city of Santa Cruz during the 1989 Loma Prieta earthquake. This included historic Pacific Garden Mall, where several people died beneath the brick rubble (Fig. 13-28). The *Unreinforced Masonry Building Law* was signed in 1986. This law required California's local governments to make an inventory of URM buildings and to develop earthquake hazard mitigation programs for those buildings by 1990. At the time of the most recent survey in 1997, only 53 percent of the 288 local agencies required seismic retrofits of these buildings; these areas included more than 78 percent of the inventoried URM buildings. Just in the seismically vulnerable metropolitan areas of the San Francisco Bay area and the Los Angeles Basin, more than 25,000 URM buildings with an average size of 10,000 square feet had been inventoried. In 1992, California established a uniform building code for earthquake hazard reduction in URM buildings with load-bearing walls. The URM law could result in an estimated $4 billion effort to reduce earthquake hazards by

Fig. 13-28 The Pacific Garden Mall, Santa Cruz, following the Loma Prieta earthquake. (Source: Nakata, J. U.S. Geological Survey.)

the beginning of the next decade. In addition to the financial burdens of strengthening the buildings, many local governments would like to exempt their historic buildings and affordable housing from the code.

Earthquake Preparedness

Since 1769, 206 known earthquakes with magnitudes estimated or measured at 6.0 or greater have occurred in California, western Nevada, and the northern part of Baja California, Mexico. Of these, 117 took place on the San Andreas fault system (see Chapter 14). Despite a greater public awareness of earthquake risk and increasingly earthquake-resistant building construction, it is not likely that California's vulnerability to earthquakes will decrease during the coming decades. As the population increases, urban areas and freeway corridors become increasingly congested, and this trend alone means that future moderate and large earthquakes will cause greater loss of life and property damage.

The U.S. Geological Survey, the California Office of Emergency Services, and local governments have prepared a series of "scenarios" that would follow a likely earthquake in the San Francisco Bay area or the greater Los Angeles area. In all cases, the scenarios predict large residential and commercial areas left without power or transportation access. Large numbers of people would be unable to obtain emergency services for 72 hours or more, and thousands could be stranded far from home by disabled transportation systems and freeways. It is important for all Californians to be prepared for earthquakes, not only at home, but also for earthquakes that might strike when they are away from home. Earthquake planning and preparedness might not eliminate all risks, but it can reduce the anxieties of people stranded far from their loved ones, pets, and households.

THE NEXT BIG ONE

It's a typical day at your college dorm. You're sitting in your room on the third floor, checking your e-mail. You only live 40 miles from home, but your parents thought that freshman year in the dorms would be a good experience for the whole family. Dad is on a business trip to Taiwan, and at this hour, Mom is probably at her job, which is an hour's drive from home. Your younger brother Bobby is at junior high school. Grandma, who lives with the family, is at home with the dog, Lucy.

Suddenly, your computer starts shaking—and so does everything else. A long minute later, you find yourself under the desk and realize that you are safe. You crawl out from under it and into a chaotic scene of books, lamps, chairs, CDs, and everything else you own. The power suddenly goes out, and the halls of the dorm fill with shouting students. People begin to evacuate the building in a fairly orderly fashion, thanks to recent drills.

Outside the dorm, you learn from someone with a portable radio that a 7.5 magnitude earthquake has struck. You immediately reach for your cell phone to call home—but it doesn't work. Three million other people are overloading all the cell phone and wireless networks. The landlines from the pay phones don't work either, and the power is still out. Panicked about your family's safety, you decide to drive home to check on them.

However, your car is somewhere in the rubble of the former parking garage.

You run though a mental list of where everyone probably is:

- Mom: At her office building, even further from home than you.
- Dad: In Taiwan, but frantic with worry.
- Bobby: Will the school take care of him?
- Grandma: Who will give her her medicine?
- Lucy: Will she run away?

This is not the earthquake your family had planned for—the one when everyone is at home together, able to grab flashlights and the emergency kit. Freeways are completely clogged, and many overpasses are down. Your only way home seems to be on foot, but your shoes are in the dorm room, and even if you could get them, it probably would take you 2 days to walk the 40 miles.

The time to plan for earthquakes is before they happen. The reading list at the end of this chapter includes a number of resources to aid in preparing for future earthquakes. In addition to these, the front sections of most telephone directories contain earthquake information on first aid and survival. The following lists have been compiled from a variety of sources.

Before the Next Earthquake

- Discuss earthquake safety with children, caretakers, and family members.
- Calmly go over possible earthquake scenarios for different times and places.
- Evaluate your home, school, and workplace for earthquake hazards and identify safe places.
- Correct or reduce hazards where possible.
- Make sure that *all* adults and teens in the household are able to turn off gas, electricity, and water.
- Plan and practice evacuation of the places where you spend time.
- Discuss reunion and communication plans with loved ones. Be sure to consider plans without cell phones.
- Plan for care of elderly or disabled family members and pets in case you are absent.
- Take a basic first aid training course.
- Store emergency supplies at home, work, and in your cars.
- Know the emergency plans of day care centers, schools, and nursing homes where your loved ones might be during an earthquake.
- Organize people in your workplace and neighborhood for emergency response and preparedness.
- Keep lists of important personal, financial, medical, and household information in safe and accessible places.

Continued.

THE NEXT BIG ONE—Cont'd

During an Earthquake

- Duck, cover, and hold if inside.
- Get into the open if outside.
- If driving, stop **if** you are in a safe place. ***Do not stop*** under bridges, power lines, or overpasses.
- If in a wheelchair, stay in it, move to a safe place, lock the wheels, and cover your head.
- Do not rush the exits in crowded public places.

Following an Earthquake

- Put on sturdy shoes if possible; get a flashlight if it is dark.
- Check for injuries and treat them appropriately.
- Make a safety check of immediate surroundings and prepare for aftershocks.
- Evacuate to a safe place; implement reunion plans if appropriate.
- Check food, water, and medical supplies.
- Do not use appliances if you suspect there is leaking gas.
- Do not use the telephone except for extreme emergencies.

COMMON EARTHQUAKE FEARS

One of the most frequently asked earthquake questions is, "Is California going to fall into the ocean?" Another common fear is that an earthquake will cause the ground to open up suddenly and swallow entire buildings. Fortunately, neither of these popular myths is in California's earthquake scenarios. Although seaside cliffs and their resident structures might indeed tumble to the sea during an earthquake-induced landslide, these are localized catastrophes. As we will see in Chapters 15 and 16, the vertical element of motion along most of California's faults is slowly causing much of the coast to rise relative to sea level. Furthermore, even the largest expected earthquake will rupture only a portion of the San Andreas system, and most of that motion will be right lateral.

SUMMARY

1. The continual movement of the Pacific and North American Plates, as well as the Gorda Plate in northwestern California, causes stress to build up in the rocks near the plate boundaries. Earthquakes occur when strain energy builds up to a level where it exceeds the strength of the rocks.
2. Most California earthquakes have fairly shallow focal depths, although deeper focus earthquakes occur in the Cascadia subduction zone. Active faults, those capable of generating earthquakes in today's tectonic settings, either display recent seismicity or offset young sediments or rocks.
3. The magnitude of an earthquake is a measure of the energy released during the earthquake. The intensity of an earthquake is a measure of the damage done by an earthquake.

4. Magnitude is a measure of an earthquake's strength and does not change from one place to another. In contrast, intensity changes from one locality to another and depends on several factors. These include the earthquake's magnitude, the distance from the epicenter to the place where intensity is measured, the nature of the soil or bedrock at different localities, and local construction standards.

5. The major hazardous earthquake events are ground rupture, which occurs when a fault ruptures to the ground surface, and ground shaking. Ground shaking generally causes greater damage because it affects areas away from the fault. Liquefaction and landsliding are hazardous geologic processes that result from ground shaking.

6. Tsunamis can occur when earthquakes with vertical displacement rupture the ocean floor. Much of the California coastline is at risk from flooding due to tsunamis generated by either local offshore earthquakes or more distant sources.

7. The assessment of earthquake risk entails the identification and mapping of active faults, followed by determination of the magnitude and frequency of earthquakes on fault segments.

8. Scientists estimate that there is a 62 percent probability of an earthquake with magnitude 6.7 or higher in the San Francisco Bay area within the next 30 years. Southern California has an estimated probability of about 85 percent within the next 30 years. Because we cannot prevent or precisely predict earthquakes, Californians must be prepared for them.

9. The Alquist-Priolo Act prohibits development within 50 feet of the surface trace of an active fault zone in California. Other laws and building codes are designed to enhance the seismic safety of schools, dams, hospitals, transportation networks, commercial buildings, and residences.

REFERENCES

General Publications

BLAIR, M.I., AND SPANGLE, W.E. 1979. *Seismic Safety and Land-Use Planning: Selected Examples From California.* U.S. Geological Survey, Professional Paper 941-B.

BOATWRIGHT, J., AND OTHERS. 1995. *Ground Shaking in San Francisco* [colored map and text illustrating predicted ground shaking in San Francisco]. U.S. Geological Survey.

BOLT, B. 1993. *Earthquakes*, 3rd ed. W.H. Freeman, 331 pp.

DVORAK, J., AND PEEK, T. 1993. Swept Away: The Deadly Power of Tsunamis. *Earth Magazine,* July, pp. 52–59.

GORE, R. 1995. Living With California's Faults. *National Geographic* 187(4):2–35.

Governor's Office of Emergency Services. *Beat the Quake: California Earthquake Preparedness* [brochure with information contacts]. California Office of Emergency Services.

JENNINGS, C.W. 1994. *The Fault Activity Map of California and Adjacent Areas.* California Division of Mines and Geology, scale 1:750,000.

———————. 1995. New fault map of California and adjacent areas. *California Geology* 48(2):31–42.

MONASTERSKY, R. 1994. Abandoning Richter: How a white lie finally caught up with seismologists. *Science News* 146:250–252.

U.S. Geological Survey. 1994. *The Next Big Earthquake in the Bay Area May Come Sooner Than You Think. Are You Prepared*? Menlo Park, CA. [Available in English, Spanish, Chinese, Braille, and Recordings for the Blind]

YANEV, P.I. 1991. *Peace of Mind in Earthquake Country.* San Francisco: Chronicle Books, 218 pp.

ZIONY, J.I., ed. 1985. *Evaluating Earthquake Hazards in the Los Angeles Region—An Earth-Science Perspective.* U.S. Geological Survey, Professional Paper 1360, 505 pp.

Earthquake Planning Scenarios, Published by the California Geological Survey

DAVIS, J., AND OTHERS. 1982. Earthquake Planning Scenario for a Magnitude 8.3 Earthquake on the San Andreas Fault in Southern California. Special Publication 60, 128 pp.

———————. 1982. Earthquake Planning Scenario for a Magnitude 8.3 Earthquake on the San Andreas Fault in the San Francisco Bay Area. Special Publication 61, 160 pp.

REICHLE, M., AND OTHERS. 1990. Planning Scenario for a Major Earthquake in the San Diego-Tijuana Metropolitan Area. Special Publication 100, 189 pp.

STEINBRUGGE, K., AND OTHERS. 1987. Earthquake Planning Scenario for a Magnitude 7.5 Earthquake on the Hayward Fault, San Francisco Bay Area. Special Publication 78, 242 pp.

TOPPOZADA, T., AND OTHERS. 1988. Planning Scenario for a Major Earthquake on the Newport-Inglewood Fault Zone. Special Publication 99, 200 pp.

———————. 1993. Planning Scenario for a Major Earthquake on the San Jacinto Fault Zone, San Bernardino Area. Special Publication 102, 208 pp.

———————. 1995. Planning Scenario in Humboldt and Del Norte Counties, California, for a Great Earthquake in the Cascadia Subduction Zone. Special Publication 115, 159 pp.

YEATS, R., 2001, *Living with Earthquakes in California.* Corvallis, OR, Oregon State University Press, 416 pp.

Web Sites and Media Resources

Surviving the Big One, video available from local libraries, Los Angeles, CA: KCET-TV.

"Killer Quake," Episode of *NOVA* about the Northridge earthquake.

U.S. Geological Survey, Earthquake Information: *http://quake.usgs.gov*

Northern California Earthquake Center: *http://quake.geo.berkeley.edu/*

Southern California Earthquake Center: *http://www.scec.org*

Seismosurfing on the Internet, University of Washington: *http://www.geophys.washington.edu/seismosurfing.html*

Association of Bay Area Governments: *http://www.abag.ca.gov/bayarea/eqmaps/eqmaps.html*

Southern California Earthquake Center, Faults of Southern California (clickable map with links to faults): *http://www.scecdc.scec.org/faultmap.html*

Largest Earthquakes in the U.S., National Earthquake Information Center, U.S. Geological Survey: *http://neic.usgs.gov/neis/eqlists/10maps_usa.html*

California Seismicity, National Earthquake Information Center, U.S. Geological Survey: *http://neic.usgs.gov/neis/states/california/*

Alquist-Priolo Earthquake Fault Zones, California Geological Survey: *http://www.consrv.ca.gov/cgs/rghm/ap/index.htm*

Pacific and Alaskan Tsunami Warning Center, National Oceanic and Atmospheric Administration: *http://wcatwc.gov/*

California Geological Survey, Probabilistic Seismic Hazard Assessment Maps: *http://www.consrv.ca.gov/cgs/rghm/psha/index.htm*

1906 Earthquake Centennial Alliance, Historic documents, 2006 events and special exhibits: *http://1906centennial.org/*

14

The San Andreas Fault System

Fig. 14-1 The Andreas and Garlock faults. The numbers show the fault segments discussed in this chapter.

SETTING OF THE SAN ANDREAS SYSTEM

The San Andreas fault system marks the boundary between the Pacific and North American Plates (see Fig. 1-4). From its southeastern end at the Mexican border, the San Andreas system follows a northwest trend for 1,350 kilometers to its northwestern end at the Mendocino Triple Junction. At the heart of the system lies the San Andreas itself (Fig. 14-1), probably the world's most famous fault. Running parallel to the San Andreas fault are several other active faults of the system (see Fig. 13-6 and Endpaper 2). These include the Rodgers Creek, Hayward, and Calaveras faults in the San Francisco Bay area, the Newport-Inglewood, Whittier, and Palos Verdes faults in the Los Angeles basin, and the Elsinore, San Jacinto, and Imperial faults in southern-most California. Some geologists and seismologists also consider the Eastern California Shear Zone in the Mojave Desert to be a part of the San Andreas system.

As outlined in Chapter 1, the direction of movement of the two plates results in relative right-lateral motion between them, and most of the faults in the San Andreas system show this same sense of movement. Along its entire length, movement takes place along one or more active faults. Here, however, the similarities along the San Andreas fault system come to an end. In some areas, movement is confined to a single, narrow fault zone, whereas in others, multiple, complex *fault strands* mark the plate boundary. At its northern end, the San Andreas system is about 100 kilometers wide, whereas in southern California, the zone encompassing the active faults of the system is 300 kilometers across.

Sections of the San Andreas fault where faulting is confined to one or more linear zones of right-lateral movement are seen where the fault is parallel to the plate boundary (Endpaper 2). These relatively simple *fault segments* include the North Coast section (the northern part of section 2 in Fig. 14-2) and the segment between San Juan Bautista and the Big Bend (section 3 in Fig. 14-2). In contrast, sections of the fault system where the San Andreas fault system is not parallel to the plate boundary are segments of complexity. The most obvious of these segments is known as the Big Bend (section 4 in Fig. 14-2), but a less pronounced bend is found near San Jose. As shown in Fig. 14-3, left-stepping bends in the San Andreas fault cause convergent motion between the two sides of the fault rather than pure right-lateral slip. The convergence results in compression between the two sides of the fault, and these fault segments show folding, faulting, and uplift associated with compressional forces (see Chapter 1). Right-stepping bends in some of the faults of the San Andreas system produce local areas of extension, referred to as *pull-apart basins*. In these areas, the crust stretches, forming a basin where young sediments can accumulate (see Fig. 1-16).

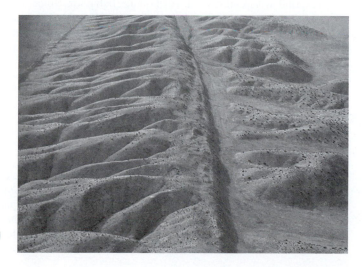

Fig. 14-2 Aerial view of the San Andreas fault, Carrizo Plain. (Source: Wallace, R. U.S. Geological Survey.)

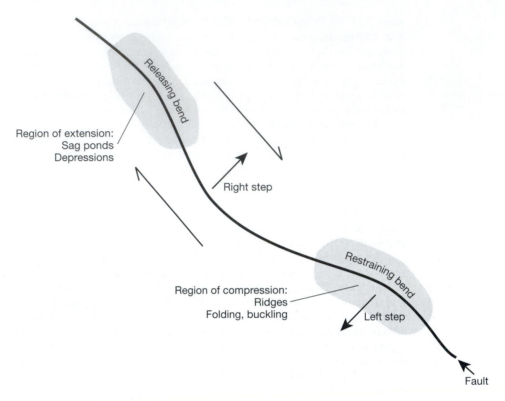

Fig. 14-3 Bends in the faults of the San Andreas system create areas of compression and extension. Bends and their effects are seen at all scales along the faults of the San Andreas system. Fig. 1-17 shows how transtension and transpression are created along transform faults.

The contrasts in the appearance of the San Andreas system along its length correspond with contrasts in earthquake behavior. Since 1769, 117 earthquakes with magnitudes greater than or equal to 6.0 have occurred on the San Andreas system (see Table 13-1 and Fig. 13-1). Some parts of the San Andreas system experienced no large earthquakes, whereas others ruptured in great earthquakes of magnitude 8.0 or greater. An understanding of each segment of the San Andreas fault and its subsidiaries is critically important for earthquake prediction.

GEOLOGIC HISTORY OF THE SAN ANDREAS SYSTEM

Fault Offset and Slip Rates

At the Skinner Ranch in Marin County, a dairy farmer's fence and his barn straddled the San Andreas fault, running almost perpendicular to the fault trace. On the morning of April 18, 1906, this fence was instantaneously separated as the 1906 earthquake ruptured the ground surface along the fault. When geologist Grove Karl Gilbert examined the area shortly after the earthquake, he measured 4.7 meters of displacement or offset of this fence (Fig. 14-4). When he estimated the 1906 offset of

Fig. 14-4 The 1906 offset of a fence at Skinner Ranch, Marin County, today part of Point Reyes National Seashore. Geologist G.K. Gilbert measured the displaced features shortly after the earthquake. The location is near the photo shown in Fig. 13-14. Broken lines show the positions of bushes, fences, and path before the earthquake. (Source: Adapted from Lawson, A. 1908.)

the fence, Gilbert assumed that no displacement had occurred from the time that the fence was built until the 1906 earthquake. Other objects sitting on the fault trace showed similar offsets: the corner of the barn's foundation, a line of raspberry bushes, and a path. All of the objects recorded the same direction of offset, with the western side of the fault moving northwest relative to the eastern side, the characteristic right-lateral motion of all movements along the San Andreas (see Fig. 1-15C). Today, some of the offsets at the old Skinner Ranch, reconstructed from Gilbert's photographs and notes, can be seen along the earthquake trail at the Point Reyes National Seashore's Visitor Center.

Since 1906, no movement has occurred along the northern part of the 1906 rupture, and the fence is still offset by 4.7 meters. If the offset were averaged for the 97 years since the earthquake, an average rate of 4.8 centimeters per year would be obtained. Of course, the locked segments of the San Andreas fault system do not move at an average pace. At Skinner Ranch, the rate of offset is normally zero, punctuated by meters of movement within seconds. However, an understanding of the long-term rate of offset is important for two reasons. First, it provides geologists with a clue to the history of motion along the plate boundary. Second, if the long-term offset rate is known, segments of a fault that lag behind the long-term rate can be identified as "overdue" for significant movement—that is, a major earthquake.

Considering that great, ground-rupturing earthquakes occur only every century to every few centuries, the average rate of 4.8 centimeters per year at Skinner Ranch between 1906 and 1996 is far too high to be a representative long-term rate. A more reliable estimate of the fault's "average" behavior would be made over a longer time period and incorporate several earthquakes.

In order to determine the long-term rate of offset, the same principles can be applied to geologic features that have been separated by an active fault. A topographic feature such as a stream channel can be offset in the same way as Skinner's fence (Fig. 14-5). However, determining the rate of offset of this natural feature involves much more uncertainty, because the original course of the stream crossing

A

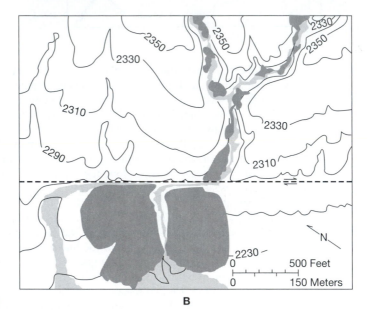

Fig. 14-5 **A**, Offset of Wallace Creek along the San Andreas fault, Carrizo Plain. **B**, Stream deposits used to calculate the offset are shaded. Contour line elevations are in feet. (Source: Wallace, R. 1990. U.S. Geological Survey, Professional Paper 1515.)

B

the fault was probably not straight. Matching up the deposits of a particular stream on either side of the fault can be difficult if several streams are flowing in the area. Finally, establishing the precise ages of the deposits or landforms can also pose problems.

Despite all of these uncertainties, geologists working in the vicinity of Skinner Ranch have used offset stream deposits to conclude that offset or slip rate has been between 2.1 and 2.7 centimeters (21 to 27 millimeters) per year during the past 1,800 years (Fig. 14-6). If all of this displacement occurred during 1906-type earthquakes, eight or nine earthquakes would be needed to create the total displacement of the stream deposits. This suggests that a great earthquake would be expected about every

Fig. 14-6 Slip rates can be calculated from offset stream deposits if the offset channels can be matched across the fault and dated. **A,** The deposits of Gravel Creek as they appeared in a trench excavated on the eastern side of the San Andreas fault near Skinner Ranch, near the location of Fig. 14-4. Similar deposits were identified and dated on the western side of the fault. **B,** The reconstructed positions of Gravel Creek, based on correlation and dating of the alluvial deposits (BP = before present). (Source: Niemi, T., and Hall, N.T. 1993. *Geology.* Geological Society of America.)

190 to 260 years along the northern San Andreas fault. At a site near Cajon Canyon, 75 kilometers east-northeast of Los Angeles, alluvial sediments offset by the San Andreas fault have been matched across the fault. After dating those offset sediments, geologists calculated an average slip rate of about 2.5 centimeters per year for this part of the San Andreas fault.

As discussed in Chapter 1, the relative motion between the Pacific and North American Plates has been calculated at about 4.9 centimeters per year during the past 3 to 4 million years. This rate is obviously much greater than the 2.5 to 3 centimeters per year documented along the San Andreas fault. Even when the slip rates along subsidiary faults of the San Andreas system are added to the San Andreas slip rate, only about 35 millimeters (3.5 centimeters) per year can be accounted for in central California. Geologists and geophysicists believe that the "missing" right-lateral plate motion is occurring on faults east of the San Andreas system, along the eastern side of the Sierra Nevada and in the Basin and Range Province (see Chapter 7).

How Old Is the San Andreas Fault? Long-Term Slip Rates

As right-lateral motion continues on the San Andreas fault system, geologic features that originally straddled the fault traces become separated by greater distances. Older geologic units are offset more than recent deposits and landforms because they have been around subjected to a longer period of displacement. Many geologists working in California actively search for rocks, faults, and other geologic features that can be matched up across the faults of the San Andreas system. These studies enable more refined estimates of slip rates to be made, even allowing geologists to see changes in the rate of slip along the system. It is also possible to determine which faults in the system are more recent breaks, and thus to detect the birth of new fault strands and the deactivation of older ones. Studies of older offset rocks and features have also enabled geologists to determine the maximum amount of displacement on the San Andreas fault.

One geologic feature that has been matched across the San Andreas fault is the Pinnacles volcanic field, located on the west side of the fault in central California (Figs. 14-7 and 14-8). The volcanic rocks at Pinnacles National Monument are a series of high-silica volcanic flows that are overlain by pyroclastic rocks, including distinctive layers of perlite, pumice tuff, and breccias (see Chapters 5 and 12). On the eastern side of the San Andreas fault, 315 kilometers southeast of the Pinnacles near Gorman, is the Neenach Volcanic Formation, where a sequence of volcanic rocks remarkably similar to those at the Pinnacles is found (Fig. 14-8). Radiometric dating has shown that the rocks were erupted in early Miocene time, about 23.5 million years ago.

Along the entire length of the main San Andreas fault, all units older than the Pinnacles-Neenach volcanic fields show about the same amount of offset, approximately 315 kilometers. The consistent amount of offset of older materials leads geologists to conclude that the modern San Andreas fault began its history of right-lateral faulting after 23.5 million years ago. Since that time, it has moved a total of 315 kilometers, or at a long-term average slip rate of 1.34 centimeters per year. As we will see later in this chapter, it is possible that the San Andreas movement started even later, which would make this rate a minimum value.

The Pelona and Orocopia schists are distinctive older rocks that have been used to calibrate the offset across the southern San Andreas fault system. These are sedimentary and igneous rocks that were metamorphosed during Mesozoic thrust faulting. Today, the schist is exposed in the Sierra Pelona, Orocopia, and Chocolate Mountains (Fig. 14-9). The axis of a fold that cuts through the Pelona schist on the western side of the San Andreas can be precisely matched to the same fold axis in the Orocopia schist on the eastern side of the fault. Today this feature is offset 220 kilometers across the southern San Andreas fault, almost 100 kilometers less than the offset of much younger features like the Pinnacles-Neenach volcanics. Geologists

Fig. 14-7 High-silica volcanic rocks at Pinnacles National Monument.

infer that this strand of the fault must be a much more recently formed break than the San Andreas to the north.

Similar studies of offset rates and amounts on other members of the San Andreas system indicate that right-lateral motion began on the ancestral faults of the

Fig. 14-8 Comparative cross sections of volcanic rocks seen at Pinnacles and Neenach volcanic fields, showing the remarkable similarity of units (MA = millions of years ago). (Source: Irwin, W.P. In Wallace, R. 1990. U.S. Geological Survey, Professional Paper 1515.)

Fig. 14-9 Pelona and Orocopia schists and their offset along the San Andreas fault in southern California. The schists are discussed further in Chapter 16. (Source: Ehlig, P. In Geological Society of America, *The Cordilleran Orogen*, Vol. G-3, p. 236.)

modern San Andreas about 28 million years ago. The birth of the transform plate boundary followed the completion of the subduction of the Farallon Plate, as discussed in Chapter 12 (see Fig. 12-2).

TOPOGRAPHY, FAULT CHARACTERISTICS, AND EARTHQUAKE HISTORY ALONG THE SAN ANDREAS SYSTEM

When viewed from an airplane, almost any part of the San Andreas fault appears on the ground as a sharp linear scar that cuts across all features of the landscape (see Figs. 14-2 and 14-10). The fault displaces ridges, valleys, and streams faster than they can be smoothed by erosion and deposition. In some areas, elongated depressions or *sag ponds* line up along the fault. Along other stretches, ridges are pushed up parallel to the fault like the diggings of giant gophers. The stream channels crossing the fault are often completely disrupted, some to the extent that they have become disconnected (see Fig. 14-5). In fact, the landforms formed in active transform fault zones are important characteristics used to identify such faults (Fig. 14-11).

The San Andreas fault system is a fundamental force shaping California's geology today. For that reason, it is important to understand the fault's history and the differences between different segments. A tour along the San Andreas fault system from Cape Mendocino to the Gulf of California highlights some characteristics of each section of the fault. Figure 14-1 shows the location of each numbered segment.

Fig. 14-10 Sharp linear breaks in the topography mark the trace of the San Andreas fault, Carrizo Plains. A sag pond, marked by white patches, can be seen along the right fault trace. (Source: Wallace, R. U.S. Geological Survey.)

Fig. 14-11 Common landforms along the San Andreas fault system. (Source: Wallace, R. 1990. U.S. Geological Survey, Professional Paper 1515.)

1. The North Coast and the Mendocino Triple Junction

Few Californians living outside of northwestern California realize that the Cape Mendocino region is one of the most seismically active parts of the state. Because of the interactions among three plates, a multitude of active faults crosses northwestern California, each capable of generating earthquakes. The Pacific, North American, and Gorda Plates meet near Cape Mendocino at the Mendocino Triple Junction (see Chapter 1). The triple junction marks the northwestern end of the San Andreas fault system; north of this boundary, the Gorda Plate is being subducted beneath North America along the Cascadia subduction zone (Fig. 14-12). Because the exact configuration of the faults that define the plate boundaries is complex, the triple junction is an area rather than a single geographic point (Fig. 14-13). Most of the Cascadia subduction zone lies offshore, as do the associated faults and folds above it. However, southern Humboldt County is unique in that many of the active folds and faults can be followed onto land. This evidence suggests that the subduction zone itself extends on land, with the trench being partly subducted beneath the Cape Mendocino region.

Earthquakes are generated along the northern San Andreas fault, the Mendocino fault, along several offshore faults within the Gorda Plate, and along thrust faults north of the triple junction within the North American Plate. Moderate earthquakes from these sources are common, shaking the residents of Eureka and surrounding areas of Humboldt County on a regular basis. Because of the low population in Humboldt County, these earthquakes have resulted in less damage than they would if they had occurred in California's urban areas. However, a potentially deadly threat is posed by another earthquake source. Recent geologic investigations indicate a potential for great earthquakes generated by rupture along the subduction zone north of Cape Mendocino. A great earthquake with a magnitude between 8.5 and 9 and could rupture the entire Cascadia subduction zone, from Cape Mendocino as far north as British Columbia. The intervals between such earthquakes are thought to be long—from several hundred to as much as a thousand years. Nevertheless, the last of these earthquakes is believed to have occurred almost 300 years ago, so significant strain has accumulated in the rocks since that time.

Fig. 14-12 Cross section of the Cascadia subduction zone. (Source: Dengler, L., and others. 1992. *California Geology*, 43[3].)

Fig. 14-13 Map showing principal faults and other features of northwestern California. Figs. 12-23 and 12-24 also show sediments in the Humboldt basin, outlined here by the dotted pattern. (Source: Adapted from Sarna-Wojcicki, A., and others. 1982. *Friends of the Pleistocene Guidebook.*)

The active faults and the earthquake history of northwestern California show the effects of both convergent and right-lateral plate motion. Active faults are numerous both on land and offshore, and Humboldt County experiences frequent earthquakes. North of Cape Mendocino, most features result from compression, whereas features produced by lateral motion are more common in the south. However, because the change in plate motions occurs over a broad area, active lateral and compressional features can be found throughout northwestern California.

Recent Earthquakes.   On April 25 and 26, 1992, three earthquakes with magnitudes between 6.0 and 7.0 shook the region of Humboldt County near Cape Mendocino. The first and largest quake, with a magnitude of 7.0, centered just north of the small town of Petrolia, caused about $48 million in damages and injured 356 people (see Table 13-1; Fig. 13-1) Geologists believe the event to be the first historic earthquake produced by thrust faulting along the Cascadia subduction zone. The earthquake triggered numerous landslides and shook houses off their foundations (Fig. 14-14); it also generated a 1-meter-high tsunami at Crescent City. The two earthquakes of April 26, with magnitudes of 6.0 and 6.5, resulted from right-lateral movement along a fault within the Gorda Plate. These events were felt as far away as San Francisco. The epicenters for the two later earthquakes were offshore, west of Cape Mendocino and north of the triple junction (Fig. 14-15). Only 8 months earlier, four large earthquakes, with magnitudes between 6.3 and 7.1, had shaken the same area. Three of the 1991 events were located in the Gorda Plate north of the triple junction. The fourth and largest earthquake, on August 17, was centered on land about 11 kilometers south of Petrolia near the town of Honeydew. The Honeydew earthquake damaged roads, bridges, and buildings and was felt in San Francisco.

Fig. 14-14 Damage caused by the 1992 earthquakes, at Ferndale, Humboldt County. See Fig. 13-1 and Table 13-1 for location and magnitude information. (Source: National Geophysical Data Center.)

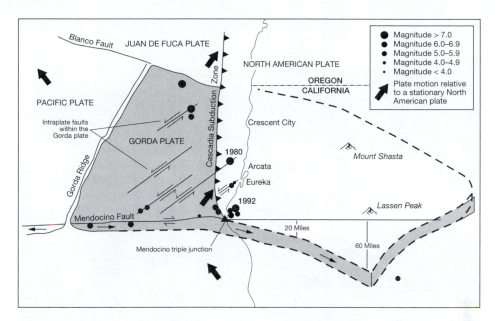

Fig. 14-15 Epicenters and plate-tectonic setting of recent earthquakes in northwestern California. (Source: Department of Geology, Humboldt State University, after Menherson, 1989.)

Along the coast near Petrolia, local fishermen noted that the tides were extremely low in the days following the April 25 and 26 earthquakes. Along the beach, bivalves and other organisms normally below the tide level were suddenly lifted high and dry, and after about 2 weeks, these communities of organisms died (Fig. 14-16). Based on the elevation of these organisms above their normal habitat, researchers concluded that between 3 and 5 feet of uplift occurred during the April 26 earthquake. Careful documentation of this earthquake-triggered uplift, called *coseismic uplift*, has provided geologists with a model for similar events preserved in the geologic record as deposits and fossils along the California coast.

The Geologic Record of Plate Convergence. At the surface above the Cascadia subduction zone is an area of intense compression where rocks, young sediments, soils, and landforms are being squeezed as the crust shortens. This area, marked by numerous thrust and reverse faults and folds, is known as a *fold-and-thrust belt* (see Fig. 14-13). As the three plates converge, compression is occurring in a northeast direction, so that the faults and the hinges of folds (*fold axes*) are oriented northwest. Sediments can be folded downward as well as upward when the crust shortens and buckles, as discussed further in Chapter 16.

One of the large folds in Humboldt County is the Freshwater syncline, a downward fold that runs northwest along the northern edge of Humboldt Bay and the mouth of the Mad River. The downstream end of the Mad River is a tidal channel or *slough* surrounded by salt marsh. Along the edges of the modern slough are layers of sediment deposited by the river during the past 1,600 years (Fig. 14-17). Two types of deposits are present within the layered section: estuarine mud layers

Fig. 14-16 Tidal flats uplifted by the 1992 earthquake, Petrolia, Humboldt County. (Source: Dengler, L. Humboldt State University.)

Fig. 14-17 Salt marshes drowned by an ancient earthquake, Clam Beach, north of the Mad River slough, Humboldt County. (Source: Carver, G., and Burke, R.M. 1982. *Friends of the Pleistocene Field Trip Guidebook.*)

formed in a tidal slough and peat layers formed by decomposed salt marsh plants. The salt marsh plants are buried upright by the overlying mud layers. At the edges of the marsh, redwood trees have been abruptly killed by burial in mud. After carefully describing the layers, geologists have concluded that the area experienced sudden subsidence (down-dropping) during earthquakes. The subsidence caused the marsh plants and nearby trees to be drowned. Mud then accumulated in the newly formed estuary until salt marshes were reestablished. Radiocarbon dating of the peat layers indicates that coseismic subsidence took place during earthquakes about 300, 1,100, and 1,600 years ago. Similar buried salt marshes found along the Oregon and Washington coasts are of about the same age. The similarity in ages along the entire length of the Cascadia subduction zone has led geologists to conclude that great earthquakes, with magnitudes as high as 9.0, have ruptured the entire zone in an earthquake similar to the 1964 Alaska earthquake. The most recent of these earthquakes might have occurred about 300 years ago. In fact, monks in Japan recorded a tsunami in January 1700, which might have been produced by the same Cascadia earthquake.

2. Northern San Andreas Fault Zone and the San Francisco Bay Area

From the Mendocino Triple Junction south to Point Arena, the San Andreas fault zone lies offshore. The Garberville and Maacama fault zones, which parallel the northernmost San Andreas, are also thought to be active, but have not slipped during historic time (Endpaper 2). At Point Arena (see Fig. 14-13), the fault emerges from the Pacific and can be traced to Fort Ross, where it runs offshore. Further south, Tomales Bay is one of the San Andreas fault's most striking landforms, separating the Point Reyes Peninsula from the "mainland" of Marin County along a narrow slash that extends south of Tomales Bay to Bolinas Bay. From Bolinas Bay the fault lies offshore, west of San Francisco, until it intersects the San Francisco Peninsula and remains on land for the rest of its length.

From Point Arena south to San Juan Bautista and Hollister, the faults of the San Andreas system cross through the San Francisco Bay area, home to 7 million people. In the same segment, movement on the San Andreas system is shared among four major, right-lateral fault zones. All four cross through urbanized areas, and all show evidence of recent, damaging earthquakes (Fig. 14-18). The San Andreas fault northwest of San Francisco was the epicentral region of the great 1906 earthquake (see Table 13-1). Along much of this segment, the San Andreas is marked by linear valleys. East of the San Andreas fault, the Rodgers Creek fault north of San Pablo Bay, the Hayward fault along the eastern margin of San Francisco Bay, and the Calaveras fault in the southeastern Bay area all display offset streams, sag ponds, and other characteristic features of active transform faults.

In addition to the "master" faults of the San Andreas system in the Bay area, several reverse faults produced by compressional forces are thought to be active. These faults are oriented roughly parallel to the San Andreas system. In the Santa Clara Valley, these faults are found along the low mountains that form the edges of the Almaden and Santa Clara valleys (Fig. 14-19). Geologists believe that the complexity of the San Andreas fault system in the Bay area is at least partly caused by the slight bend in the orientation of the fault in the Santa Cruz Mountains.

Fig. 14-18 Faults of the San Andreas system in the San Francisco Bay area. (Source: Wahrhaftig, C., and Sloan, D. 1989. *Geology of San Francisco and Vicinity*. Washington, DC: American Geophysical Union, Field Trip Guidebook T105.)

Fig. 14-19 Active reverse faults in Santa Clara County are associated with the San Andreas fault system. (Source: McLaughlin, R. U.S. Geological Survey.)

The 1989 Loma Prieta Earthquake. On the warm afternoon of October 17, 1989, 62,000 baseball fans filled Candlestick Park to watch the San Francisco Giants and the Oakland Athletics play the third game of the World Series. Thousands more had left work early to watch the contest at home, leaving the freeways between San Francisco and Oakland emptier than normal on a Tuesday at 5 P.M. It would prove a fateful coincidence that saved lives. At 5:04 P.M., the stands rocked and shook, and a cheer went up from the crowd. Two thousand feet overhead, the Goodyear blimp lost its TV link to the stadium. Within seconds, the fans fell silent, realizing that this was no swell of foot-stomping. As power to the stadium lights was lost, the sky darkened. Above, the blimp operator saw a flash as transformers exploded under a nearby highway. During the next hour, portable radios and the blimp transmitted the news: Part of the Bay Bridge had fallen, fires were ablaze, power was lost, and people were

Fig. 14-20 Diagram and map showing the rupture zone of the 1989 Loma Prieta earthquake. (Source: Rymer, M. U.S. Geological Survey.)

dead. Remarkably, there was no panic in Candlestick Park that night, as fans left the stadium not knowing what damage they would find at home. It was the first time that a Major League Baseball game was canceled due to an earthquake.

The Loma Prieta earthquake, named for a peak near its epicenter, was the most damaging earthquake on the San Andreas system since the great earthquake of 1906. (The 1994 Northridge earthquake, discussed in Chapter 13, did not occur on a right-lateral fault). The Loma Prieta earthquake left 62 people dead, 3,757 injured, and more than 12,000 homeless. Estimated losses from the earthquake were between $6 and $10 billion. The lives of most of the 6 million people living in the Bay area were affected for months after the quake. It took almost a decade to repair all of the damage from the Loma Prieta earthquake. The earthquake had a moment magnitude of 6.9 (see Table 13-1) and was centered in the southern Santa Cruz Mountains along a remote stretch of the San Andreas fault. The ruptured segment was approximately 40 kilometers long, and both right lateral and vertical movement occurred (Fig. 14-20). Unlike the 1906 event, this earthquake did not rupture the ground surface along the fault, although cracks formed because of landsliding and severe shaking along ridges near the epicenter (Fig. 14-21).

As would be expected, communities near the epicenter were hard-hit by the 1989 earthquake. Unreinforced brick buildings crumbled and wooden structures were knocked off their foundations (Fig. 14-22). In some towns, almost every house lost its chimney. Almost 100 kilometers north of the epicenter, unexpected devastation and

Fig. 14-21 Ground cracking near the epicenter of the Loma Prieta earthquake; Summit Road south of Highway 17, Santa Cruz Mountains. (Source: Wilshire, H. U.S. Geological Survey.)

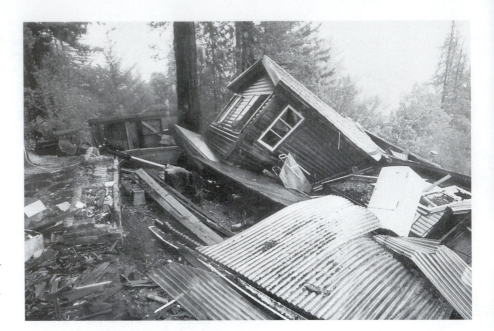

Fig. 14-22 House thrown from its foundation near the epicenter of the 1989 earthquake. (Source: Nakata, J. U.S. Geological Survey.)

Fig. 14-23 Collapsed apartments, Marina District, San Francisco, following the 1989 earthquake. (Source: Meyer, C. U.S. Geological Survey.)

loss of life also struck. In the Marina District of San Francisco, entire apartment buildings collapsed (Fig. 14-23), and several burned in a major fire. Ironically, the disastrous ground shaking in the Marina District can be traced back to the 1906 earthquake (Fig. 14-24). To celebrate the city's recovery from the great earthquake, San

SAN FRANCISCO BAY

Fig. 14-24 **A,** Map showing the extent of filled land and the materials in the Marina District. **B,** The severity of ground shaking recorded on bedrock (*MAS*), dune sand (*PUC*), and filled land (*LMS*) during the 1989 earthquake. (Source: U.S. Geological Survey.)

Francisco held an exhibition in 1913 and created the exhibition grounds by filling in a lagoon on the edge of San Francisco Bay (Fig. 14-25). Some of the artificial fill was actually rubble cleaned up after the 1906 earthquake. When the artificially filled land failed during the 1989 earthquake, liquefaction brought up old debris from 1906 along with the wet, sandy fill.

Across the Bay in Oakland, the top of a double-decked section of Interstate 880 collapsed in sections like dominoes, crushing 41 commuters beneath. If traffic had

Fig. 14-25 A view of San Francisco looking toward the Golden Gate, circa1870. The arrow shows the lagoon that was later filled in to become San Francisco's Marina District.

been as heavy as normal, hundreds more could have died there. The heavy damage in this area was the result of design defects. One end of the collapsed section was anchored in soft sediments that amplified the ground shaking.

The 1906 San Francisco Earthquake.   San Francisco in 1906 was a prosperous city of 400,000 inhabitants. It was a major center for world trade and banking, as well as a city of beautiful hills overlooking the ocean and San Francisco Bay. But at 5:12 A.M. on April 18, a great earthquake brought disaster to San Francisco and the surrounding areas. Centered on the San Andreas fault west of

Fig. 14-26 Golden Gate Avenue in San Francisco, 1906, after the earthquake but before the fire that destroyed many of the buildings. (Source: Branner Collection.)

the city, the earthquake registered at a magnitude of about 7.8 (see Table 13-1). It was felt as far away as Los Angeles and central Nevada. Many residents died in their beds as unreinforced masonry buildings collapsed on them; many others were killed by falling debris as they ran outside (Fig. 14-26). Enormous fires swept through the city unchecked, because water mains had broken, leaving fire hoses completely useless. Firefighters eventually controlled the fires by dynamiting entire blocks to make fire breaks. As many as 2,000 people are believed to have died in San Francisco, and 189 people were killed in surrounding areas (Fig. 14-27). Afterward, people lived in tent camps for months while soldiers kept order in the ruined city.

The 1906 earthquake is one of the world's most famous disasters, partly because of the lessons learned by geologists, seismologists, and engineers who studied its effects. Geologist Andrew Lawson compiled a comprehensive report published in 1908 detailing the earthquake phenomena observed throughout the area. One important finding was the correlation between geologic materials and the intensity of ground shaking in different parts of San Francisco (Fig. 14-28). The severe damage in areas that had been reclaimed by filling in San Francisco Bay was a pattern that was repeated in 1989 in the Marina District (see Fig. 14-24).

Geologists measured and photographed surface offset along the rupture zone, carefully documenting the fact that the entire northwestern 430 kilometers of the San Andreas fault had ruptured at once. The southern end of the 1906 rupture zone was near the mission at San Juan Bautista, where a wall collapsed. At the time, the evidence of lateral offset was puzzling—the plate tectonics theory would not be accepted until 60 or more years later. However, after studying the 1906 earthquake, seismologist H.F. Reid developed the model that is still the basis for our understanding of earthquakes (see Chapter 13).

Fig. 14-27 Santa Rosa City Hall after the 1906 earthquake. (Source: Branner Collection.)

Fig. 14-28 A 1908 map showing the severity of ground shaking during the 1906 earthquake. Compare this map with that shown in Fig. 13-23. (Source: U.S. Geological Survey, Professional Paper 941-A.)

3. Central California: The Creeping Segment and Parkfield Area

Between San Juan Bautista and Parkfield (Endpapers 2 and 3), the San Andreas system behaves very differently from its neighboring segments. This central segment experiences numerous small earthquakes, usually with magnitudes less than 4.0 (Fig. 14-29). The Calaveras fault and the southern Hayward fault east of San Francisco Bay show a similar pattern. Seismologists believe that strain energy is being released in continuous, small amounts, so that the likelihood of a great earthquake on these seismically active segments is small. However, the 1984 magnitude 6.2 earthquake on the Calaveras fault and the magnitude 7.0 (estimated) 1868 Hayward fault earthquake are proof that damaging moderate earthquakes are possible.

Relatively constant, slow displacement causes *fault creep* along the central segment of the San Andreas. The continuous offset displaces sidewalks, curbs, and other manmade features along the fault (Fig. 14-30 and Color Plates 19 and 20). In the town of Hayward in the San Francisco Bay area to the north, cracks, buckles, and small offsets produced by creep on the Hayward fault can be traced through the

Fig. 14-29 Seismicity of the central San Andreas fault system showing creeping segments of the Hayward, Calaveras, and San Andreas faults. The creeping segments are those with solid clusters of earthquake epicenters along the faults. (Source: Wallace, R. 1990. U.S. Geological Survey, Professional Paper 1515.)

downtown area (Fig. 14-31). The average rate of creep on the San Andreas fault southeast of Hollister is about 1 centimeter per year. Following the 1989 Loma Prieta earthquake north of the creeping segment of the San Andreas fault, a "burst" of several millimeters of creep was recorded.

At the southern end of the creeping segment near the town of Parkfield, a 30-kilometer-long stretch of the San Andreas fault experienced magnitude 6.0 earthquakes at 22-year average intervals between 1881 and 1966. This regular pattern led seismologists to predict an earthquake for Parkfield in the late 1980s, but the event has not occurred as of 2003.

Fig. 14-30 An aerial photograph showing the San Andreas fault in Cienega Valley, south of Hollister. The main fault trace runs from left to right, just below the center of the photo, and through the Cienega Winery buildings. (See Color Plates 19 and 20 for a closer view of the offsets.)

Fig. 14-31 Trace of the Hayward fault through downtown Hayward, Alameda County. (Source: Hirschfield, S. 1987. *Field Guide to the Hayward Fault. Hayward, California*. Hayward: California State University Hayward.)

4. Carrizo Plain and the Fort Tejon Earthquake

The unpopulated grasslands of the Carrizo Plain provide excellent opportunities to view the effects of the San Andreas fault on the landscape. Along this straight, relatively simple segment of the fault, one can view the offset streams, compressional

Fig. 14-32 Map showing the fault rupture produced by the 1857 Fort Tejon earthquake. (Source: Sieh, K. *Seismological Society of America Bulletin,* Vol. 5.)

ridges, linear valleys, and sag ponds that characterize a transform fault on land (see Figs. 14-2, 14-5, and 14-10). On a clear day, travelers flying commercial airliners between the San Francisco Bay area and the greater Los Angeles area are often treated to a bird's-eye view of this section of the San Andreas fault. Knowledgeable pilots frequently point out the fault to passengers as they fly over it.

The 1857 Fort Tejon Earthquake.   On January 9, 1857, an enormous earthquake ruptured the San Andreas fault, from Parkfield, through the Big Bend segment, and southeast at least to Wrightwood, a total distance of at least 360 kilometers (Fig. 14-32). Fort Tejon, a military post at the southernmost end of the Carrizo Plain, was one of a few population centers near the epicenter. There the ground shook for 1 to 3 minutes. Hundreds of antelope stampeded in terror, and water sloshed out of a sag pond along the fault and stranded fish to die on dry land. One eyewitness, a cowboy working in the southern Carrizo Plain, reported that a round sheep corral was instantly bent into an S shape by right-lateral slip of several meters. The "earthquake crack" along the fault trace could be seen for years by travelers and early settlers.

The earthquake produced as much as 9 meters of offset in the Carrizo Plain and 3 to 4 meters in the Mojave Desert. The average displacement was between 4.5 and 4.8 meters. Seismologists estimate the magnitude of the Fort Tejon earthquake as slightly larger than the 1906 earthquake, at an approximate magnitude of 8.0 (Table 13-1). Since 1857, this segment of the San Andreas has remained locked.

5. The Big Bend: The San Andreas Fault System in the Transverse Ranges and the San Gabriel and San Bernardino Mountains

From its junction with the Garlock fault, the San Andreas makes a marked bend to the southeast for about 120 kilometers (Fig. 14-33). This segment of the fault is appropriately referred to as the Big Bend. The complexity found along all bending segments of the fault is maximized along the Big Bend. Because of the significant component of compression in this region, rocks are actively being squeezed and uplifted (Fig. 14-34). As a result of the compression, spectacular mountain ranges have been thrown up along the margins of the fault—the San Gabriel, San Bernardino, and San Jacinto ranges have peaks that reach more than 3,000 meters in elevation. The steep mountain slopes have shed enormous quantities of debris that are spread across the range fronts in large alluvial fans (see Chapter 6, Fig. 6-12). The fans, increasingly occupied by high-density housing and commercial developments, are in many places cut by active range-front faults of the San Andreas system. Many of the range-front faults are reverse faults that are caused by the compression in this area. Reverse faulting and folding in the Transverse Ranges and the Los Angeles basin are discussed in more detail in Chapter 16, and the Garlock fault is discussed later in this chapter.

The right-lateral motion along the San Andreas system is taken up by a number of fault strands in the Big Bend segment. The San Andreas fault itself defines the eastern end of the system, and it is presently the most active fault strand. As discussed at the beginning of this chapter, Mesozoic rocks are offset about 220 kilometers across

1 San Clemente Fault		
2 Palos Verdes Fault		
3 Rose Canyon Fault		
4 Newport-Inglewood Fault	**22** West Calico Fault	
5 Whittier Fault	**23** Emerson Fault	
6 Santa Cruz Fault	**24** Camprock Fault	
7 Malibu Coast Fault	**25** Lockhart Fault	
8 Santa Monica Fault	**26** Lenwood Fault	
9 Raymond Hill Fault	**27** Old Woman Springs	
10 Sierra Madre Fault	**28** Helendale Fault	**40** Rinconada Fault
11 Elsinore Fault	**29** Sierra Frontal Fault	**41** San Juan Fault
12 Superstition Mtn. Fault	**30** San Andreas Fault	**42** Ozena Fault
13 Superstition Hills Fault	**31** Harper Fault	**43** Santa Ynez Fault
14 Imperial Fault	**32** Blackwater Fault	**44** Big Pine Fault
15 Banning Fault	**33** Garlock Fault	**45** Pine Mountain Fault
16 San Jacinto Fault	**34** So. Death Valley Fault	**46** San Cayetano Fault
17 Pinto Mountain Fault	**35** Panamint Valley Fault	**47** San Gabriel Fault
18 Blue Cut Fault	**36** Sierra Nevada Fault	**48** Arroyo Parida Fault
19 Ludlow Fault	**37** Kern Front Fault	**49** Oakridge Fault
20 Pisgah Fault	**38** White Wolf Fault	**50** Santa Susana Fault
21 Calico Fault	**39** Pleito Fault	**51** North Frontal Fault

Fig. 14-33 Major faults of the San Andreas system in southern California. (Source: Working Group on California Earthquake Probabilities.)

the southern San Andreas fault. In the northern part of this segment, the San Gabriel fault can be traced through the eastern Transverse Ranges and through the western San Gabriel Mountains near Pasadena. In the Ridge Basin, the Mint Canyon Formation, a 12- to 13-million-year-old sedimentary unit, contains cobbles of unusual volcanic rocks (Fig. 14-35). Well to the southeast in the Chocolate Mountains, the source for these volcanic clasts can be found. These relatively young rocks are offset as much as the older rocks across the San Gabriel and San Andreas faults, about 220 kilometers. This indicates that the faults did not begin movement until after the Mint Canyon Formation was deposited 12 to 14 million years ago.

Geologists have estimated that the San Gabriel fault has offset the Pelona schist and other older rocks about 50 kilometers. Today the San Gabriel fault is inactive. Sediments younger than about 3 million years old show no displacement across the fault, and it is seismically quiet. Geologists believe that the San Andreas

Fig. 14-34   Folded and faulted rocks in the San Andreas fault zone, San Bernardino County. (Source: Tarbuck, E. Central Illinois University.)

fault to the east is presently accommodating the slip formerly taken up by the San Gabriel fault.

South of the San Gabriel Mountains, near Cajon Pass and San Bernardino, the San Andreas fault systems plays into several complex strands (see Fig. 14-33). The San Andreas fault zone consists of three subparallel strands that bend eastward along the San Bernardino Mountains: the Mission Creek, San Gorgonio Pass, and Banning (also called the southern San Andreas) faults. The San Jacinto fault zone trends southeast from Cajon Pass, and, further west, the Elsinore fault parallels the San Jacinto's trend. The San Jacinto and Elsinore faults are discussed in the following section.

Pallet Creek and Earthquake Frequency.   Along the bank of Pallet Creek, about 55 kilometers northeast of Los Angeles, lies a marsh that has been cut by a strand of the San Andreas fault. Deposits here record amazing evidence of past earthquakes on the San Andreas fault in southern California. At least 12 earthquakes have broken the sediments at Pallett Creek during the past 1,700 years at this site, giving geologists the best information about earthquake history along the San Andreas (Figs. 14-36, 14-37, and 14-38). Clusters of earthquakes, two or three within 100 to 200 years, are followed by 200 to 300 years without large earthquakes (see Fig. 14-38). The average recurrence interval for this segment of the fault is about 132 years, and 146 years have passed since the great Fort Tejon earthquake of 1857. The established recurrence times between earthquakes are one major reason for assigning a high probability of a large earthquake on the southern San Andreas within the next 30 years.

Fig. 14-35 **A,** Present location of the Mint Canyon Formation, a late Cenozoic sedimentary unit containing volcanic clasts, and its volcanic source location. **B,** The reconstructed drainage pattern before the units were offset along the San Andreas system. (Source: California Division of Mines and Geology. From Crowell, J., ed. 1975. *The San Andreas Fault in Southern California.* Special Report 118.)

Fig. 14-36 Offset sedimentary layers, including peat beds, at Pallett Creek. (Source: Sieh, K. 1990. In Wallace, R. 1990. U.S. Geological Survey, Professional Paper 1515.)

Recent Earthquakes. Three major earthquakes associated with this part of the San Andreas system—Long Beach in 1933, San Fernando in 1971, and Northridge in 1994—were described in Chapter 13. Although none of these events was centered on the San Andreas fault itself, all were the indirect result of the same transform motion, and all caused major damage in the greater Los Angeles area. This

Fig. 14-37 The San Andreas fault zone exposed in the banks of Swarthout Creek near Wrightwood; view to the northwest. The fault ruptures are marked by tapes, with the date of the earthquake that produced each break shown at the top of the tape. To the left (southwest) of the fault zone, deposits consist of alternating dark peat beds and light debris-flow deposits, similar to the deposits at Pallet Creek. Massive gravelly sediment from nearby hillslopes appears on the right side of the faults. (Source: Fumal, T. U.S. Geological Survey.)

segment of the San Andreas fault was part of the rupture zone during the 1857 Fort Tejon earthquake just discussed. At a site near Wrightwood, geologists have found evidence of earlier fault rupture in an 1812 earthquake (see Table 13-1). There the rings of trees along the San Andreas fault show evidence of simultaneous damage,

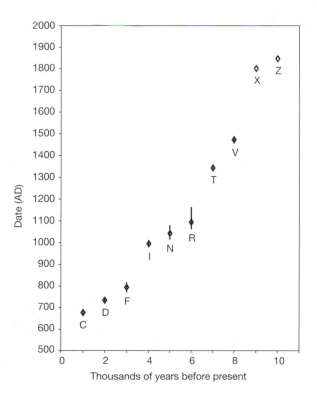

Fig. 14-38 Timing of earthquakes at Pallett Creek. (Source: Sieh, K., and others. 1989. *Journal of Geophysical Research.*)

and geologists are able to date the event by radiocarbon dating of the damaged section of the wood cores.

Since 1857, the segment of the San Andreas fault along the Big Bend north of Cajon Pass has been seismically quiet. South of Cajon Pass, the 1948 Desert Hot Springs earthquake (ML 6.0) ruptured the Mission Creek fault, and the 1986 North Palm Springs earthquake (ML 5.6) ruptured the Banning (southern San Andreas) fault. The area near San Gorgonio Pass, at the southern end of the Big Bend, shows southern California's highest level of background seismicity.

Fig. 14-39 Epicenter of the 1987 Whittier Narrows earthquake. W= Whittier Fault. (Source: Weber, F. 1987. *California Geology,* 40[12].)

Whittier Narrows, 1987. Two earthquakes within the first 4 days of October 1987 shook residents of the northern Los Angeles, particularly those living in Whittier. The larger of the two events had a magnitude of 5.9 and was centered 16 kilometers east of the Los Angeles Civic Center (Fig. 14-39). Together, the two earthquakes damaged 10,500 structures and caused at least $358 million in damages. They were the first damaging earthquakes in the Los Angeles area since the San Fernando earthquake of 1971 (discussed in Chapter 13).

Like the 1971 San Fernando and 1994 Northridge earthquakes, the Whittier Narrows earthquake occurred on a fault that is only subtly expressed at the Earth's surface. The southeastern Whittier fault was known to offset 2,000-year-old sediments, and further southeast, the Whittier fault joins the active Elsinore fault. However, in the vicinity of Whittier, no evidence was found of a recently active strand, and it was not marked as a Special Studies zone. However, recent studies have shown that beneath the surface, a blind thrust fault does offset young sedimentary units (Fig. 14-40). The epicenters of the main shock and aftershocks of the Whittier Narrows earthquake appear to have occurred on this fault. Geologists now agree that the Whittier Narrows earthquake probably occurred on a hidden and potentially deadly fault beneath Los Angeles (Fig. 14-41; see Chapters 13 and 16).

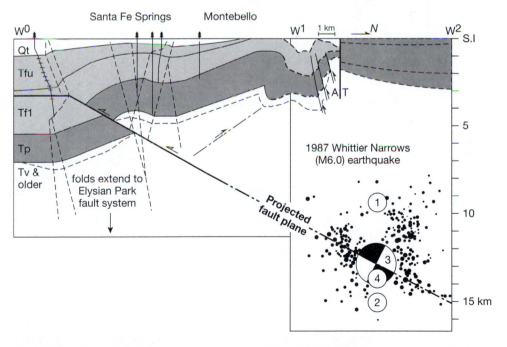

Fig. 14-40 Cross section showing the epicenter of the 1987 Whittier Narrows earthquake (see Table 13-1) and a proposed blind fault beneath Los Angeles. The vertical lines show oil wells near Santa Fe Springs. Qt = Quaternary terraces; Tfu, Tfl, Tp, Tv = late Cenozoic formations. (Source: Shaw, C., and Shearer, P. Harvard University.)

Fig. 14-41 A block diagram showing the orientation of the proposed blind thrust fault beneath Los Angeles. (Source: Shaw, C., and Shearer, P. Harvard University.)

6. The Southern San Andreas, the Salton Trough, and the Imperial Valley

The Salton Trough, a long desert valley that contains both the Imperial Valley and the Salton Sea, is the most striking feature of the southern San Andreas system (Fig. 14-42; see also Fig. 10-11). Bordered on the northeast by the San Andreas fault and on the southwest by the San Jacinto fault zone, the Salton Trough is about 390 kilometers long. The Salton Sea, discussed in Chapter 10, lies about 72 meters below sea level. Along the margins of the modern lake are shorelines of Pleistocene Lake Cahuilla, one of the desert lakes discussed in Chapter 6.

The transform plate boundary in the region of the Salton Trough shows some features characteristic of the divergent plate boundary immediately to the south in Baja California, Mexico. The trough is a pull-apart basin that formed during the past 4 million years as the Peninsular Ranges pulled away from North America (Fig. 14-43). The spreading has not yet caused a midocean ridge to appear, as it has further south. However, rhyolite and obsidian domes at Salton Buttes, on the southeastern side of the Salton Sea, appeared about 16,000 years ago. Within the high-silica volcanic rocks, geologists have found inclusions of the type of basalt erupted from midocean ridges; the same type of basalt also appears as dikes within the domes. This indicates that midocean ridge magma is interacting with the continental crust beneath the Salton Trough and that rifting is extending northward from the Gulf of California. High measured heat flow in the Salton Trough also supports the idea that rifting is

Fig. 14-42 Map showing the major faults in the Salton Trough. A satellite image of the same region can be seen in Fig. 10-11. (Source: Adapted from U.S. Geological Survey.)

active in the region. The geothermal gradient is high enough to produce several geothermal reservoirs; some of these have been tapped for geothermal energy.

Continued right-lateral motion on the faults bordering the Salton Trough causes the basin to increase in length. At the same time, vertical motion on the Brawley fault, the northern Imperial fault, and other faults (Fig. 14-42) is causing the basin to deepen. As the center of the Salton Trough subsides in response to extension, more than 5 kilometers of Pliocene and Pleistocene sediments have accumulated. Geophysical data suggest that at depth, hot fluids escaping from the nearby magma have metamorphosed these young sediments.

On the flanks of the Salton Trough there are low mountains where sediments of late Cenozoic age have been uplifted and exposed. One widespread unit of Pliocene age is the Imperial Formation, a sequence of sandstone, shale, and limestone deposited

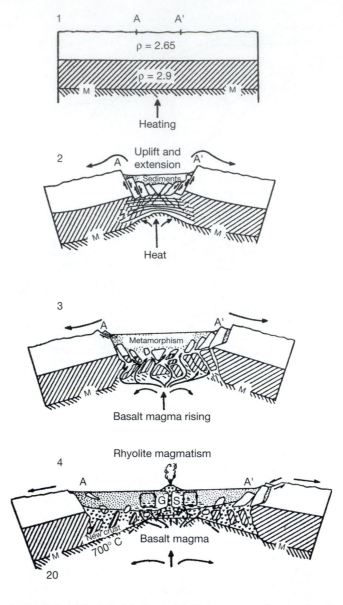

Fig. 14-43 Cross sectional model showing the rifting of the crust beneath the Imperial Valley. (Source: Elders, W., and others. 1972. *Science* 8[4056].)

in the Salton Trough. Marine microfossils, as well as snails, clams, and oyster beds indicate that an ocean embayment existed in the region. Northwest of the Salton Sea, the mostly nonmarine Palm Spring Formation also contains some beds deposited in a marine environment. Nonmarine sediments of Pleistocene and Pliocene age include both fine-grained lake deposits and conglomerate deposited in alluvial fans.

Active Faults and Earthquake History. The earthquake history of the Salton Trough region is stamped with the signature of the San Andreas fault. Right-lateral motion has been documented for earthquakes on three of the four major faults in the area during the century; only the San Andreas fault proper has been aseismic in historic time. The Imperial fault, which roughly bisects the Salton Trough south of the Salton Sea, has been the most active member of the system, generating two sizable earthquakes during the past 50 years. In 1940, a magnitude 6.9 event ruptured the entire 60-kilometer length of the fault. After this earthquake, known as the Imperial Valley earthquake, formerly straight rows of orange trees along the California–Mexico border were displaced by 6 meters of right-lateral offset (Fig. 14-44). The Imperial Valley earthquake was an important event in the development of California building codes, because it was the first event to be recorded by strong-motion sensors located near the fault. In 1979, a magnitude 6.4 earthquake ruptured about 30 kilometers of the fault, producing less than a meter of right lateral displacement. This earthquake damaged the main municipal building in nearby El Centro. Seismicity continues to be high in the Brawley seismic zone along the Brawley fault, which parallels the Imperial fault.

The San Jacinto fault is actually a zone of right-lateral faults that form the southwestern border of the Salton Trough. At least 10 earthquakes with magnitudes between 6.0 and 6.6 have occurred along the San Jacinto system since 1890, more moderate events than on any other southern California fault zone. In 1968, the magnitude 6.5 Borrego Mountain earthquake resulted in 0.4 meters of right lateral offset. An earthquake with a magnitude of 6.6 occurred in November 1987 on the Superstition Hills fault. This earthquake, which was preceded by a magnitude 6.2 foreshock on a nearby left-lateral fault, was followed by a number of aftershocks that increased the right-lateral offset along the 1987 fault trace (Fig. 14-45).

The southern San Andreas fault, which forms the northeastern edge of the Salton Trough (see Fig. 14-42), has had no earthquakes in historic time, although its southern end showed minor sympathetic slip during recent earthquakes on the Imperial and San Jacinto faults. Geologists estimate that this segment has experienced four ground-rupturing earthquakes since 1000 ad, with the last event occurring about 300 years ago. These earthquakes have displaced sediments of ancient Lake Cahuilla.

Fig. 14-44 Offset rows of orange trees in the Imperial Valley east of Calexico after the 1940 Imperial Valley earthquake (see Table 13-1).

Fig. 14-45 Surface rupture produced by the 1987 Superstition Hills earthquake. Note canteen at left for scale. (Source: Rymer, M. U.S. Geological Survey.)

The southern San Andreas, Imperial, and San Jacinto faults all exhibit fault creep of a few millimeters per year.

West of the Salton Trough, the Elsinore fault is another northwest-trending member of the southern San Andreas system. The Elsinore fault cuts through the plutonic crystalline rocks of the Santa Ana Mountains. To the northwest in the Los Angeles basin, the Elsinore fault splits into two strands, the Whittier fault and the Chino fault. In the Peninsular Ranges, the Elsinore fault has a *slip rate* of about 5 millimeters per year. The northern part of the fault is more active, with only small earthquakes documented on the southern Elsinore fault. A magnitude 6.0 event on the central Elsinore fault was recorded in 1910.

Together, the faults of the southernmost San Andreas system account for right-lateral displacement of about 45 to 50 millimeters per year between the North American and Pacific Plates. The San Jacinto fault accounts for as much as 12 millimeters per year, and the southern San Andreas fault south of Cajon Pass between 25 and 30 millimeters per year. Further west, slip on the southern Elsinore fault is about 5 millimeters per year, and the Newport-Inglewood and Palos Verdes faults 0.6 and 3 millimeters per year, respectively.

7. The Garlock Fault and the Eastern California Shear Zone

The Garlock fault trends east-northeast from Tejon Pass for 250 kilometers, forming the northern border of the Mojave Desert province (Fig. 14-46). It is an unusual active fault for California because its motion is left-lateral. No significant earthquakes have occurred on the Garlock fault during historic time, but the western part of the fault exhibits low-level seismicity and aseismic creep. After mapping and studying the displacement of young alluvial deposits in trenches across the fault, geologists estimate that the average slip rate is about 7 millimeters per year, with between 200 and 3,000 years between earthquakes. About 8 meters of left-lateral offset have occurred on the western Garlock fault during the past 10,000 years. Along the eastern

Fig. 14-46 High-altitude U-2 photo of the Garlock fault. (Source: U.S. Air Force and Clark, M. U.S. Geological Survey.)

Garlock fault, offset lake shorelines and other geomorphic features suggest that earthquakes have produced slip of as much as 7 meters during the past 10,000 years. As discussed in Chapter 18, geologists believe that the Garlock fault has had a long history of activity, dating back to about 20 million years ago, when it resulted from extension.

A zone of active faults known as the Eastern California Shear Zone runs parallel to the San Andreas fault through the central Mojave Desert (Fig. 14-47). Many of the faults are young, normal faults, but right-lateral motion also occurs on several, including the Johnson Valley, Landers, Homestead Valley, Emerson, and Camprock faults. The pattern of activity on these faults, which form a north-trending zone (see Fig. 14-47, inset map), indicates that they could correctly be considered part of the boundary between the Pacific and North American Plates. Geologists currently refer to these faults collectively as the Eastern California Shear Zone. Several moderate earthquakes have ruptured the faults of the Eastern California Shear Zone, including the 1975 Galway Lake (Mw 5.0), the 1979 Homestead Valley (ML 5.6), and the April 1992 Joshua Tree (Mw 6.1) events (Fig. 14-47). The Joshua Tree earthquake is now considered a foreshock of the 1992 Landers earthquake.

Landers and Big Bear Earthquakes of 1992.

Two earthquakes on separate faults shook the southern California desert on June 28, 1992. The Landers earthquake, which ruptured parts of the Eastern California Shear Zone, occurred at about 5 A.M. and measured 7.5 (see Table 13-1). Because it was centered in a sparsely populated area of the Mojave Desert, the Landers earthquake caused only one fatality. It ruptured the ground surface along an 80-kilometer stretch across the desert floor. The rupture occurred along parts of four previously mapped faults (Fig. 14-48). Both right-lateral displacement of up to 6 meters and vertical offset of as much as 1.2 meters

Fig. 14-47 Active faults in the Mojave Desert. The inset map shows the Eastern California Shear Zone in southern California in the white area. (Sources: Inset: Hopson, R. 1998. *California Geology* 51[6]; Map: Dokka, R., and others. 1979. Geological Society of America field trip guidebook.)

Fig. 14-48 Map showing earthquake epicenters of the 1999 Hector Mine, 1992 Landers, and 1992 Big Bear earthquakes and their aftershocks. (Source: Southern California Earthquake Center.)

were observed by geologists (Fig. 14-49). Three hours later, a magnitude 6.5 earthquake took place on a buried, northeast-oriented fault near Big Bear Lake. This earthquake took place on a left-lateral fault, but no surface rupture was found. Seismologists believe that the Landers earthquake might have triggered the Big Bear event. Many seismologists are also concerned that the Landers and Big Bear earthquakes have brought the nearby segment of the San Andreas fault closer to failure.

Hector Mine Earthquake of 1999. California's most recent significant earthquake occurred on October 16, 1999, about 50 kilometers north of the town of Joshua Tree (Fig. 14-48). Called the Hector Mine earthquake, it produced as much as 5.2 meters of right lateral offset along about 41 kilometers of the Lavic Lake fault. (See Fig. 13-13.) The earthquake occurred in an area even less populated than the Landers quake, and thus caused little damage. However, the Landers and Hector Mine earthquakes are reminders that not all of the transform motion between the Pacific and North American Plates takes place on the San Andreas system, and they confirm the existence of the Eastern California Shear Zone.

Fig. 14-49 Vertical fault scarp along the Johnson Valley fault produced by the 1992 Landers earthquake. (Source: Rymer, M. U.S. Geological Survey.)

SUMMARY

1. The San Andreas fault system marks the transform boundary between the Pacific and North American Plates. The faults of the San Andreas system show an average of about 2.5 to 3 centimeters of offset per year, accounting for about 50 to 60 percent of the total movement across the plate boundary.

2. Parts of the San Andreas system consist of multiple fault strands, whereas other segments display a single fault trace. The fault is most complex in the region of the Big Bend, where a significant component of compression is created by the bend.

3. Right-lateral motion began on the San Andreas transform during late Cenozoic time, most likely about 28 million years ago. The maximum amount of displacement by the fault is about 315 kilometers; this offset is best documented by 23.5-million-year-old volcanic rocks at Pinnacles and Neenach.

4. The traces of the San Andreas stream disrupt the land surface, producing sag ponds, offset streams, linear valleys, and scarps.

5. In northwestern California, subduction of the Gorda Plate creates active faults as a result of convergent as well as right-lateral motion.

6. In the San Francisco Bay area, multiple right-lateral faults make up the San Andreas system: These include the Rodgers Creek, Hayward, and Calaveras faults. Both the 1906 San Francisco and 1989 Loma Prieta earthquakes occurred on the San Andreas fault.

7. In central California, the southern Hayward and Calaveras faults, as well as the San Andreas fault south of San Juan Bautista, are creeping segments. The San

Andreas fault south of its junction with the Calaveras fault is a single, straight break through central California.

8. The 1857 Fort Tejon earthquake, with an estimated magnitude of 8.0, is the largest historic event on the San Andreas fault. It produced up to 9 meters of offset and ruptured about 360 kilometers of the fault's surface trace.

9. The Big Bend region of the San Andreas system is characterized by complex, multiple, active faults. These include the San Andreas, San Gorgonio Pass, Mission Creek, Newport-Inglewood, Banning, San Jacinto, and Elsinore faults. The San Gabriel fault is an inactive member of the San Andreas system along this segment. The paleoseismic record at Pallett Creek indicates that the Big Bend ruptures with a large-magnitude earthquake about every 132 years.

10. The 1971 San Fernando, 1987 Whittier Narrows, and 1994 Northridge earthquakes occurred on blind thrust faults under Los Angeles. Numerous blind thrust faults are the result of compression in the region of the Big Bend.

11. South of the Big Bend in the Salton Trough, the San Andreas system includes the southern San Andreas, Imperial, Elsinore, and San Jacinto faults. The Salton Trough is a pull-apart basin formed as the result of divergent motion at the southern end of the transform boundary.

12. The Garlock fault is an active left-lateral fault running almost perpendicular to the San Andreas fault. South of the Garlock fault, active faults of the Eastern California Shear Zone produced the 1992 Landers and 1999 Hector Mine earthquakes, both of which produced right-lateral offset.

REFERENCES

Additional references are listed at the end of Chapter 13.

California Division of Mines and Geology. 1993. The Landers–Big Bear earthquakes. California Geology 46(1).

DENGLER, L., CARVER, G., AND MCPHERSON, R. 1992. Cape Mendocino. California Geology 45(2):40–51.

EISMAN, D.B. 1972. Dust from Antelope. California Geology 25(8):171–173.

HUNTER, C. 2000. Hector Mine earthquake. California Geology 52(1):12–15.

IACOPI, R. 1971. Earthquake Country. Menlo Park, CA: Lane, 160 pp.

JENNINGS, C.W. 1994. Fault Map of California. California Division of Mines and Geology, scale 1:750,000 (Endpaper 2).

PETERSON, M.D., AND WESNOUSKY, S.G. 1994. Fault slip rates and earthquake histories for active faults in southern California. Bulletin of the Seismological Society of America 84(5):1608–1649.

WALLACE, R.E., ed. 1990. The San Andreas Fault System, California. U.S. Geological Survey, Professional Paper 1515, 283 pp.

WEBER, H.F., JR. 1987. Whittier Narrows earthquakes. California Geology 40(12):275–281.

Web Sites and Media Resources

Faults, Past Earthquakes, and Geology, U.S. Geological Survey: *http://quake.usgs.gov/info/hist-geo.html*

Includes links to 1906 earthquake information and a virtual field trip to the Hayward fault zone.

Interactive fault map of Southern California, Southern California Earthquake Center: *http://www.scecdc.scec.org/faultmap.html*

Field trip guide to the Calaveras and San Andreas faults, Hollister

Harden, D.R., In U.S. Geological Survey, Online Bulletin 2188: *http://geopubs.wr.usgs.gov/bulletin/b2188/*

Paleoseismology of the Hayward fault, U.S. Geological Survey: *http://quake.usgs.gov/research/geology/paleoseis/index.html*

Wallace Creek Field Trip guide, Southern California Earthquake Center: *http://www.scec.org/wallacecreek/*

SCOTT W. STARRATT

CHAPTER

15

The California Coast

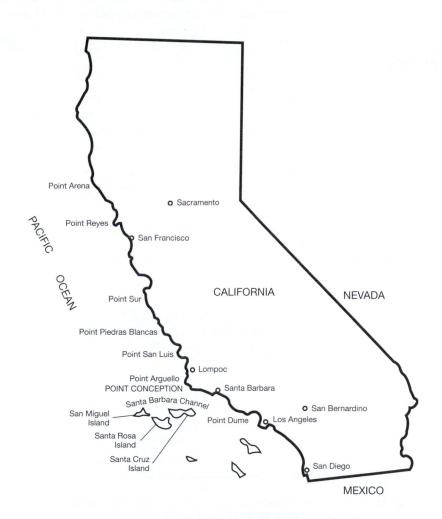

Fig. 15-1 The California coastline. Note the change in the coast's orientation at Point Conception.

The western edge of the North American continent meets the Pacific Ocean along 1,800 kilometers of California coastline, a varied landscape where change is constant. Tectonics, waves, and wind combine forces to create beautiful scenery and a wealth of geologic features along the coastal zone. California's beaches are one of its most valuable resources, enjoyed by residents and tourists alike for recreation and relaxation.

The California coastline can be broadly divided into two distinct sections north and south of Point Conception (Fig. 15-1). Here, at the northern end of the Transverse Ranges, the coastline changes direction from its overall northwesterly trend. The southern coast heads eastward and then gradually returns to its northwesterly direction. As a result, the coast south of Point Conception is broadly curved. The shift in the orientation of the coast brings warmer water currents to southern California's coastal areas (see Fig. 15-1). Because of its orientation and the presence of a broad offshore shelf, the southern coast is also somewhat protected from winter storm waves, which usually approach the shore from the northwest.

WAVES

With the exception of tsunamis (see Chapter 13), all ocean waves are generated by winds blowing over the water surface. Waves radiate outward from storm areas, traveling hundreds or thousands of kilometers on the open ocean without losing much energy. During winter, most waves that approach the California coast are generated by storms in the northern Pacific Ocean near the Gulf of Alaska. These waves approach the coast from the northwest. During late summer and fall, storms off Baja California and the Mexican mainland generate waves that approach the coast from the south.

In the open ocean, the surface water moves in circular orbits to transmit the waves, leaving the ocean bottom unaffected. When waves reach the shallow water at

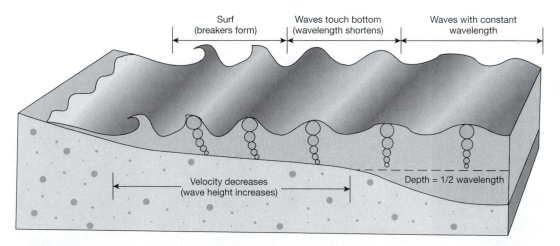

Fig. 15-2 The circular motion of water in ocean waves. (Source: Tarbuck, E., and Lutgens, F. 1993. *The Earth,* 4th ed. Upper Saddle River, NJ: Prentice Hall.)

the coastline, they begin to "feel" the bottom (Fig. 15-2). At this depth, the wave slows down, and the bottom of the wave slows down the most. The top of the wave overruns the base, and the wave breaks. The exact depth at which waves break depends on the size of the wave: Large waves will break further from the shore, in deeper water, than will small waves. The zone of breaking waves along a coastline, the *surf zone*, is constantly subjected to wave attack.

The continental shelf is relatively narrow along most of the coast north of Point Conception. As a result, the ocean bottom drops off quickly from the beach into deep water. Even large waves arrive close to the shoreline before breaking. Wave energy along unprotected coastlines is high because of the steep, narrow shelf. Wave energy is particularly high on straight beaches that face northwest, because they directly face incoming winter waves. The open beach at Point Reyes National Seashore, unprotected by any nearby land masses, is one of California's highest energy beaches (Fig. 15-3).

Rocky points that protrude into the ocean are attacked by waves more fiercely than embayments in the shoreline. A wave front is slowed down as it approaches the rocky headlands, but continues into the embayment for a greater distance before breaking. As a result, the wave fronts are bent, or *refracted* (Fig. 15-4). As the protruding points are vigorously eroded by waves, the protected embayments tend to fill in with beach sediment. Over time, the irregularities in the coastline are removed and the coast tends to straighten (Fig. 15-5).

California's famous surfing beaches result from a combination of coastline geometry, offshore topography, or bathymetry, and the wave conditions at the sites. Three of USA Today's 2003 top surfing beaches in the United States are in California: Black's Beach at Torrey Pines Park in San Diego (see Fig. 15-21); Trestles in the city of San Clemente; and Malibu Beach in Malibu, west of Los Angeles. Further north, Mavericks in San Mateo County is a world-famous spot for enormous surfing waves (Fig. 15-6). A zone of offshore rocks along a fault zone combines with seasonal waves to produce waves that can only be handled by the most expert surfers. The swells at Mavericks can reach 25 to 30 feet, a height that would measure over 40 feet if measured at the breaking edge.

Fig. 15-3 Waves breaking at Point Reyes National Seashore.

Fig. 15-4 Wave refraction around rocky point.

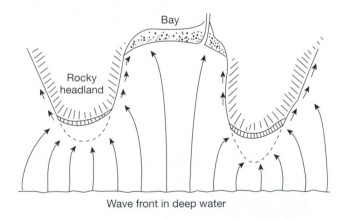

Bay

Rocky headland

Wave front in deep water

Bay-mouth bar

Longshore transport

Fig. 15-5 Diagrams illustrating how coastlines straighten over time, as headlands are eroded and bays filled in by wave refraction and longshore transport. (Source: Bascom, W. 1980. *Waves and Beaches.* Garden City, NY: Anchor Books, p. 15.)

Fig. 15-6 Mavericks Surfing Beach near Pillar Point, San Mateo County. (Source: Montgomery, D.)

BEACHES AND LONGSHORE TRANSPORT

The southern California coast is particularly famous for its long, smooth stretches of sand. Most of California's beaches are narrow strips of sand at the base of sea cliffs (Fig. 15-7), but some, like those at the mouth of the Santa Clara River near Ventura, are backed by coastal lowlands (discussed later). Although most California beaches are composed of sand, gravel beaches can be found in many areas (Fig. 15-8). Some gravel beaches form in areas where the nearby sea cliffs contain conglomerate, as is the case just north of Encinitas in southern California. Beach sediments—gravel as

Fig. 15-7 Narrow sandy beach and steep cliffs at Point Conception. Note the wave-cut platform in the surf zone, where vertical beds of the Monterey Formation can be seen. (Source: Sarna-Wojcicki, A. U.S. Geological Survey.)

Fig. 15-8 A gravelly pocket beach in northern California (Source: Photo by author.)

well as sand—are typically well rounded and sorted by the tumbling action of the waves (see Chapter 2).

A beach is a narrow strip of sand parallel to the coastline. Above the water level is the beach berm, where sand has been carried to shore by waves. Below the water, out to about 10 meters of water depth, are elongated piles of sand parallel to the beach, called ***offshore bars*** (Fig. 15-9). Both the beach berm and the offshore bars are part of the beach system, and the sand on a beach is in almost constant motion. Evidence of this can be seen by comparing the same beach in winter and in summer. During the winter storm season, large waves pound the beach. The waves carry much of the sand to the offshore bars. The beach berm becomes narrower. During summer, the beach widens because the smaller summer waves are unable to carry sand from the beach (Fig. 15-10).

Most beach sand is supplied from rivers that empty into the ocean. Eroding sea cliffs supply a lesser amount, although they might be important local sources. From the river mouths, sand is carried along the coast by ***longshore currents***, which result

Fig. 15-9 Profiles of a typical beach showing the seasonal movement of sand. During winter, storm waves move much of the sand to offshore bars. (Source: Bascom, W. 1980. *Waves and Beaches.* Garden City, NY: Anchor Books, p. 251.)

A

B

Fig. 15-10 **A**, Winter beach and **B**, summer beach, Santa Cruz. The photos were taken at low tide, when the wave-cut platform is exposed in the surf zone. The houses sit on a marine terrace. (Source: **A**, Bradley, W. **B**, Photo by author.)

from the interaction between approaching waves and the coastline. Waves approach the coast from the direction of wave-generating winds. In California, the largest waves originate from winter storms in the northern Pacific, and they approach the coast from the northwest. As they approach the coastline, the waves are refracted, but

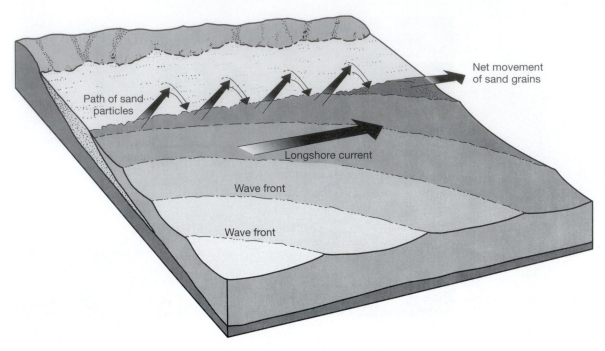

Fig. 15-11   Movement of sand along a beach by longshore currents. The longshore currents are created by obliquely breaking waves. These currents transport sand along the beach and in the surf zone. (Source: Tarbuck, E., and Lutgens, F. 1993. *The Earth,* 4th ed. Upper Saddle River, NJ: Prentice Hall.)

they generally arrive at the surf zone at an angle to the beach (see Fig. 15-3). When the waves break, the water runs up the beach face at right angles to the line of approach (Fig. 15-11). The water returns down the beach face at right angles to the shoreline. Unless the wave fronts and the shoreline are exactly parallel, the water is displaced a small distance parallel to the shore (see Fig. 15-11). Sand moving with the waves is also carried parallel to the shore; this type of sand movement is known as ***longshore transport***.

Longshore currents operating in the shallow waters adjacent to California's beaches effectively distribute sand along the coast. Because the largest waves approach the California coast from the northwest, the overall direction of longshore transport is generally from north to south. North of Point Conception (see Fig. 15-1), the overall direction of sand transport is from north to south because the coastline faces southwest in most areas. South of Point Conception, sand is carried east, and then southeast because of the bend in the coastline (Fig. 15-12). Oceanographers have confirmed the direction of longshore transport by measuring the movement of sand in the surf zone. They have also identified the sources of many beach sands by determining their mineral content and matching it with that of rocks found in the drainage basins of coastal streams.

The local direction of longshore transport might be complex, because transport depends on the geometry of the coastline as well as the direction of wave approach.

Fig. 15-12   Sand movement along southern California beaches. The arrows show the pathway of sand carried southeastward by the longshore currents. The numbers adjacent to the arrows show the five major beach segments along the southern California coast. At the southeast end of each segment, longshore transport is interrupted by submarine canyons and fans (see Chapter 12); the fans are shown by the striped areas. (Source: Emery, K.O. 1960. *The Sea off Southern California.* New York: Wiley.)

For example, in Monterey Bay, longshore currents move both north and south from protruding headlands (Fig. 15-13). Summer and autumn storms south of California can also generate sizable waves that approach the coast from the south.

Sand enters the southern California coastal zone from six major rivers (see Fig. 15-12). Longshore currents carry the sand east and southeast from the river deltas. The beaches grow wider going south and southeast as sand builds up. Then, where the longshore currents reach the head of a submarine canyon, the beaches

Fig. 15-13 The general pattern of longshore currents in Monterey Bay. (See Fig. 12-6 for the bathymetry of Monterey Bay.)

abruptly become much narrower. The submarine canyon heads are so close to the shore that they intercept the sand moving along the beaches and transport it seaward into deep water. (See Chapter 12 for information about submarine canyons and fans.)

From the example of southern California, it is clear that beach environments are closely linked with each other and with coastal streams. Any structure that interrupts the longshore current acts like a dam to halt the flow of sand along the coast. Beaches down-current of such a structure are cut off from a sand supply and are quickly eroded by waves. The breakwaters at the Santa Barbara harbor, at Ventura and Port Hueneme, and at Santa Monica all interrupt the longshore transport of sand. It has been necessary to artificially replenish the sand on "sand-starved" beaches down-current of these structures, which is a costly process. Damming of southern California's rivers for flood control has also diminished the supply of sand to the beaches.

In many places along the California coast, the longshore currents are naturally interrupted by a change in the coastline. For example, the coastline might bend to form an embayment. The longshore current might build a ridge of sand beyond the

Fig. 15-14 Limantour Spit north of San Francisco. The Coast Ranges of Marin County can be seen in the background. (Source: Sarna-Wojcicki, A. U.S. Geological Survey.)

bend, forming a *spit* that projects into the bay (Fig. 15-14; see also Fig. 15-5). A sand spit can provide a natural harbor, protecting an embayment from storm waves, as in the case of the Coronado spit at San Diego harbor and the large spits crossing Humboldt Bay. Spits can grow until they completely seal off an embayment from the open ocean. This is a frequent occurrence at the mouths of northern California rivers during summer. The sand bars are broken during the first winter storms, when the river runoff is sufficient to break through them.

Sand dunes are found in several places along the California coast. High winds, patchy vegetation, and abundant loose sand at these sites provide ideal conditions for dune formation (see Chapter 6). In some areas, dune fields cover large areas and extend several kilometers inland. The largest dune fields in coastal California are the Monterey Bay dunes south of the Salinas River, Nipomo Dunes north of the Santa Maria River, and the dune fields at San Diego Bay.

EMERGENT COASTS

Much of California's coast is steep and rocky. Here, instead of long, straight, sandy beaches, the shoreline is composed of small pocket beaches tucked between rocky points (see Fig. 15-8). Along many rocky shores, wave erosion has isolated protruding points from the mainland, forming *sea stacks* (Fig. 15-15 and Color Plate 28). In a few areas, including Anacapa Island and Natural Bridges near Santa Cruz, waves have carved caves, tunnels, and arches into the rocks (see Color Plate 29). Steep cliffs are found along rocky coastlines (see Color Plate 2), and landslides are a common occurrence, as discussed later in this chapter and in Chapter 19.

The features along California's rocky coasts are the result of tectonic uplift. As a result of ongoing uplift, much of the coastline is rising relative to sea level. Geologists classify this type of coast as *emergent*; that is, it is emerging from the ocean. Even though worldwide sea level is high today (see Chapter 8 and Chapter 12), most of the coastline is emergent because tectonic uplift has outpaced the postglacial sea level rise.

Fig. 15-15 Sea stacks along the Humboldt County coast near Trinidad. Trinidad Head is the large rock in the center of the background. (Source: Photo by author.)

BIG SUR

Fig. 15-16 Wind-swept granitic rocks along the coast at Big Sur. (Source: Skapinsky, S. San José State University.)

This verse expresses the austere, enduring beauty of coastline formations like Big Sur (Fig. 15-16):

> This gray rock, standing tall
>
> On the headland, where the seawind
>
> Lets no tree grow,
>
> Earthquake-proved, and signatured
>
> By ages of storms: on its peak
>
> A falcon has perched.

Source: "Rock and Hawk," by Robinson Jeffers. From Hunt, T., ed. 1989. The Collected Poetry of Robinson Jeffers, Vol. 1, 1920-1928. Stanford, CA: Stanford University Press.

Marine Terraces

Marine terraces are one of the most striking features of emergent coasts (Fig. 15-17). At some localities—San Clemente Island, Palos Verdes, or Santa Cruz, for example—multiple terraces are cut against the hillslopes, their levels rising from the shoreline like flights of stairs. Terraces are conspicuous along most of California's coast; only where the coast is low-lying are they completely absent.

Marine terraces are ancient marine shorelines, or strandlines. They consist of a relatively flat ***wave-cut platform***, eroded by waves in the surf zone, and a cover of beach sand (Fig. 15-18; see also Figs. 15-7 and 15-10). Fossils, including microfossils and mollusk species like clams, might be preserved in the terrace deposits. Many geologists believe that the most extensive wave-cut platforms are formed during times when worldwide sea levels are high. The line where the platform meets the steep sea cliff behind it is called the shoreline angle. As tectonic uplift causes the coastline to rise from the sea, new platforms are cut at lower levels, and the former platforms are lifted above sea level to become terraces (Fig. 15-19).

Because they represent former sea-level positions, marine terraces provide excellent indicators of uplift rates. Using radiometric techniques to date fossils in the marine terraces, it is possible to determine the rate of uplift since the terrace formed. Geologists have calculated the uplift rates along the California coast using the widespread 125,000-year-old terrace cut during the last global high stand of sea level. At most places, the rates are 0.1 to 0.4 meters per 1,000 years. Along most of California, the 125,000-year-old terrace is found at 12.5 to 50 meters elevation. However, the rates are much greater in areas of rapid uplift. Both the north coast near Cape Mendocino (see Chapter 14) and the western Transverse Ranges near Ventura (see Chapter 16) are rising rapidly. This is not surprising, considering that these are the regions most strongly affected by compression. At Ventura, a fold

Fig. 15-17 Marine terraces near Rincon Point, Ventura County. (Source: Sarna-Wojcicki, A. U.S. Geological Survey.)

Fig. 15-18   An uplifted wave-cut platform is overlain by sandy beach sediments. The platform is cut on dark-colored sedimentary rocks in the lower cliffs. Fitzgerald Marine Preserve, San Mateo County. (Source: Photo by author.)

Fig. 15-19   Schematic diagram illustrating marine terraces and sea cliffs along an emergent coast. (Source: LaJoie, K. 1972. *Friends of the Pleistocene Field Trip Guidebook.*)

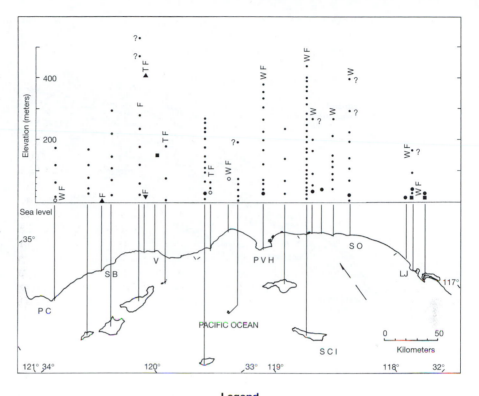

Fig. 15-20 Schematic profiles showing the elevations of marine terraces along the southern California coast. The map in the lower portion of the figure shows the locations of the terrace profiles. The symbols in the profiles portray the elevations and ages of terraces at each site. Note that the lower terraces are youngest. Places where terraces are warped (W), faulted (F), or tilted (T) are noted by letters on the profiles. (Source: LaJoie, K., and others. 1979. *Quaternary Terraces and Crustal Deformation in Southern California*. Geological Society of America Field Trip Guidebook.)

Legend

Map names		Terrace Ages	
P C	Point Conception	●	Undated
S B	Santa Barbara	○ ●	120,000
V	Ventura	■	85,000
P V H	Palos Verdes Hills	▲	45,000
S C I	San Clemente Island	▼	5000
S O	San Onofre		
LJ	La Jolla		

is lifting marine terraces at an impressive rate of 13 meters per 1,000 years. There, a 40,000-year-old terrace is found 400 meters above sea level (see Fig. 15-17; see also Chapter 16).

Terraces also record any tilting, folding, and faulting that have occurred since they formed. The shoreline angles can be thought of as geologic "level lines" (Fig. 15-20). Wherever the shoreline angles are warped, tectonic activity must have deformed them. Geologists have used California's marine terraces extensively when evaluating the long-term seismic safety of nuclear reactors, refinery facilities, and other critical projects.

In the San Diego area, ancient beach ridges—linear piles of sand up to 30 meters high—are preserved on most of the 14 marine terrace levels (Fig. 15-21). The ridges are probably the ancient equivalent of the sand bars and spits that enclose Mission Bay and San Diego harbor today.

Fig. 15-21 Ancient beach ridges along the San Diego County coast. The cross section shows the elevations of beach ridges at Torrey Pines. San Onofre Bluff is shown as locality SO in Fig. 15-20. (Source: Lajoie, K. 1992. In *Nonglacial Quaternary of the U.S.* Geological Society of America, Decade of North American Geology, Vol. K-2.)

ESTUARIES, COASTAL WETLANDS, AND SUBMERGED COASTS

Although most of California's coastline has been tectonically uplifted during the past few million years, some areas of the coast have sunk relative to sea level. In these areas, known as *submerged coastlines*, sea level has risen sufficiently to inundate the lower parts of the landscape. In many of these areas, tectonic uplift has not

Fig. 15-22 Coastal wetlands and estuary at the mouth of the Mad River, Humboldt County. (Source: Dengler, L. Humboldt State University.)

been rapid enough to elevate the river valleys. As a result, tidal marshes, lagoons, or estuaries have formed at the mouths of rivers (Fig. 15-22). San Francisco Bay is an excellent example of an estuary formed by the "drowning" of a river valley (see Chapter 12 and Color Plate 17). Coastal wetlands have also developed in areas that have been tectonically down-dropped. As we will see in Chapter 16, down-warped areas are found in regions of compression. Two areas where coastal wetlands are found in tectonic depressions are along the Cascadia subduction zone (see Fig. 15-22) and in the Transverse Ranges near Carpinteria (Fig. 15-23).

Estuaries are important breeding and feeding areas for a variety of animal species, including seabirds and a variety of fish. Estuaries are particularly important nurseries for juvenile salmon and steelhead, fish species that spend part of their life cycle in fresh water and part in the open ocean. After spawning in rivers, juvenile fish increase their chances of survival by feeding and growing in the estuary before entering the open ocean to face their adult habitat. Two of California's protected National Marine Sanctuaries have been established in estuaries and their surrounding wetlands. Elkhorn Slough is located in Moss Landing along the central coast of Monterey Bay (see Fig. 15-13), and the Tijuana Slough forms the Mexican border at the mouth of the Tijuana River. Despite the untreated sewage that flows down the river from Tijuana, the slough and the surrounding marshes are a rich habitat for many species. In contrast, the filled, paved, and diked wetlands of Mission Bay in San Diego or San Francisco Bay are human habitats created with much greater impact on other species.

Fig. 15-23 A portion of the 1944 Carpinteria topographic map showing the coastal wetlands near Carpinteria. (Source: U.S. Geological Survey.)

COASTAL EROSION

Tremendous energy is focused on a narrow strip of rock and sediment, as waves pound ceaselessly against the shoreline for years. Waves generated by large storms are particularly effective agents of erosion. For example, at Redondo Beach, in 1953, observers witnessed large storm waves throw 5-centimeter rocks more than 30 meters into the air. Coastal erosion is a natural process along all shorelines, particularly during storms. Along emergent coasts, waves undermine the lower parts of sea cliffs, causing large blocks to fall into the ocean. Rocky points are eroded to form sea stacks, which are eventually removed by the waves. Over time, the entire shoreline retreats inland.

Rates of coastal erosion are determined by several factors, including the orientation of the coast; the height of storm waves; and the erodibility of the rocks along the shore. Natural erosion rates can be increased by human activities that interrupt the longshore transport of sand. One example of this is found at the Half Moon Bay Harbor in San Mateo County. Half Moon Bay's distinctive curved shape results from the refraction of waves around Pillar Point and the distribution of sand southward along the coast (Fig. 15-24). Between 1959 and 1961, a breakwater was constructed

Fig. 15-24 Progressive coastal erosion and shoreline retreat in Half Moon Bay near El Granada, San Mateo County. (Source: LaJoie, K. 1972. Modified from *Friends of the Pleistocene Field Trip Guidebook*.)

to protect the harbor from summer waves approaching from the southwest. The breakwater disrupted the southward flow of sand, with the predictable result that the waves south of the breakwater are actively eating away at the cliffs to replenish the system with sediment. Between 1962 and 1985, the coast retreated by as much as 2.5 meters per year, a fourfold increase over the prebreakwater rates. Portions of roads, including State Highway 1, disappeared with the coastal area.

The town of Encinitas north of San Diego has lost several buildings, streets, and railroad lines to coastal erosion. During the winter of 1941, large storm waves along the base of the cliffs and heavy rains on the hills above combined forces. A relatively new four-story temple was tossed into the ocean as the cliffs collapsed. In 1977, similar storms toppled miles of cliffs and destroyed the train tracks along the terrace.

Coastal erosion poses problems because people build supposedly permanent structures along mobile coastlines. The U.S. Army Corps of Engineers has estimated that 86 percent of California's coast is experiencing significant erosion. Along 124 kilometers of coastline, coastal erosion is threatening houses, roads, and other structures. Erosion is typically episodic, with major cliff retreat, landsliding, and sand removal taking place during large storms and El Niño winters (Fig. 15-25). Many years might pass between cliff-eating storms, fading the threat of coastal erosion in the memories of communities and developers. In some of California's coastal counties, the California Coastal Zone Commissions and other civic agencies are attempting to restrict coastal development to reduce the risk posed to property by future coastal erosion. However, because much of the coastline is already developed, coastal erosion will continue to cause property damage and loss.

A

B

Fig. 15-25 A, The remains of seven homes hanging at the edge of a cliff in Pacifica, CA, that crumbled under the battering of relentless El Nino storms in the winter of 1998. (AP Photo/Adam Turner). **B**, Work crews cleaning up after the collapse of homes in Pacifica, CA (Photo by Monty Hampton, U.S. Geological Survey, Menlo Park, CA).

SUMMARY

1. The California coastline north of Point Conception is characterized by a narrow continental shelf, rocky headlands, and pocket beaches. Southern California beaches, oriented east–west and protected by a broader continental shelf, are longer and straighter.
2. Waves are generated by storm winds at sea, and most commonly approach the California coast from the northwest. During the winter months, larger storm waves move beach sand into offshore bars, whereas sand accumulates on the beaches during summer.

3. Sand is distributed from the mouths of California's rivers to the beaches by long-shore currents. Most longshore transport is from north to south in California, but complex patterns develop from variation in the coastline's orientation and seasonal variations in wave approach. Human activities can interrupt longshore transport, resulting in coastal erosion or beach starvation.
4. Much of California's coastline is being uplifted at a faster rate than the postglacial rise in sea level, resulting in an emergent coastline. Emergent coasts are characterized by rocky shores, steep beaches, cliffs, wave-cut platforms, and marine terraces. Marine terraces are useful indicators of faulting and folding along the coastline.
5. Submerged or drowned coasts are present at the mouths of rivers or at localities where the coastline has been tectonically down-dropped. Estuaries at the mouths of drowned rivers are important habitats for aquatic species and birds.
6. Coastal erosion poses a continuing threat to developments along much of the California coastline.

REFERENCES

BASCOM, W. 1980. Waves and Beaches. Garden City, NY: Anchor Books, 366 pp.

California Coastal Commission. 1994. California Coastal Resource Guide, 2nd ed. Berkeley: University of California Press, 384 pp.

DESANTIS, M. 1985. California Currents: An Exploration of the Ocean's Pleasures, Mysteries, and Dilemmas. Novato, CA: Presidio Press, 238 pp.

EMERY, K.O. 1960. The Sea off Southern California. New York: Wiley, 364 pp.

GRIGGS, G., and SAVOY, L. 1985. Living With the California Coast. Durham, NC: Duke University Press, 415 pp.

KAUFMAN, W., and PILKEY, O. 1979. The Beaches Are Moving. Garden City, NY: Anchor Press, 326 pp.

Web Sites and Media Resources

The Beach, A River of Sand, 1965, Encyclopedia Britannica Educational Film, 20 minutes.

California Coastal Commission: *http://www.coastal.ca.gov/*

California Shore and Beach Preservation Association: *http://www.csbpa.org/*

USA Today's 2003 Beach Guide: California Beaches: *http://www.usatoday.com/travel/vacations/beach/california.htm*

Big Sur Beaches: *http://www.bigsurcalifornia.org/beaches.html*

Cabrillo National Monument, National Park Service: *http://www.nps.gov/cabr/home.html*

Channel Islands National Park, National Park Service: *http://www.nps.gov/chis/*

Coastal Erosion and El Niño U.S. Geological Survey: *http://walrus.wr.usgs.gov/elnino/*

National Marine Sanctuaries, National Oceanic and Atmospheric Administration: *http://www.sanctuaries.nos.noaa.gov*

Point Reyes National Seashore, National Park Service: *http://www.nps.gov/pore/*

U.S. Geological Survey, San Francisco Bay and Delta: *http://sfbay.wr.usgs.gov/*

The Transverse Ranges, the Los Angeles Basin, and the Offshore Islands: Compression and Rapid Change

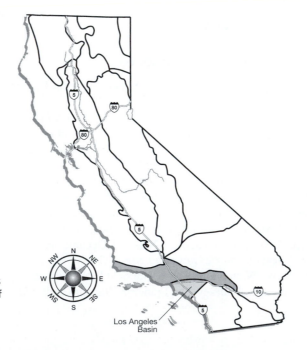

Fig. 16-1 Location of the Peninsular Ranges Province, the Los Angeles basin, and the location of the offshore islands. The offshore islands are also shown in Endpaper 3.

Los Angeles Basin

Transverse Ranges, Los Angeles Basin, and Offshore Islands Highlights

Major geographic feature:	Transverse Ranges Mountains, Offshore Islands, Los Angeles basin
Named mountain ranges (west to east):	Santa Ynez, San Rafael, Santa Susana, Santa Monica, San Gabriel, San Bernardino, Little San Bernardino, Orocopia
Major valleys and basins:	Ventura, Soledad, San Fernando, Los Angeles
Offshore islands (north to south):	Channel Islands: Anacapa, Santa Cruz, Santa Rosa; San Nicolas, Santa Barbara, Santa Catalina, San Clemente
Selected high points:	White Ledge Peak, Santa Ynez Mts.: 1,414 m (4,640 ft.) Mt. Pinos, Pine Mts.: 2,692 m (8,831 ft.) Mt. San Antonio, San Gabriel Mts.: 3,068 m (10,064 ft.) Cucamonga Peak, San Gabriel Mts.: 2,700 m (8,859 ft.) San Gorgonio Mt., San Bernardino Mts.: 3,505 m (11,501 ft.) San Bernardino Peak, San Bernardino Mts.: 3,246 m (10,649 ft.)
Predominant rock types:	Cenozoic marine and nonmarine sedimentary, Mesozoic marine sedimentary and metasedimentary, Mesozoic granitic plutonic, Precambrian igneous and metamorphic
Plate tectonic events:	Mesozoic subduction and accretion, mid-Cenozoic extension and rotation, late Cenozoic transform motion and transpression
Other geologic events:	Sea level changes
Famous faults:	San Andreas, Santa Ynez, Pine Mountain, San Cayetano, Oak Ridge, Ventura, Malibu Coast, San Gabriel, Sierra Madre, Cucamonga, San Jacinto, Holly wood, Santa Monica, Raymond, Palos Verdes, Newport-Inglewood, Puente Hills, Whittier, unnamed blind thrust faults
Resources:	Petroleum, diatomite
National Parks:	Channel Islands National Park, Santa Monica Mountains National Recreation Area
State Beaches (north to south):	Gaviota, Refugi, El Capitan, Carpenteria, Emma Wood, San Buenaventura, McGrath, Point Mugu, Leo Carillo, Robert Meyer, Point Dume, Malibuagoon, Bolsa Chica, Huntington, Crystal Cove, Corona del Mar, Doheny, San Clemente
Other sites of interest:	Page Museum (La Brea Tar Pits)

Even a quick look at a California map (Endpaper 3) reveals how the Transverse Ranges got their name: They run transverse, or crosswise, to almost all of California's other mountains and valleys. Several rugged mountain ranges make up the Transverse Ranges (Fig. 16-2). The highest peaks in the ranges are San Gorgonio Peak in the San

Fig. 16-2 The major mountain ranges of southern California. (Source: California Division of Mines and Geology, Bulletin 170.)

Bernardino Mountains (3,500 meters) and San Antonio Peak in the San Gabriel Mountains (3,072 meters). On a clear day, the Transverse Ranges form an imposing fringe around the flat valleys that are sandwiched against them (Fig. 16-3). The largest of these valleys, the San Fernando Valley and the Los Angeles basin, are home to many of California's residents (Fig. 16-4). Although the Los Angeles basin has traditionally been considered part of the Peninsular Ranges geomorphic province (see Chapter 17), its geologic history is closely tied to that of the Transverse Ranges, so it is logical to include it in this chapter.

Fig. 16-3 A view of the San Bernardino Mountains and the San Bernardino basin, looking northeast. The San Andreas fault runs along the base of the mountains. (Source: U.S. Geological Survey, Southern California Areal Mapping Project.)

Fig. 16-4 Landsat satellite image of the Los Angeles basin. The Los Angeles basin and the San Fernando Valley to the north are the smooth plains, and mountain ranges ring the valleys. (Source: Image processing by Rymer, M. U.S. Geological Survey.)

Like the mountains, this part of the California coastline also runs east–west (see Chapter 15). A person standing at the ocean's edge to watch the waves in Santa Barbara or Ventura is actually facing south. Eight islands and numerous smaller rocks are found off the southern California coast. The four northern islands, called the Channel Islands, form a chain at the southern boundary of the Santa Barbara Channel, whereas the four southern islands are a more scattered group lying southwest of Los Angeles and Long Beach (Endpaper 3). Five of the islands—San Miguel, Santa Rosa, Santa Cruz, Anacapa, and Santa Barbara—make up Channel Islands National Park, established in 1980. Santa Catalina Island is more developed and a very popular summer resort, and San Clemente Island is a military reservation. Numerous smaller sea stacks are scattered throughout the islands. The area offshore is referred to as the Southern California Borderland, and like the Los Angeles basin, it has a geologic history closely tied to that of the Transverse Ranges Province.

Today, compression is the geologic process creating the distinctive features of the Transverse Ranges Province. Because of the Big Bend in the San Andreas fault, the Pacific and North American Plates converge in this region (Fig. 16-5; see Chapter 14).

Fig. 16-5 The Big Bend in the San Andreas fault causes compression in the Transverse Ranges Province. (Source: Southern California Earthquake Center, *Reviewing the Basics, The Master Fault: http://www.scecdc.scec.org/eqabc.html.*)

As a result, the entire region is being squeezed together, with the maximum compression oriented in a north–south direction. The mountains, the basins, and many of the faults crossing the province are all expressions of this squeezing. Most trend east–west, perpendicular to the direction of maximum compression. As the crust is compressed, it is being squeezed and thickened. This process is expressed at the surface, where some areas are being uplifted to form mountains, as others are being pushed down to form basins.

The Transverse Ranges are one of the Earth's most rapidly uplifting areas. The steep, rugged mountains along the Big Bend are some of the evidence of rapid uplift. The stepped marine terraces near Ventura and Palos Verdes (see Chapter 15) provide further proof of rapid uplift. The terraces are rising by as much as 5 to 10 millimeters per year. Dramatic uplift takes place during earthquakes. During the 1994 Northridge earthquake, the Santa Susana Mountains rose 70 centimeters, and during the 1971 San Fernando earthquake, the same mountains rose about 2 meters.

REVERSE FAULTS AND FOLDS

Not surprisingly, many of the faults in the Transverse Ranges Province are the type produced by compressional forces. When blocks of rocks are compressed, they buckle together, with one block sliding over the other (Fig. 16-6; see also Fig. 1-15B). Both vertical movement and shortening of the crust occur across the resultant fault, known as a ***reverse fault***. If a reverse fault is only slightly inclined, it is called a ***thrust***

Reverse

Oblique-slip reverse
(component of strike-slip
and reverse slip)

Thrust
(low-angle reverse)

Buried reverse fault

Fig. 16-6 Block diagrams of reverse and thrust faults. (Source: Adapted from Keller, E., and Pinter, N. 1996. *Active Tectonics.* Upper Saddle River, NJ: Prentice Hall.)

Fig. 16-7 View to the north across the Ventura fault in Ventura. The steep rise in the road shows the surface expression of the fault. A trench excavated by geologists across the fault trace is marked by the linear mound of sediment (from the backhoe excavation) on the right side of the photo. (Source: Sarna-Wojcicki, A. U.S. Geological Survey.)

fault. Earthquakes on reverse faults can generate strong vertical motion as well as horizontal motion.

Many of southern California's reverse and thrust faults have not ruptured the sediments or rocks at the surface. In many cases, the ground surface is buckled where the fault projects to the surface (Fig. 16-7), but some buried faults have no surface expression. Geologists have recently recognized that these buried, blind thrust or reverse faults pose significant hazards. The 1994 Northridge, 1987 Whittier Narrows, and 1971 San Fernando earthquakes were all centered on reverse faults buried beneath the Los Angeles basin (Fig. 16-8; see also Chapter 13 and Chapter 14). The 1983 Coalinga earthquake (magnitude 6.5) was also centered on a blind fault, one of several bordering the western edge of the San Joaquin Valley.

When rocks and sediments are compressed, they can also be crumpled together without rupturing. When this happens, *folds* are created (Fig. 16-9). It is important to realize that folding and reverse faulting can both be produced by compression (as in northwestern California's fold and thrust belt, discussed in Chapter 14), and that both phenomena result in crustal shortening. A fold created by an upward buckling of rock or sediment is an *anticline*, whereas a downward-buckled fold is a *syncline* (see Fig. 16-9). In many regions where rocks are folded, multiple anticlines and synclines are aligned with their hinge lines, or fold axes, perpendicular to the direction of maximum compression. The east–west alignment of reverse faults and fold axes in the Transverse Ranges is good evidence that both were created by north–south compression.

Southern California's *basins* are regions where the crust has been tectonically depressed in the center and pushed up on all sides, forming a low-lying bowl or basin. Southern California's basins are the result of down-buckling of the crust, just as many of the hills and mountains have resulted from up-buckling. The basins are

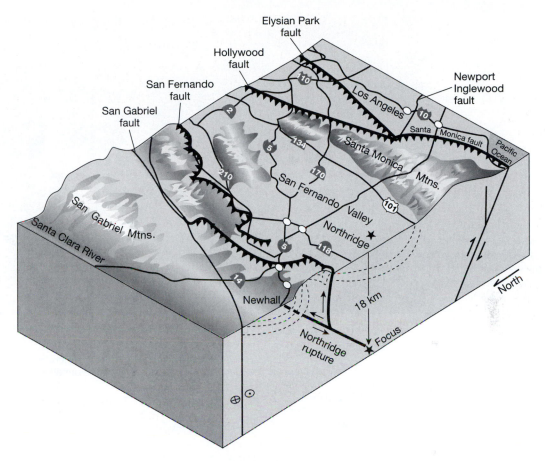

Fig. 16-8 A cross sectional view of the Northridge area showing reverse faults beneath the San Fernando Valley and the northern Los Angeles basin. (Source: Williams, P. Lawrence Berkeley Laboratory. In Learning from Los Angeles, *Earth Magazine*, September 1994.)

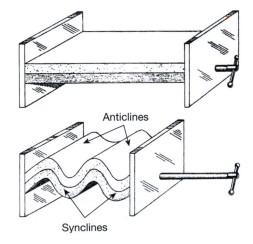

Fig. 16-9 An anticline and a syncline. The vise shows the direction of maximum compression. (Source: Coch, N., and Ludman, A. 1991. *Physical Geology.* New York: Macmillan.)

Fig. 16-10 The Ventura Avenue anticline near Ventura. The location of the anticline is shown in Fig. 16-16. The view is toward the west-northwest along the axis of the fold. (Source: Sarna-Wojcicki, A. U.S. Geological Survey.)

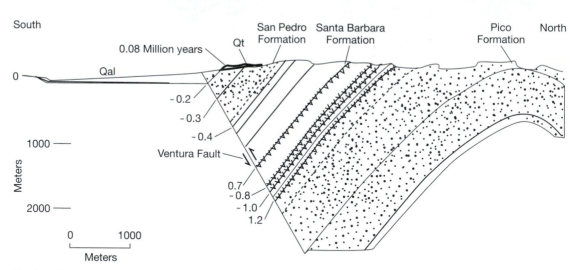

Fig. 16-11 A cross section showing the Ventura fault (see Fig. 16-7) and the southern limb of the Ventura anticline. Volcanic ashes used to date the sedimentary layers are shown by Vs in the diagram. The ages of units are shown along the left of the cross section in millions of years. (Source: Sarna-Wojcicki, A., and Yerkes, R. U.S. Geological Survey.)

typically surrounded by mountain ranges or lower hills, regions where the crust is tectonically uplifted. For example, the Santa Ynez, Topatopa, and Santa Monica Mountains are large, complex anticlines. The borders between basins and mountains are typically active faults—either reverse faults or transform faults of the San

Andreas system (see Fig. 16-8). The correlation between the topography and geological structures is strong evidence that tectonic forces are active.

Some folds in the Transverse Ranges appear to be growing rapidly. For example, the Ventura Avenue anticline, just north of the town of Ventura, has tilted very young sedimentary layers, only a few hundred thousand years old, by as much as 30 degrees (Figs. 16-10 and 16-11). Repeated surveys indicate that the valley of the Ventura River, near the axis of the anticline, rose about 0.5 centimeters per year between 1920 and 1968. Even the surface of California Route 33 is buckled slightly where it crosses the fold. Further south, the 200,000-year-old San Pedro sand is uplifted in the Palos Verdes and San Joaquin Hills and buckled beneath the area between them (Fig. 16-12). The total difference in elevation of the once-flat formation is now tens of meters.

The Los Angeles basin and the Transverse Ranges Province, including the Channel Islands, are crisscrossed by reverse faults and folds. Trending east-west, these faults and folds intersect the northwest-trending right-lateral faults of the San Andreas system, like the Newport-Inglewood fault zone (see Endpaper 2; Fig. 14-33). Not all of the Transverse Range's reverse faults are buried, and along some, rugged mountain ranges have been pushed up. For example, at the northern edge of the

Fig. 16-12 Diagram showing a syncline in the south Los Angeles basin. Very young sedimentary units are uplifted in the San Joaquin Hills and the San Pedro Hills, at the ends of the cross section. Beneath the southern Los Angeles basin, in the center of the cross section, the same units are down-warped below sea level. (Source: Lajoie, K., and others. 1979. Geological Society of America, field trip guide.)

Fig. 16-13 A portion of the 1900 Pasadena topographic map near the present community of Monterey Park. Because the map was made before extensive urbanization of the Los Angeles basin, one can clearly see the low hills that mark the surfaces of young folds and blind faults. Today, Interstate 10, the San Bernardino Freeway, runs east–west along the western part of the railroad line shown in the northern part of this map running through Shorb. A photograph of the Monterey Park Hill, taken after the 1933 Long Beach earthquake, is shown in Fig. 13-25B.

province, the Santa Ynez fault runs along the northern slope of the Santa Ynez Mountains (see Fig. 14-33). Further south, the San Cayetano and Oak Ridge faults, two steep reverse faults dipping in opposite directions, have sandwiched and down-dropped the Santa Clara Valley between the Topatopa Mountains on the north and the Santa Susana Mountains on the south (see Fig. 16-16). The Sierra Madre fault zone forms the southern boundary of the San Gabriel Mountains. At the southern edge of the Transverse Ranges, the Malibu Coast, Santa Monica, and Raymond Hills reverse faults all extend to the surface, where they cut across young sediments (see Figure 14-33).

Other fault and fold zones are marked by low hills crossing southern California's basins—for example, the Puente Hills, the San Pedro Hills, and the Repetto Hills. In the urbanized basins, subtle ramps along streets mark the surfaces of folds and reverse faults. On older maps, these changes in slope sometimes appear as aligned marshy areas or small hills (Fig. 16-13). At some localities in southern California, folding at or near the surface could be the expression of reverse faulting in the deeper layers beneath. The less consolidated sediments above a buried reverse fault might be buckled into an anticline. Geologists believe that many of southern California's faults are still growing upward and have not yet broken the younger sedimentary layers near the surface. Recent earthquakes on blind reverse faults, most notably the 1994 Northridge and 1987 Whittier Narrows earthquakes discussed in Chapter 13, prompted geologists to reexamine southern California's folds in light of their connection to buried faults (see Figs. 14-40 and 14-41).

LATE CENOZOIC BASINS OF THE TRANSVERSE RANGES PROVINCE

Another dramatic effect of compression in the Transverse Ranges is the formation and rapid filling of deep basins during the past 4 million years. As the San Andreas fault system evolved and the Big Bend formed, the basins subsided and began to fill with sediment (Fig. 16-14). About 4 million years ago, the rate of subsidence (or depression) of the basins increased, as did the rate of uplift in the surrounding mountains.

The young basins of the Transverse Ranges are in various stages of being filled with sediment. The Santa Cruz and San Nicolas offshore basins are entirely below sea level. In contrast, the Los Angeles basin is filled to the brim, so that its surface lies above sea level. Beneath its relatively flat surface lies a complex depression containing up to 5,500 meters of sediments deposited during the past 4 million years (Fig. 16-15) and another 3,600 meters of older Miocene sediment beneath that.

The Los Angeles basin is also broken by faults and folds. Large fault zones slice the basin into four well-defined blocks. On the west, the Palos Verdes fault zone separates the uplifted Palos Verdes Hills from the lower Long Beach–Wilmington area. The Newport-Inglewood fault zone, a part of the San Andreas system, runs northwest through the basin (see Figs. 16-15 and 14-33). Further east, the Whittier-Elsinore fault cuts across the basin, separating the Puente Hills and the San Gabriel Valley from Los Angeles. Anticlines and buried faults cross through the basin within these major blocks.

Fig. 16-14 Deep basins of southern California. The numbers show the maximum thickness (in kilometers) of sediments filling the basins during the past 4 million years. (Source: Yeats, R. 1991. In *The Cordilleran Orogen.* Geological Society of America, Decade of North American Geology, Vol. G-3, p. 336.)

Ventura Basin

The Ventura basin is sandwiched between the Santa Ynez and the Santa Monica Mountains, as if caught between the jaws of a giant vise. As the two ranges move together, sediments caught in the trough are being deformed and pushed up into mountains (Fig. 16-16). The western Ventura basin, beneath the Santa Barbara Channel, lies below sea level west of the city of Ventura. Further east, the surface of the Ventura basin forms the Santa Clara Valley and the Oxnard Plain. Beneath those smooth plains lie up to 10,000 meters of sediment, 7,300 meters of it deposited during the past 4 million years. The Ventura is the deepest of all the basins of the Transverse Ranges, and it contains the world's thickest section of Pleistocene marine deposits.

The Ventura basin began to form in early Pliocene time, about 4 to 5 million years ago. Using both the types of sediment being deposited in the basin and the microfossils, mostly foraminifera, geologists can reconstruct the changes in water depth during the basin's history. At first, the Ventura basin was subsiding faster than it was filling with sediment. As a result, the water depth increased

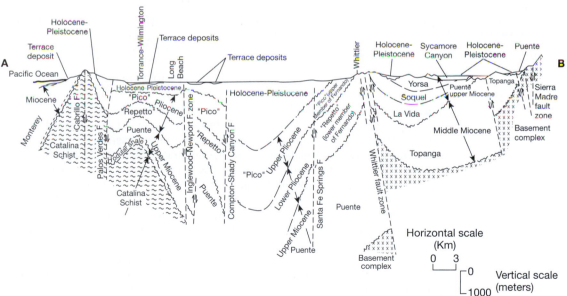

Fig. 16-15   Above, map, and below, cross section of the Los Angeles basin. The shaded areas on the map show oil fields. Fig. 14-40 shows the epicenter of the 1987 Whittier earthquake and a proposed blind thrust fault beneath the same area. (Source: Map, Yeats, R. 1991. In *The Cordilleran Orogen*. Geological Society of America, Decade of North American Geology, Vol. G-3, p. 339; cross section, California Oil and Gas Report. California Division of Mines and Geology, Publication TR12, Vol. 2.)

Fig. 16-16   Map of the Ventura basin area (Source: Yeats, R. 1991. In *The Cordilleran Orogen.* Geological Society of America, Decade of North American Geology, Vol. G-3, p. 335.)

(Fig. 16-17). The sediments and fossils found in the older Ventura basin formations are typical of deep marine conditions. During the time that the Pico Formation was deposited, the basin was receiving sediments from deep-water turbidity currents (Fig. 16-18). Then, between about 3.1 and 1.8 million years ago, 2,600 meters of sediment were deposited in the basin. During this period, the rate of sedimentation was greater than the rate of tectonic subsidence, so the depth of water in the basin decreased. The evidence of shallower marine waters is clear: The fossils represent organisms that inhabit shallower water, and the sediments were deposited in a more near-shore environment. By the time the Santa Barbara and San Pedro formations were deposited, between about 1.2 and 0.5 million years ago, shallow marine conditions prevailed in the central Ventura basin (Fig. 16-19).

Between about 200,000 and 400,000 years ago, the central Ventura basin emerged above sea level and rivers began to deposit nonmarine sediment. The river sediment contains distinctive granitic pebbles and cobbles derived from the San Gabriel Mountains to the east. Reverse faulting and folding continued during and after sedimentation in the Ventura basin. The older basin sediments are disrupted by numerous faults and folds, like the Ventura Avenue anticline discussed earlier.

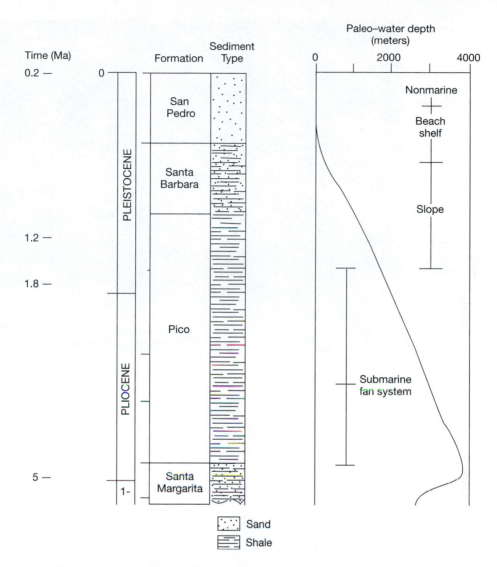

Fig. 16-17 The graph on the right shows the ancient water depths during the filling of the Ventura basin, based on the species of foraminifera found in the sedimentary formations. (Source: Ingle, J. 1981. *IGC Field Trip Guidebook*.)

Continued compression is squeezing the deposits, uplifting them so that they are exposed in the hills surrounding the Santa Clara Valley (Fig. 16-20). Deposition in the eastern Ventura basin ceased earlier than in areas to the west. Folded, faulted, and uplifted Ventura basin sediments can now be seen in the Santa Susana Mountains. These sediments are now being eroded and redeposited on the Oxnard Plain and in the Santa Barbara Channel to the west.

Fig. 16-18 Turbidites in the Pico Formation near Pitas Point, Santa Barbara County. As shown in Fig. 18-17, the Pico Formation represents deep-water deposition in a submarine fan. (Source: Sarna-Wojcicki, A. U.S. Geological Survey.)

Fig. 16-19 Shallow-water fossils in the Santa Barbara Formation near Ventura. During the time that these sediments were deposited, the sedimentary basin had filled almost to sea level (see Fig. 16-17). (Source: Sarna-Wojcicki, A. U.S. Geological Survey.)

Fig. 16-20   Uplifted sedimentary rocks of the Ventura basin, near Rincon Point, Santa Barbara County. (Source: Photo by author.)

MONTEREY FORMATION

One of California's most widespread and famous sedimentary formations is the Miocene Monterey Formation (also called the Modelo Formation in southern California by some authors). Today, it is exposed throughout coastal California, from Point Arena south to Dana Point, as well as on the offshore islands (Fig. 16-21). It occupies as much as 100,000 square kilometers of offshore and onland California. Because it was the source rock for much of California's petroleum, geologists have taken a particular interest in the conditions that created the Monterey Formation.

The Monterey Formation accumulated in marine basins along the California coast between about 23 and 12 million years ago. The environment of deposition was similar to today's southern California borderland or the Gulf of California. Like the sediments in the Ventura basin, the Monterey Formation's sediments reflect deposition in a variety of marine conditions. During the time of deep-water deposition, when water depths ranged from 1,000 to 2,000 meters, fine-grained shale and siliceous sediment accumulated by the settling of fine sediment from the overlying waters. During this time, the basins were not receiving large amounts of sediment from the coastal areas.

Much of the silica-rich sediment in the Monterey Formation is **diatomite**, a fine-grained sediment composed of the skeletal remains of marine diatoms. Diatoms are single-celled marine plankton. Today diatoms flourish in offshore waters where upwelling of deep water provides large amounts of nutrients. When the diatoms die, their siliceous cells settle to the bottoms of the marine basins. Today, diatomaceous sediments are accumulating in deep ocean basins where little or no land-derived sediment is deposited. These conditions apparently characterized the Miocene basins during the time the Monterey diatomite formed, beginning about 12 to 13 million years ago. Throughout coastal California, one can find distinctive, white outcrops of diatomite (Figs. 16-22 and 16-23). Some of the diatomite has been altered to harder siliceous rocks—chert and porcelanite. Diatomite can also form in lakes, and beds of diatomite can be seen in the sediments that accumulated in many of California's Pleistocene lakes (see Chapter 6).

Pure diatomite has many commercial uses, including many types of filters, absorbents, and insulation. Diatomite filters are used in brewing, sugar refining, and water treatment. The diatom skeletons are ideal for these purposes because they are hollow and of variable shapes; diatomite filters contain about 90 percent open space and 10 percent solids. California supplies more than 60 percent of the world's diatomite, and much of that is mined from the Monterey and Sisquoc Formations near Lompoc in Santa Barbara County.

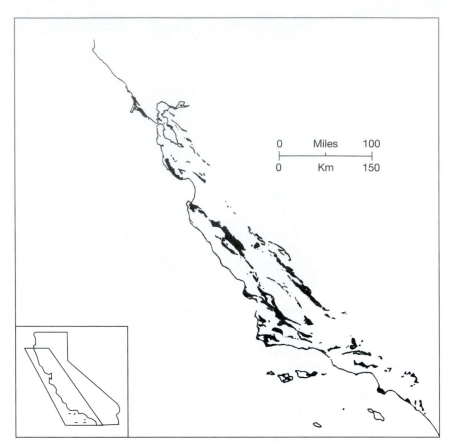

Fig. 16-21 The distribution of Monterey Formation and similar siliceous rocks. (Source: Pisciotto, K., and Garrison, R. 1981. *The Monterey Formation and Related Siliceous Rocks of California.* Society of Economic Paleontologists and Mineralogists, p. 97.)

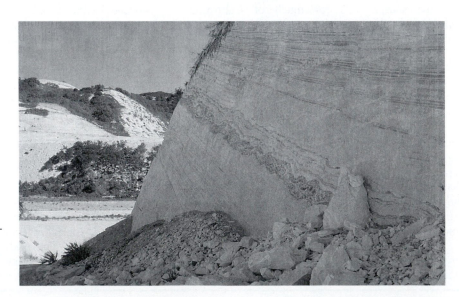

Fig. 16-22 Diatomite beds in the Monterey Formation, Lompoc, Santa Barbara County. (Source: Sarna-Wojcicki, A. U.S. Geological Survey.)

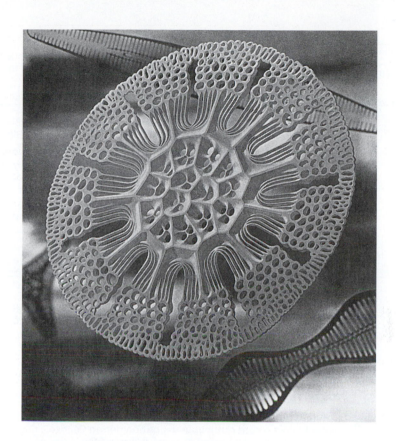

Fig. 16-23 Microphotograph of diatoms in the Monterey Formation. (Source: Grefco Corporation.)

EARLY CENOZOIC AND LATE MESOZOIC ROCKS

Marine sedimentary rocks of late Mesozoic and early Cenozoic age are present in the Santa Ynez, Topatopa, and Santa Monica Mountains and south of the Los Angeles basin in the Santa Ana Mountains (see Fig. 16-2). Like the younger sedimentary rocks in the region, they have been uplifted, folded, and faulted during the recent period of compression. Careful mapping and correlation have enabled geologists to reconstruct the original depositional sequences of sediments from these deformed rocks. Based on the similarities between sequences in different ranges, geologists concluded that the marine sediments accumulated along the continental margin before it was sliced into separate basins. The Cretaceous and early Cenozoic marine sedimentary rocks are also similar to those seen in the Great Valley Sequence to the north, suggesting that all of these rocks formed in the forearc basin along the western margin of North America before the San Andreas transform developed (see Chapter 11 and Chapter 18).

A thick sequence of Mesozoic and early Cenozoic sedimentary rocks can be viewed in the Santa Ynez Mountains north of Santa Barbara (Fig. 16-24). The light-colored and resistant Matilija Sandstone forms the most rugged slopes in the area. Another, younger sandstone, the Coldwater Sandstone, also forms rugged slopes along the southern Santa Ynez Mountains. Similarly resistant sandstone forms layered cliff exposures in the San Emigdio Mountains (Fig. 16-25). Less resistant, more thinly bedded shale units such as the Eocene Cozy Dell Formation are found on less prominent slopes.

During the Oligocene Epoch, these parts of the Transverse Ranges Province emerged from the sea, as indicated by the presence of nonmarine sedimentary rocks

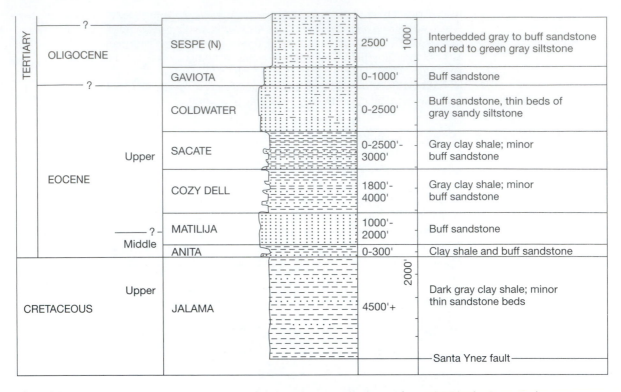

TERTIARY	OLIGOCENE		SESPE (N)		2500'	Interbedded gray to buff sandstone and red to green gray siltstone
		?	GAVIOTA		0-1000'	Buff sandstone
	EOCENE	Upper	COLDWATER		0-2500'	Buff sandstone, thin beds of gray sandy siltstone
			SACATE		0-2500'-3000'	Gray clay shale; minor buff sandstone
			COZY DELL		1800'-4000'	Gray clay shale; minor buff sandstone
		Middle ?	MATILIJA		1000'-2000'	Buff sandstone
			ANITA		0-300'	Clay shale and buff sandstone
CRETACEOUS	Upper		JALAMA		4500'+	Dark gray clay shale; minor thin sandstone beds

Santa Ynez fault

Fig. 16-24 A simplified stratigraphic section of early Cenozoic sedimentary formations in the Santa Barbara area. Rocks in the western Transverse Ranges record continuous deposition during Cenozoic time. Note that thicknesses are given in feet. (Source: California Division of Mines and Geology, Bulletin 186.)

Fig. 16-25 Rugged cliffs underlain by the Eocene San Emigdio Formation, San Emigdio Mountains north of Los Angeles, Kern County. (Source: Brabb, E. U.S. Geological Survey.)

of that age. These distinctive rocks are the maroon, red, and green Sespe Formation, which contains a mixture of sandstone, siltstone, and conglomerate. The sediments of the Sespe Formation and the Vasquez Formation to the east were deposited in ancient alluvial fans along the flanks of mountains that have long since disappeared from the landscape. Marine conditions returned to the region during Miocene time and persisted until the most recent emergence began during Pliocene time. Some areas, like the Ventura basin, emerged from the sea only a few hundred thousand years ago.

MIOCENE EXTENSION AND ROTATION

Late Mesozoic and early Cenozoic sedimentary rocks are absent in the Los Angeles basin east of the Newport-Inglewood fault and in the offshore Santa Monica, San Pedro, and Catalina Basins (Fig. 16-26). In those areas, younger Miocene formations directly overlie the Catalina schist. The Catalina schist, which is exposed on Catalina Island and in the Palos Verdes Hills, is a fine-grained metamorphic rock of Mesozoic age. Like the Franciscan complex (see Chapter 12), it contains blueschist minerals indicative of metamorphism during subduction, probably at depths of 20 to

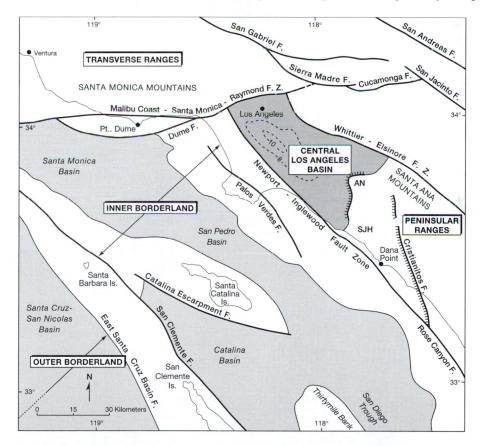

Fig. 16-26 Major basins formed by Miocene extension in the Inner Borderland and Los Angeles basin. (Source: Crouch, J., and Suppe, J. 1993. *Geological Society of America Bulletin*, Vol. 105, p. 1416.)

Fig. 16-27 A schematic cross section showing the tectonic setting of the volcanoes in the Santa Monica Mountains during Miocene time. Crustal extension created steep normal faults and low-angle detachment faults, exposing the Mesozoic Catalina schist in highly extended areas. See Chapter 7 for further discussion of detachment faults. (Source: Adapted from Crouch, J., and Suppe, J. 1993. *Geological Society of America Bulletin*, Vol. 105, p. 1422.)

45 kilometers. Also like the Franciscan, its parent rocks were marine sedimentary and volcanic rocks. In fact, geologists believe that the Catalina schist is actually Franciscan basement rock that was shoved, or ***underplated***, beneath the Transverse Ranges during Mesozoic subduction and accretion.

Geologists believe that the deposition of Miocene rocks directly onto Catalina schist is evidence that sedimentary basins did not exist in those areas until Miocene time. The oldest sedimentary rocks deposited on the schist date the opening of the basins, and these are Miocene sedimentary rocks, commonly the San Onofre breccia, which is discussed in Chapter 17. This implies that the central Los Angeles basin, the Inner Borderland basins, and the Santa Maria basin north of the Transverse Ranges were created by crustal extension during Miocene time (Fig. 16-27). Extension created both steep normal faults and low-angle detachment faults. Rocks from the lower crust—the Catalina schist—were exposed in highly extended areas.

During Miocene time, today's Santa Monica Mountains were also the site of numerous active volcanoes, as indicated by thick sequences of submarine Miocene volcanic rocks (Figs. 16-27 and 16-28). The active volcanoes extended to the northern

A

B

Fig. 16-28 **A**, Outcrop, and **B**, Closeup of Miocene pillow basalts in the Santa Monica Mountains. (Source: Wentworth, C. U.S. Geological Survey.)

Channel Islands, as discussed later. The volcanic activity was another consequence of Miocene activity (see Fig. 16-27).

One of the more dramatic conclusions geologists have made about southern California geology in recent years is that the western Transverse Ranges Province has been rotated more than 90 degrees in a clockwise direction during Miocene time. The conclusion is based on several different lines of geologic evidence. First, rock types, sedimentary features, and boundaries of early Cenozoic rocks can be matched up with those in the Peninsular Ranges. The second line of evidence that supports the rotation model comes from studies of the Earth's magnetic field as recorded in these Transverse Ranges rocks. The direction of magnetic north appears in the proper position only if these older rocks are rotated back to their restored positions.

Recent work by geologists has provided a tectonic explanation that links the anomalous aspects of southern California's Cenozoic geology. Clockwise rotation of the western Transverse Ranges is thought to have caused the Miocene extension and opening of the basins. The newly formed San Andreas transform is the likely cause of the rotation. Geologists hypothesize that a fragment of North America was partially captured by the northwest-moving Pacific Plate. Clockwise rotation resulted because the eastern edge of this plate fragment remained attached to North America as the western edge swung to the northwest (see Chapter 18 for figures and further discussion).

BASEMENT ROCKS

A look at the California geologic map (Endpaper 1) reveals that many of southern California's mountain ranges contain large areas of granitic and metamorphic rocks, brought to the surface by recent uplift and faulting. Of course, older basement rocks also underlie the Cenozoic basins of the Transverse Ranges Province. Geologists have identified four distinct basement blocks in the mountains of southern California: the Peninsular Ranges, San Gabriel Mountains lower plate, San Gabriel Mountains upper plate, and San Bernardino Mountains (Fig. 16-29).

The basement rocks of the Peninsular Ranges block in the vicinity of Riverside (see Fig. 16-29) are Mesozoic granitic rocks and associated metamorphic rocks similar to the rocks of the Sierra Nevada batholith. As discussed earlier, the Catalina schist, the equivalent of the Franciscan complex, is the basement rock for the Los Angeles basin west of the Newport-Inglewood fault zone and the Inner Borderland. The boundary between the oceanic basement represented by the Catalina schist and continental basement represented by the Peninsular Ranges block is not completely defined at this time.

San Gabriel Mountains

Two very different types and ages of basement rocks are found in the San Gabriel mountains. The Pelona schist and very similar rocks called by different names—including the Orocopia schist and the Rand schist—are distinctive metamorphic rocks found in several ranges of southern California (Fig. 16-30B). Radiometric dating indicates that most of the sedimentary parents of the schists were deposited in late Cretaceous time, about 70 million years ago. Like the Catalina schist and the Franciscan

Fig. 16-29   The major basement rocks of southern California. (Source: Southern California Areal Mapping Project, U.S. Geological Survey: *http://scamp.wr.usgs.gov*.)

complex, they were marine sedimentary rocks subjected to high-pressure conditions during subduction. The schists were thrust under North America during Mesozoic subduction and accretion, and some geologists believe that they underlie a large portion of southwestern North America. Because they are unusual, distinctive rocks (Fig. 16-31), the schists also provide an important record of long-term offset along the San Andreas fault system (see Fig. 16-30).

Some of California's oldest known rocks can be found in the San Gabriel Mountains, where they are part of the upper plate basement rocks. The ancient rocks include the Mendenhall gneiss and related rocks that are up to 1.6 billion years old. A younger body of plutonic rock, dating to about 1.3 billion years ago, underlies about 250 square kilometers of the western San Gabriel Mountains and can be seen along the Angeles Forest Highway. The pluton contains anorthosite, syenite, and gabbro, and the rocks display examples of cumulate textures that result from crystal accumulation at the bottom of a magma chamber, similar to textures of rocks of the Klamath Mountains described in Chapter 9. The upper plate San Gabriel basement block also includes Mesozoic granitic rocks that intrude the ancient rocks and are similar to the rocks of the Sierra Nevada batholith.

Both types of San Gabriel basement rocks are involved in the Vincent Thrust, a late Mesozoic to early Cenozoic low-angle fault in the San Gabriel Mountains (see Fig.16-29). The thrust fault placed the pre-Cambrian basement rocks on top of much younger Mesozoic basement rocks (Fig. 16-32). The placement of the two basement types by the thrust faulting is the reason for their names: The younger, but structurally

Fig. 16-30 The Pelona and related schists **A**, before right-lateral offset along the San Andreas fault, and **B**, in their present areas of exposure. (Source: Adapted from Jacobson, C., and others. 2000. *Geology,* Vol. 28, p. 220.)

Fig. 16-31 A polished sample of Pelona schist from the San Bernardino Mountains. (Source: Morton, D. U.S. Geological Survey.)

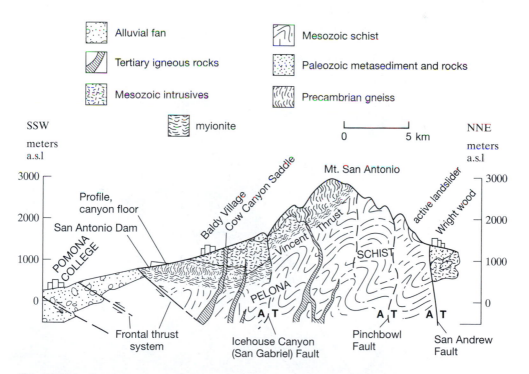

Fig. 16-32 Schematic cross section across the San Gabriel Mountains between Pomona College and Wrightwood. Note that the Vincent Thrust fault places ancient Precambrian metamorphic rocks over Mesozoic Pelona schist. (Source: Hazlett, R. Pomona College.)

lower Pelona schist block is the lower plate, and the older, but structurally higher pre-Cambrian block is the upper plate. This complexity caused by faulting violates the "normal" order of things, as outlined by the principles of superposition in Chapter 3.

San Bernardino Mountains

The most widespread older rocks of the San Bernardino Mountains are similar to those of the Mojave Desert and Basin and Ranges Provinces (see Chapter 7). They include metasedimentary rocks that originally formed on the passive margin of western North American between about 900 and 300 million years ago. In the San Bernardino Mountains, the original sedimentary rocks were metamorphosed to marble, schist, gneiss, and quartzite during later deep burial, faulting, and folding. Granitic rocks similar to those of the Sierran batholith intruded into the older rocks during Mesozoic time, as occurred in the provinces of southeastern California. The San Bernardino Mountains basement block also contains ancient igneous and metamorphic rocks similar to those in the San Gabriel Mountains. These rocks, up to 1.7 billion years old and similar to those found in the interior of the North American continent, are also present in the San Bernardino Mountains (see Chapter 18).

OFFSHORE ISLANDS

Southern California's offshore islands and their surrounding waters are an important ecological and recreational refuge where visitors can study geology and coastal processes, as well as plants, animals, and habitats (Figs. 16-33 and 16-34).

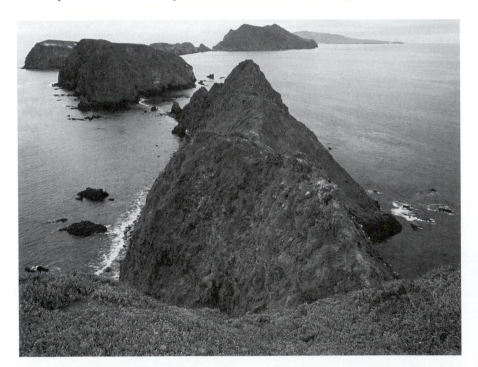

Fig. 16-33 Anacapa Island is divided into three islets, East, Middle, and West Islands. Santa Cruz Island is in the distance, Channel Islands National Park, California. (Source: Corbis.)

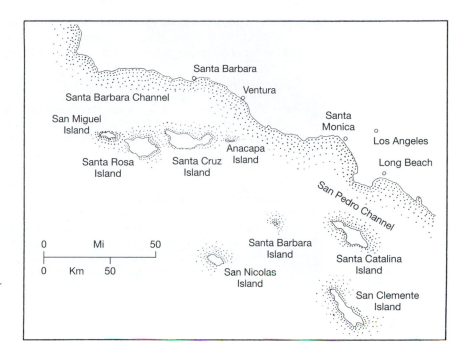

Fig. 16-34 Map of southern California's offshore islands and the Santa Barbara channel. (Source: Moon Publications, Chico, CA.)

Juan Rodriquez Cabrillo, a Portuguese explorer who traveled up the California coast in 1542 and 1543, is thought to be buried on San Miguel Island. Spanish explorer Sebastian Vizcaino also visited the islands in 1602. He named Santa Barbara, San Clemente, and Santa Catalina Islands. Both parties found well-established groups of Native Americans living on the islands and making frequent trading trips to the mainland. About 17 kilometers north of Avalon, on Santa Catalina Island, native Californians quarried steatite, an impure variety of talc. They used the steatite to make distinctive bowls, which they traded on the mainland.

The Channel Islands were isolated from each other about 12,000 to 10,000 years ago when sea level rose as the world's climate warmed and ice sheets melted (see Chapter 12). The rising sea level isolated the plant and animal communities of the islands. Many species of plants and animals are found only on the offshore islands, and some are found only on a single island (see the box on exiled elephants of the Channel Islands in Chapter 3). Prior to the sea level rise, the northern islands were connected, forming a single large island named Santarosae by geologists.

Marine terraces are prominent on the offshore islands, particularly on San Clemente, which might have the most complete sequence of marine terraces in California (Fig. 16-35). Twenty-five terrace levels are preserved on San Clemente Island. The highest terrace, about 580 meters above sea level, is about 2.8 million years old. As discussed in Chapter 15, the terraces attest to the tectonic uplift of the islands. On Santa Catalina Island, terraces are not nearly as prominent as on the other islands, but narrow terrace benches do mark former shorelines.

The four northern Channel Islands are the western extension of the Santa Monica Mountains. Many of the rocks, folds, and faults of the Santa Monica Mountains can be matched up with those seen on the islands. For example, the Santa Monica fault

Fig. 16-35 At least eight marine terraces can be seen on this aerial photograph of San Clemente Island near Eel Point. The terrace edges make wavy lines parallel to the shoreline. For scale, the long runway is approximately 1.6 kilometers long. (Source: NOAA, 1971.)

runs along the south side of the Santa Monica Mountains, continuing west along the south side of the northern Channel Islands (see Endpaper 2). The Catalina schist, a distinctive rock discussed earlier, is found on and named for Catalina Island, but it is not found on the other offshore islands.

The northern Channel Islands are composed mainly of volcanic rocks. Pillow basalt, indicating submarine lava flows, is found on Santa Barbara Island. On Anacapa Island, andesitic breccia and tuff are present, in addition to basalt. Volcanic rocks are also present on San Clemente, Santa Catalina, and Santa Cruz islands. The volcanic rocks on the islands are of Miocene age, similar to thick sequences of volcanic rocks found in the Santa Monica Mountains. Most are characteristic of rocks produced by submarine eruptions.

CALIFORNIA OIL

On August 17, 1769, Gaspar de Portola's expedition encountered a group of Chumash Indians who were caulking their sea canoe with tar, or **brea**, collected from a nearby pool. Portola's men named the site Carpinteria (carpenter's shop). Carpinteria, located along the Ventura County coast, is one of several places in the western Transverse Ranges where tar oozes from the ground in natural seeps (Fig. 16-36). Much later, in 1892, prospector Edward Doheny would look out of the window of his Los Angeles boarding house and notice a wagon loaded with chunks of brea. Southern Californians used it to waterproof their roofs and fence posts, and later to lubricate machinery. On a brilliant hunch, Doheny figured that the brea oozing from the ground might be a signal of oil riches beneath. He and partner Charles Canfield leased a lot east of Los Angeles, dug a 155-foot-deep shaft by hand, deepened it to 460 feet with a sharpened eucalyptus drill, and struck oil. Thus began the big oil boom in the heart of Los Angeles.

Fig. 16-36 Natural tar seeps along Highway 150 near Ojai, Santa Barbara County. (Source: Photo by author.)

The world's first oil well had been drilled in Pennsylvania in 1859, starting the systematic search for petroleum. In the early 1860s, prospectors drove tunnels into the side of Sulphur Mountain near Ojai to extract oil from sand there. In 1866, the first commercial oil well in southern California was completed in Ventura County. In the 1870s, wells were drilled near Newhall in Los Angeles County, one of them yielding 30 barrels per day from a depth of 300 feet. This well, designated Pico Number 4, was later deepened, and its production increased to 150 barrels per day. As a result, California's first oil pipeline was built. The pipeline carried oil from Pico Canyon 5 miles to a refinery near Newhall. Offshore wells were drilled from piers at Summerland east of Santa Barbara as early as 1896.

> The Doheny discovery inaugurated an oil boom in the heart of downtown Los Angeles. During the next five years derricks "as thick as the holes in a pepper box" sprouted along a narrow two-mile residential strip. Gardens, palm trees, and even homes disappeared as steam-powered rigs and primitive hand-driven drills probed the earth. Most found oil at shallow depths and produced a paltry ten daily barrels. But by the turn of the century, thousands of wells in the Los Angeles field yielded almost two million barrels a year.*

By 1900, more than 1,000 oil companies were drilling in California, producing about 12,000 barrels per day. The first major oil boom in southern California began when geologists realized the connection between anticlines and oil fields, and by 1930, most of California's major oil fields associated with surface faults or folds had been discovered. During the 1920s, California's oil production reached a peak of 850,000 barrels per day, much of it from the Long Beach and Huntington Beach fields. Wells pumped oil from beneath the streets of Los Angeles, Wilmington, Huntington Beach, and other cities (Fig. 16-37). In 1914, a publishing company offered a 25- by 100-foot plot of land in Huntington Beach to buyers of its encyclopedia. When the Huntington Beach oil field was discovered under the "encyclopedia lots," those who had kept up on their property taxes got sizable monthly royalty checks.

---

*Source: Tygiel, J. 1994. *The Great Los Angeles Swindle.* New York: Oxford University Press, p. 21.

Fig. 16-37 Early oil fields in the Signal Hills near Los Angeles. (Source: Shell Oil Company.)

The early oil boom brought a wave of fortune-seekers to southern California, and with them came scandal, swindle, and speculation. The oil boom overlapped with the land and water wars (see Chapter 10) and the early movie-making boom, making Los Angeles a colorful place during the early 20th century.

TODAY'S OIL PRODUCTION

In 1994, California ranked fourth among U.S. oil-producing states, after Texas, Alaska, and Louisiana. A total of 344.5 million barrels of oil were produced from 214 oil fields, half of them in Kern County (Fig. 16-38). These numbers seem large compared to the

Fig. 16-38 Pumping oil wells in Taft, Kern County. (Source: Photo by author.)

Fig. 16-39 Offshore oil platform in the Santa Barbara Channel. (Source: U.S. Geological Survey.)

TABLE 16-1   SELECTED CALIFORNIA OIL FIELDS

Field	Year Discovered	Cumulative Production (Millions of Barrels)	Number of Producing Wells in 1994
Wilmington	1932	2,436,452	1,456
Midway-Sunset	1894	2,238,723	10,330
Kern River	1899	1,474,280	7,381
Huntington Beach	1920	1,089,370	443
Elk Hills	1911	1,054,609	1,037
South Belridge	1911	983,345	4,519
Ventura	1919	930,252	419
Long Beach	1921	921,009	297
Coalinga	1890	820,223	1,922
Hondo[1]	1969	139,368	28
Point Arguello[1]	1981	76,320	34

[1]Offshore field

oil production during California's early oil years, but even this amount of oil is very small compared with the United States' annual use of about 7 billion barrels. During the 1990s, new wells drilled offshore along the southern California coast accounted for an increasing amount of California's production (Table 16-1 and Fig. 16-39).

FORMATION OF PETROLEUM POOLS

California's geologic history during Cenozoic time has supplied the ideal conditions for the formation of oil and natural gas resources. The basins of the Transverse Ranges, including offshore basins and the Los Angeles basin, have the exact combination of rock types and geologic structures necessary for the development of petroleum fields, as outlined in the following sections.

Source Rocks

Petroleum, which includes oil and natural gas, is a combination of **hydrocarbons**, chemical compounds of hydrogen and carbon. Studies of the hydrocarbon compounds found in petroleum have shown that they are derived from organic, or living, matter. Many people share the misconception that this living matter is the skeletons of dinosaurs, but the prime source of petroleum is marine organisms, particularly plankton like the diatoms found in the Monterey Formation. The Monterey Formation has been considered a prime source for California's hydrocarbons. The siliceous sedimentary rocks in the Monterey Formation contain between 1 and 5 percent organic carbon, and the phosphate-rich rocks in the Monterey contain 5 to 20 percent organic carbon. Other Cenozoic marine sedimentary formations containing organic material are also suitable sources for hydrocarbons.

Heating

The next necessary step in forming a petroleum pool is to heat the source rocks sufficiently to form hydrocarbon compounds. Heating occurs as sediments are buried by younger deposits. At depths of about 500 meters, heat and pressure are great enough to transform organic matter into hydrocarbon compounds. The ideal temperatures for hydrocarbon formation are between 50° and 80°C. If the rocks are heated to higher temperatures, which would result from deep burial, igneous activity, or subduction, the hydrocarbons are converted to other organic compounds or lost from the rocks. Rocks that have been heated sufficiently to convert organic matter to hydrocarbon compounds are known as oil shales.

Migration to Reservoir Rocks

For petroleum to concentrate in a pool, the hydrocarbon compounds must next migrate from the source rock into rocks that can become saturated with petroleum. The rocks or sediments capable of carrying and holding large amounts of petroleum are the same materials that make good aquifers. That is, reservoir rocks must be permeable. Porous rocks like sandstone and highly fractured rocks make suitable reservoir rocks. A petroleum-saturated reservoir rock is referred to as a pool, but it is not an underground lake of oil. Like an aquifer, a petroleum reservoir is a rock or sediment that has its pore space saturated with oil, natural gas, or both. Many of southern California's reservoir rocks are late Cenozoic sandstone formations, but fractured Monterey shale can also be a reservoir for significant petroleum.

Traps

To accumulate as a pool or reservoir of petroleum, the hydrocarbon compounds must be trapped in the reservoir. Oil geologists estimate that more than 90 percent of all the petroleum formed escapes to the Earth's surface, where it oxidizes and escapes into the atmosphere or water. Geological traps can be formed in a variety of ways. Faults and anticlines are two common types of **structural traps** found in California. If a reservoir rock, such as a permeable sandstone, is arched into an anticline *and* sandwiched between less permeable rocks, such as shale, petroleum is trapped at the crest of the fold. Because oil is lighter than water, and natural gas is lighter than oil, a layered pool often forms (Fig. 16-40). Faults also form effective structural traps. The traps in many of California's oil fields are formed by a combination of reverse faults and folds (Fig. 16-41). Petroleum traps can also form where permeable reservoir

A simple fold trap

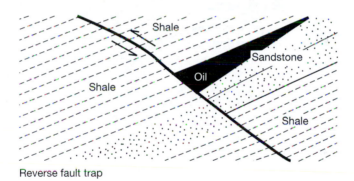

Reverse fault trap

Fig. 16-40 Diagrams of petroleum traps. (Source: Craig, J. and others. *Resources of the Earth.* Upper Saddle River, NJ: Prentice Hall.)

Anticline over blind reverse fault

Fig. 16-41 Cross section of the central Ventura basin, showing the oil fields associated with the Ventura anticline on the south and San Cayetano fault on the north side. See Fig. 16-16 for location. Note that the scale is in feet. (Source: Bailey, E., and Jahns, R. 1954. *Geology of Southern California.* California Division of Mines and Geology, Bulletin 170.)

rocks form lenses within less permeable rocks; known as ***stratigraphic traps***, these are also common in southern California.

Many of southern California's blind reverse and thrust faults are expressed as anticlines at the surface (see Chapter 13). Because petroleum is trapped in many of these structures, oil wells commonly tap into the anticlines above the blind faults. Although petroleum geologists long recognized that the folds are the surface expression of buried faults, they did not recognize the earthquake potential of those faults until the 1987 Whittier Narrows earthquake.

ENVIRONMENTAL CONCERNS

Because tar naturally seeps from the ground at several sites beneath the Santa Barbara Channel and along the southern California coast, oil on southern California's beaches can be a natural phenomenon. The 1969 Santa Barbara oil spill, however, was the result of a blowout at an offshore oil well in the Santa Barbara Channel. The well had not been properly capped, and the pressurized oil and gas literally exploded into the sea. The accident blackened Santa Barbara's beaches and wildlife with crude oil, provoking public outcries for greater environmental protection. Since then, the oil industry has had no major accidents along the coast or on the offshore platforms.

Ground subsidence following removal of oil is a major concern, particularly in coastal areas. In the period between 1933 and 1953, the ground above the Wilmington oil field sank as much as 6 meters. Parts of central Long Beach subsided 25 centimeters.

Groundwater contamination associated with oil-field operations is also a concern in some parts of the region.

During the 1994 Northridge earthquake, 19 oil fields in northern Los Angeles and eastern Ventura County sustained some damage. At the Aliso Canyon oil field, closest to the epicenter, a string of well casing 1,400 feet below the surface was damaged. Further south, a well in a residential area of Hermosa Beach, which had already been leaking small amounts of oil at the surface, began to leak significant amounts of oil. Damage at the other sites was confined to pipelines, tanks, roads, and power lines. For a short time after the earthquake, greater flow rates were noted at a number of sites in southern California.

It is geology and not coincidence that has brought beaches and oil together in southern California. Both onshore and along the continental borderland, the region is dotted with producing oil fields, pipelines, refineries, storage facilities, and tanker ports. However, the same areas are southern California's most valuable recreation areas and wildlife habitat. Protection of those resources makes environmentally sound management of petroleum operations an important priority. In addition, the likelihood of future earthquakes in the region poses additional risks from petroleum-related accidents.

SUMMARY

1. The oldest rocks in the region are gneiss, schist, and plutonic rocks found in today's San Gabriel and San Bernardino Mountains. The oldest plutonic rocks are about 1.7 billion years old and both they and older metamorphic rocks that they intrude are thought to be related to North America's ancient core.

2. Metasedimentary rocks in the San Bernardino Mountains were initially formed by deposition along the passive continental margin of western North America during late Proterozoic and all of Paleozoic time. The passive margin also extended through today's Basin and Range, Mojave Desert, and eastern Sierra Nevada and Peninsular Ranges (see Chapter 7). In the San Bernardino Mountains, Mesozoic tectonic events metamorphosed the sedimentary rocks to marble, quartzite, and schist that are exposed today.

3. During Mesozoic time, the western edge of the North American Plate, including the area that is today's southern California became a subduction zone. Along a magmatic arc that extended along California from the Sierra Nevada (see Chapter 8) to today's Baja Peninsula, Mesozoic plutons intruded the older rocks of the Transverse Ranges. During the same period, marine sediments accumulated in a forearc basin (see Chapter 11). Today these are represented by late Mesozoic and early Cenozoic sedimentary rocks, similar to the Great Valley Sequence, that are found in the northern Transverse Ranges and in the Santa Ana Mountains.

4. The Catalina schist, Pelona schist, and related Mesozoic formations that were metamorphosed during subduction form the basement rocks in the western part of the area. These rocks were underplated beneath North America during accretion and subduction and are equivalent to parts of the Franciscan complex in the Coast Ranges.

5. The Pacific and North American Plates came into direct contact about 27 million years ago, creating the San Andreas transform boundary (see Chapter 12). During the early history of the San Andreas system, in Miocene time, the western Transverse Ranges were partly captured by the northwest-moving Pacific Plate. Because the eastern part of the block remained fixed to North America, the western Transverse Ranges rotated more than 90 degrees.

6. The Miocene rotation caused extension and created new sedimentary basins in today's central Los Angeles basin, Inner Borderland, and Santa Maria basin. Deposition of the San Onofre breccia, which was derived from mountains containing the Catalina schist (see Chapter 17), signaled the creation of these basins. Miocene volcanic activity in the area of today's Santa Monica Mountains and Channel Islands was also a result of crustal extension.

7. The Monterey Formation was deposited in Miocene time throughout the province, when fine-grained shale and siliceous sediment accumulated in deep marine, fault-bordered basins. The Monterey Formation was the source rock for much of California's petroleum, and today diatomite is mined from many Monterey Formation outcrops.

8. Because the plate boundary shifted inland and created the Big Bend in the San Andreas fault, the region has undergone compression during the past 4 million years. Numerous active thrust faults and folds result from the compression. In addition, fault-bordered basins have been formed by down-warping. Some of these basins, including the Los Angeles basin, have been rapidly filled by sediment shed from actively uplifting mountains. Other offshore basins remain below sea level. Actively uplifting mountains and marine terraces (Chapter 15) have also resulted from compression.

9. The present San Andreas system in the Transverse Ranges and Los Angeles basin includes a number of active faults with right-lateral offset: the San Andreas, San Gorgonio Pass, Mission Creek, Newport-Inglewood, Banning, San Jacinto, and Elsinore faults (see Chapter 13 and Chapter 14).

10. Petroleum is an important economic resource associated with the late Cenozoic sedimentary rocks of the region's sedimentary basins. The petroleum pools are associated with reverse fault and fold traps. Petroleum is currently extracted from offshore and onland pools, and the region's oil extraction has been a colorful part of its history.

REFERENCES

CROUCH, J.K., AND SUPPE, J. 1993. Late Cenozoic tectonic evolution of the Los Angeles basin and inner California borderland: A model for core complex-like crustal extension. *Geological Society of America Bulletin* 105:1415–1434.

DAVIDSON, K. 1994. Learning from Los Angeles. *Earth Magazine*, September, pp. 40–48.

FRITSCHE, E., COLBURN, I., RAMIREZ, P., AND WEIGAND, P. 2001. Transverse/Peninsular Ranges connections—Evidence for the incredible Miocene rotation. In *Geologic Excursions in Southwestern California*. Santa Fe Springs, CA: Pacific Section of the Society for Sedimentary Geology, pp. 101–146.

GORE, R. 1995. Living with California's faults. *National Geographic*, April, pp. 2–35.

JAHNS, R., ed. 1954. *Geology of Southern California*. California Division of Mines and Geology, Bulletin 170.

MORTON, D.M., AND YERKES, R.F., eds. 1987. *Recent Reverse Faulting in the Transverse Ranges*. U.S. Geological Survey, Professional Paper 1339, 203 pp.

SHARP, R.P., AND GLAZNER, A.F. 1993. *Geology Underfoot in Southern California*. Missoula, MT: Mountain Press, 224 pp.

STEIN, R.S., AND YEATS, R.S. 1989. Hidden earthquakes. *Scientific American*, June, pp. 48–57.

TYGIEL, J. 1994. *The Great Los Angeles Swindle: Oil, Stocks, and Scandal During the Roaring Twenties*. New York: Oxford University Press, 398 pp.

Web Sites and Media Resources

Atwater, Tanya, *Southern California: 20 MA to present,* video animation and related text: *http://www.geol.ucsb.edu/faculty/atwater*

Channel Islands National Park: *http://www.nps.gov/chis*

Local Geology, Pomona College: *http://www.geology.pomona.edu/local_geology.html*

Probing the Los Angeles Basin—Insights Into Ground-Water Resources and Earthquake Hazards, U.S. Geological Survey, Fact Sheet 86-02: *http://geopubs.wr.usgs.gov/factsheet/fs086-02/*

Southern California Areal Mapping Program, U.S. Geological Survey: *http://scamp.wr.usgs.gov/scamp/html/scgeo.html*

CHAPTER

17

The Peninsular Ranges

Fig. 17-1 Location of the Peninsular Ranges Province in California.

Los Angeles Basin

Peninsular Ranges Highlights

Major geographic feature: Peninsular Ranges Mountains
Named mountain ranges (north to south): Santa Ana, San Jacinto, Santa Rosa, Agua Tibia, Palomar, San Diego (Laguna)
Other major landscape elements: Coronado Spit, Mission Bay, San Diego Bay, San Diego River, Tijuana River, Anza-Borrego Desert

Selected high points:	Santiago Peak, Santa Ana Mts.: 1,733 m (5,687 ft.) San Jacinto Peak, San Jacinto Mts.: 3,296 m (10,804 ft.) Mt. Palomar: 1,871 m (6,140 ft.) Cuyamaca Peak, San Diego Mts.: 1,985 m (6,512 ft.)
Predominant rock types:	Mesozoic granitic and mafic plutonic rocks (Peninsular Ranges batholith); Mesozoic volcanic rocks; Mesozoic marine sedimentary rocks; late Cenozoic sediments
Plate tectonic events:	Mesozoic subduction, accretion, arc magmatism, and forearc sedimentation; Cenozoic transform faulting
Famous faults (west to east):	Newport-Inglewood/Santa Rosa, Elsinore, San Jacinto, Banning, San Andreas, Imperial (see Chapter 13 and Chapter 14)
Resources:	Gemstones, water, sand, and gravel
National Parks:	Cabrillo National Monument
State Beaches (north to south):	San Onofre, Cardiff, Leucadia, San Elijo, Carlsbad, Torrey Pines, Silver Strand, Tijuana River National Estuarine Research Reserve, Border Field
Other State Parks of interest:	Cuyamaca Ranch, Palomar Mountain

SETTING

The Peninsular Ranges Province is dominated by a series of northwest-oriented mountain ranges extending from the Baja California peninsula north to the Transverse Ranges (Fig. 17-1). The mountain ranges are higher in the northern part of the province than near the Mexican border. Like the Sierra Nevada, the Peninsular Ranges are steeper on the eastern side than on the west. The eastern San Jacinto Mountains drop as much as 3,000 meters into the Coachella Valley near Palm Springs. Only the northern 240 kilometers of the Peninsular Ranges lie in California; the ranges extend south of the border for about 1,200 kilometers, forming the long crest of the Baja Peninsula.

Because it comes at the end of this tour of California's geologic highlights, this discussion of the Peninsular Ranges might seem brief to the reader who turns here first. Many aspects of Peninsular Ranges geology have been discussed in earlier chapters. The Los Angeles basin and the southern offshore islands are often considered part of the Peninsular Ranges Province, although they were discussed in Chapter 16. The Colorado Desert and the Salton Trough, which form the eastern edge of the province, are discussed in Chapter 6. The major active faults of the Peninsular Ranges, including the offshore Santa Rosa fault, and the Elsinore and San Jacinto fault zones, are parts of the San Andreas system discussed in Chapter 13 and Chapter 14. Water issues, including the formation of the Salton Sea and San Diego's recent water disputes, are covered in Chapter 10, and coastal processes are covered in Chapter 15.

THE PENINSULAR RANGES BATHOLITH

The similarity between the Peninsular Ranges and the Sierra Nevada extends beneath the landscape. A look at the geologic map (Endpaper 1) reveals that Mesozoic plutonic

Fig. 17-2 Map showing the extent of the Peninsular Ranges batholith in southern California and Baja California, Mexico. (Source: Jahns, R. 1954. *Geology of Southern California.* California Division of Mines and Geology, Bulletin 140, p. 30.)

rocks make up much of the Peninsular Ranges Province. Both the Peninsular Ranges batholith (Fig. 17-2) and the Sierra Nevada batholith (see Chapter 8) are part of a great chain of Mesozoic granitic rock that extends along much of western North America, from Alaska to the southern tip of Baja California, Mexico. Like the Sierran batholith, the Peninsular Ranges batholith represents the roots of a magmatic arc formed as a result of active subduction along western North America. The age of the Peninsular Ranges batholith is also similar to the Sierra Nevada. Plutonic rocks in the batholith crystallized between 140 and 80 million years ago during Cretaceous time, with the largest volume of magma intruded about 100 million years ago.

Geologists have found systematic variations in the types of plutonic rocks within the Peninsular Ranges batholith. On the western side of the batholith, more mafic plutonic rocks, including gabbro, are found along with granitic rocks. Because of its composition, the dark-colored gabbro is sometimes (incorrectly) called black granite, which confuses those who have learned that granite is light-colored because mafic minerals are rare (see Table 2-1 and Chapter 8). In San Diego County, gabbro is quarried for use as building stone. Geologists believe that the basement rocks beneath the western batholith are oceanic lithosphere. In contrast, plutonic rocks in the eastern Peninsular Ranges batholith are more silica-rich granodiorite and *tonalite* (Fig. 17-3; also see Figure 8-12). The relative abundance of rare earth elements sharply differs in the two areas, indicating that the magmas formed from different source rocks. The chemical composition of the rocks suggests that the basement beneath the eastern batholith is continental lithosphere.

In general, the plutons in the eastern batholith are younger than those to the west. The rocks of the western batholith are between about 140 and 105 million years old, whereas those from the eastern belt are dated between about 99 and 92 million years old. The plutons of the eastern Peninsular Ranges batholith, which are

Fig. 17-3 Igneous rocks of the eastern Peninsular Ranges batholith (lighter toned) and associated high-grade metamorphic rocks (darker foliated rocks), Mission Santa Ysabel, Highway 79. (Source: Metzler, C. Mira Costa Community College.)

Fig. 17-4 The eastern and western parts of the Peninsular Ranges batholith, the Julian schist, and the Santiago Peak Volcanics. (Source: Adapted from Todd, V. U.S. Geological Survey.)

referred to as the La Posta suite, are exposed over a very large area and represent a huge volume of magma emplacement during a geologically short time period.

The boundary between the eastern and western belts of plutonic rocks is a north to northwest-trending zone near the center of the batholith (Fig. 17-4). Geologists do not agree about how and when the eastern and western batholith were placed next to each other. However, they do agree that the western mafic belt represents the roots of oceanic volcanic arcs, as discussed in the section about the Santiago Peak volcanic rocks. One current theory suggests that the northern part of the western batholith developed on oceanic basement after the terrane had been accreted to the continent, whereas the southern part of the western batholith, which is now exposed south of the border, formed before that oceanic lithosphere was accreted. The juxtaposition of the two parts of the batholith occurred about 115 to 108 million years ago, before the plutons of the eastern batholith were emplaced.

Santiago Peak Volcanic Arc

The Santiago Peak Volcanics form a discontinuous belt on the western edge of the Peninsular Ranges batholith (see Fig. 17-4). They consist mostly of andesitic and silicic flows and pyroclastic material, as well as sedimentary rocks rich in volcanic fragments. They can be seen east of Del Mar, in the Santa Ana Mountains. Marine and plant fossils of late Jurassic age have been found in sedimentary rocks of the sequence, and radiometric ages for the volcanic rocks range from 130 to 120 million years ago. The likely source of the Santiago Peak Volcanics was an andesitic volcanic arc at or near the continental margin. Geologists interpret the Santiago Peak Volcanics as the volcanic arc that formed above the plutonic rocks of the western Peninsular Ranges batholith in California.

GEMS OF SAN DIEGO COUNTY

One interesting feature of the Peninsular Ranges batholith is a belt of gem-bearing *pegmatite* dikes found in the mountains northeast of San Diego. Pegmatites are coarse-grained, silica-rich plutonic rocks with crystals that grow to unusually large sizes because of the large amount of water vapor and boron present in the magmas. Because boron and water do not fit easily into the crystalline structure of earlier formed silicate minerals, the last Peninsular Ranges batholith magmas to crystallize were enriched in these substances. The dikes were intruded into fractures in older rocks, similar to the formation of quartz veins in the Sierra Nevada Province (see Chapter 8).

Most pegmatite dikes in San Diego County are a few meters thick, although some are as thick as 30 meters. Near the center of the dikes there are pocket zones containing crystals, some of which are unusually large and beautiful (Fig. 17-5). The

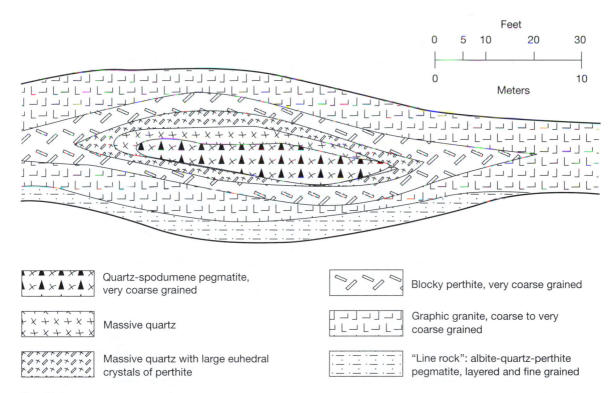

Feet

0   5   10   20   30

0   10
Meters

Quartz-spodumene pegmatite, very coarse grained

Massive quartz

Massive quartz with large euhedral crystals of perthite

Blocky perthite, very coarse grained

Graphic granite, coarse to very coarse grained

"Line rock": albite-quartz-perthite pegmatite, layered and fine grained

Fig. 17-5 The mineral zones in a typical vein of pegmatite. (Source: Jahns, R. 1977. Gem pegmatites in San Diego County. *California Geology* 30(2):44.)

pegmatite dikes of San Diego County occur in a 30-kilometer-long zone, mostly within the gabbro and tonalite plutons of the Peninsular Ranges batholith. The dikes crystallized in fractures in the older plutonic rocks, late in the formation of the batholith, about 90 to 100 million years ago. Many of the pegmatite dikes are well exposed on the hillsides. They stand out as light-colored ledges because of their resistance to erosion.

The most common minerals in pegmatite dikes are quartz, feldspar, and large, platy muscovite mica crystals. However, the uncommon chemistry of pegmatites gives rise to 100 different minerals, some rare and very beautiful. The most famous gems of the Pala area are the pink, green, and "watermelon" (pink and green) *tourmaline* crystals (Fig. 17-6). Beautiful specimens of many other gems have also been mined from the area, including black tourmaline, kunzite, beryl, topaz, and garnet.

Native Americans in the Pala area were the first to collect tourmaline crystals, which have been found in a number of burial sites. In the late 1800s, prospectors

Fig. 17-6 Watermelon tourmaline crystals from the Pala District, San Diego County. (Source: Van Pelt, H., and Van Pelt, E., for Pala International.)

began to excavate crystals, and between 1900 and 1912, mining flourished. Many of the gems were shipped to China and other overseas localities. Crystals have been mined from three areas in San Diego County—the Pala, Rincon, and Mesa Grande districts. At the Stewart mine, lepidolite was mined for its lithium content until the late 1920s, when the lepidolite zones were mined out. More recently, gem tourmaline, kunzite, and beryl, including its blue variety, aquamarine, have been mined from underground operations at the Stewart and Tourmaline mines. The gem mines are not open to the public, but examples of many of the crystals can be viewed in museums, including the California Division of Mines and Geology's mineral museum in Mariposa.

PREBATHOLITHIC ROCKS OF THE PENINSULAR RANGES

A complex assortment of rocks older than the batholith can be found in the Peninsular Ranges. The intrusion of the batholith, as well as later tectonic events, resulted in the rocks being sheared and metamorphosed. However, careful mapping and detailed studies of rock composition have led geologists to group these prebatholithic rocks into two general zones. Like the metamorphic belts in the Sierra Nevada province, the zones are elongate and trend generally northwest (Fig. 17-7).

Rocks of the Paleozoic Carbonate Platform

The oldest zone of prebatholithic rocks in the Peninsular Ranges can be matched with similar rocks seen throughout southeastern California. Rocks in this zone are the metamorphosed remnants of the stable carbonate platform that existed along western North America prior to Mesozoic time (see Chapter 7). In the Peninsular Ranges, these rocks are well exposed in the San Jacinto and Santa Rosa Mountains, west of Borrego Valley, and in the Coyote Mountains. Like the roof pendants in the Sierra Nevada, the original sedimentary rocks have been metamorphosed to marble and schist.

Triassic and Jurassic Marine Sedimentary Rocks

East of the Santiago Peak Volcanics, a zone of metasedimentary rock is preserved within the central part of the batholith. Geologists recognize the metamorphosed sandstone and shale in this zone as turbidite sequences originally deposited on submarine fans (see Chapter 12). The turbidites must have accumulated in one or more ocean basins, but their exact origins are unknown. The metasedimentary rocks include the Julian schist (see Fig. 17-3) and the Bedford Canyon Formation of the northern Santa Ana Mountains.

The Bedford Canyon Formation is a thick sequence of sandstone and mudstone turbidites of Jurassic age. It is exposed in the Santa Ana Mountains, including at Temecula Gorge in the Santa Margarita Ecological Reserve. The Julian schist is a metasedimentary formation of Triassic age, named for the town of Julian. It was metamorphosed during intrusion of the eastern Peninsular Ranges

Fig. 17-7 A generalized map showing belts of prebatholith metamorphic rocks and the Santiago Peak Volcanics in the Peninsular Ranges. On this map, plutonic rocks and younger rocks and sediments are not shown. (Source: Adapted from Jahns, R. Metamorphic and tectonic evolution of the northern Peninsular Ranges batholith. In Ernst, W., ed. *Metamorphism and Crustal Evolution of the Western United States.*)

Zone

1 ■ Santiago peak volcanics   Metavolcanic rocks, age unknown

2 Metamorphosed submarine fan deposits

Transitional between zones 2 and 3

3 Rocks of the carbonate platform

batholith during Cretaceous time. The Julian schist contains gold-bearing quartz veins, which triggered a small gold rush in the late 1800s and led to the founding of the town of Julian. The Julian schist can be seen at Volcan Mountain Preserve north of Julian and at Cuyamaca Rancho State Park.

CRETACEOUS SEDIMENTARY ROCKS OF THE SAN DIEGO AREA

Sedimentary rocks of Cretaceous age are exposed along the uplifted cliffs of the San Diego County coastline and in road cuts in the greater San Diego area. These rocks were deposited on the older plutonic rocks of the Peninsular Ranges batholith and on the Santiago Peak Volcanic sequence. Because the San Diego coastal area is relatively unaffected by faults of the San Andreas system, the Cretaceous rocks have been less

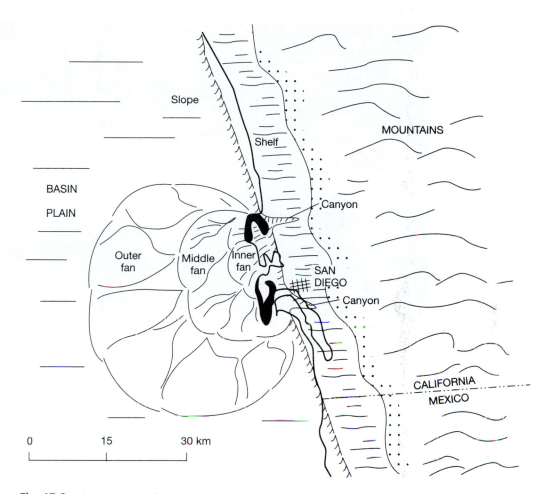

Fig. 17-8 The submarine fan that existed in the vicinity of present-day San Diego during Cretaceous time, when the Point Loma and Cabrillo formations were deposited. (Source: Nilsen, T. 1979. Geological Society of America, Field Trip Guidebook.)

disturbed by recent folding and faulting than rocks in coastal areas to the north. This has enabled geologists to reconstruct their sedimentary history from the outcrops exposed today along the coast.

The Point Loma and Cabrillo formations were deposited by turbidity currents, as indicated by the presence of graded bedding and other features typical of turbidites (see Chapter 12). Some units within the turbidites contain coarse-grained conglomerate deposited in channels near the continental margin (Fig. 17-8). Other units contain only fine-grained sediments, indicating deposition on parts of the submarine fan that were further from the edge of the continent (Fig. 17-9). Conglomerate- and sandstone-filled

Fig. 17-9 Sedimentary rocks of the Cretaceous Point Loma Formation, San Diego. The disrupted layer (above the keys) within this turbidite sequence shows numerous burrows. (Source: Nilsen, T.)

channels are well exposed along the cliffs at Tourmaline Surfing Park north of Mission Bay (Fig. 17-10). In general, the sequence is similar to the Great Valley Sequence and represents deposition in a forearc basin.

SAN ONOFRE BRECCIA

One of the most unusual rocks seen along the coast of the Peninsular Ranges is the San Onofre *breccia* (Fig. 17-11). A sedimentary breccia is a mixture of large, angular clasts in a matrix of finer sediments (see Table 2-3). The San Onofre breccia can be seen at Dana Point near the town of San Juan Capistrano, where it forms the resistant headland, and it is also exposed north of the point along Aliso Beach. It also is found further north in the Transverse Ranges province, including localities in the Santa Monica Mountains and the offshore islands to the west.

The San Onofre breccia is unusual because of the composition of the large clasts. They are black schist, greenstone, blueschist, and quartz—rocks unlike any of the older bedrock in the area. Applying the principles of the rock cycle, it is clear that a sedimentary breccia must have a source area that contains older rocks of the same type as those in the breccia. Because the fragments in the breccia are large and angular, that source area must have been fairly close and mountainous. The rock types are very similar to those found in the Franciscan complex in central and northern California (see Chapter 12), but a closer source is the Catalina schist found on Catalina Island (see Chapter 16). It is therefore likely that a mountainous land mass existed just west of the present southern California coastline. Geologists have

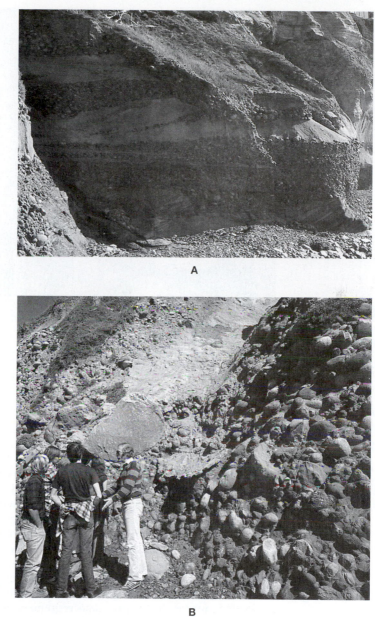

A

B

Fig. 17-10 **A**, Conglomerate of the Cretaceous Cabrillo Formation exposed along the beach cliffs north of Tourmaline Surfing Park, San Diego. **B**, Close-up of the Cabrillo Formation showing very large boulders in the conglomerate. (Source: Nilsen, T.)

speculated that an offshore fault created the mountains along a steep scarp, shedding the angular fragments that are seen today as the breccia. The process of forming the sediments would be similar to the formation of fault scarps and fans discussed in Chapter 6.

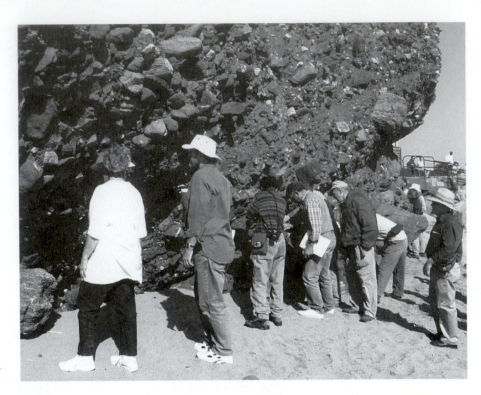

Fig. 17-11 The San Onofre breccia at Dana Point. (Source: Fritsche, A.E. California State University, Northridge.)

THE SAN DIEGO COASTLINE

On September 28, 1542, Spanish explorer Juan Rodríguez Cabrillo landed at what is now called San Diego Bay (Fig. 17-12), marking the first time that a European expedition had set foot on the west coast of what is now the United States. Cabrillo named the area San Miguel, but it was renamed San Diego by explorer Sebastian Vizcaino in 1602. Vizcaino also named the Coronado peninsula, or spit, after four islands (Las Yslas Coronadas) that he had sailed past before entering Coronado Bay. Today the city of San Diego, California's second largest after Los Angeles, occupies the areas around San Diego Bay and Mission Bay.

The bays and spits of San Diego create one of the world's best natural harbors, home to numerous naval facilities and other industries. San Diego's spectacular beaches and surrounding hills, combined with its climate, make it one of the major U.S. tourist destinations and retirement areas. As outlined in Chapter 15, longshore currents play a major role in sculpting the long, straight beaches north of San Diego. Marine terraces are also prominent along many coastlines north of San Diego (Fig. 17-13).

As is true in much of California, natural hazards create risks as well as beautiful landscapes in the San Diego area. Scientists are increasingly concerned about coastal erosion (see Chapter 15) and the tsunami threat posed by active offshore faults, particularly the Santa Rosa fault (Fig. 17-14; see also Chapter 13).

Fig. 17-12 The San Diego area from the Space Shuttle, October 1989. The rocky headland at La Jolla (LJ), Point Loma (PL), and Coronado Spit (CS) are labeled. San Diego Bay is the large embayment between Point Loma and Coronado Spit, and Mission Bay is the smaller embayment south of La Jolla. Note the wave refraction offshore and the contrast between the urbanized coastal plain and the rugged topography on the older rocks of the Peninsular Ranges. (Source: NASA.)

Fig. 17-13 Tilted gravels and overlying marine terrace, San Diego. (Source: Sarna-Wojcicki, A. U.S. Geological Survey.)

Fig. 17-14 Map of active faults in the San Diego area, including offshore faults that could pose a tsunami risk to coastal areas. (Source: Abbot, P. 1989. *Environment Southwest*. San Diego Natural History Museum, Report 524.)

ON THE BORDER

Every year, 50 million people cross the U.S. border to enter Tijuana, Mexico from San Ysidro at the world's busiest international crossing. Several kilometers east of San Ysidro, others make the crossing at Otay Mesa. The Tijuana River flows through the city of Tijuana, one of Mexico's fastest growing cities, on its journey to the Pacific Ocean north of the border. Today flood-control levees protect low-lying parts of the city from flooding, but floodplain areas on both sides of the border have been inundated in the past. In 1891, the Tijuana overflowed its banks, destroying homes and businesses on both sides of the border. People tied themselves to trees to avoid being swept away by the floodwaters.

The landscape between Tijuana and San Diego is a series of uplifted mesas cut by steep canyons. From the sparsely vegetated mesa tops, one has a broad view, but the canyons are concealed and chaparral covered. Many have come to the United States through these rugged canyons, often at great risk, to seek a better life.

The mesas along the border near San Diego are uplifted marine and river terraces similar to those discussed in Chapter 15. South of the border, travelers to Ensenada along Mexican Highway 1 have frequent opportunities to view uplifted terrace gravel and sand. Like the natural landscape features, many sedimentary formations are continuous across the border. As we have seen elsewhere in California, geologic features fail to recognize the significance of recently created, artificial human boundaries.

SUMMARY

1. Like other parts of southeastern California, the eastern Peninsular Ranges were the site of a marine carbonate platform during Paleozoic time, when the western edge of North America was a passive continental margin (see Chapter 7).
2. The western edge of North America became an active margin during Mesozoic time, when terrane accretion and magma emplacement were active (see Chapter 8). Deep marine environments were present in the Peninsular Ranges during early Mesozoic time, producing the Julian schist, Bedford Canyon Formation, and other marine sedimentary rocks deposited by turbidity currents.
3. During early Cretaceous time, a magmatic arc, which is represented today by the plutonic rocks of the western Peninsular Ranges batholith and the volcanic rocks of the Santiago Peak Volcanics, was intruded along the western edge of the North American Plate.
4. Later during Cretaceous time, between about 99 and 92 million years ago, the plutons of the eastern batholith were emplaced, after the terranes containing the western batholith had been accreted to the North American Plate.
5. Gem-bearing pegmatites, containing tourmaline and other gem-quality minerals, formed during the late stages of crystallization of the batholith.
6. During the same period that the magmatic arc was active, Cretaceous marine sediments were deposited in a forearc basin. The sediments were deposited by turbidity currents and represent both deep and shallow-water environments.
7. Marine sedimentation continued into Cenozoic time. Cenozoic sedimentary rocks include the Miocene San Onofre breccia, an unusual formation containing large clasts shed from a highland that lay west of the coast.
8. Late Cenozoic uplift of the coast has created marine terraces in the San Diego area.
9. A number of active faults of the San Andreas system run northwest–southeast through the Peninsular Ranges Province. These faults control the orientations of mountain ranges and valleys and pose a significant earthquake threat to the area. Offshore faults, particularly the Santa Rosa fault, could pose a significant tsunami threat to the San Diego area (see Chapter 13).
10. Uplift, longshore transport, postglacial sea level rise, and coastal erosion have produced the dynamic San Diego coastline.

REFERENCES

SHARP, R., AND GLAZNER, A. 1993. *Geology Underfoot in Southern California*. Missoula, MT: Mountain Press, 224 pp.

SMITH, J. 1997. Gem pegmatites in San Diego County. *California Geology* 30: 43-44.

WALAWENDER, M., AND HANAN, B., eds. 1991. *Geological Excursions in Southern California and Mexico: Geological Society of America Annual Meeting Guidebook*. San Diego, CA: Department of Geological Sciences, San Diego State University, 503 pp.

Web Sites and Media Resources

Abbot, P. "The Rise and Fall of San Diego," an episode of the television series *Written in Stone*. San Diego State University: *http://www-rohan.sdsu.edu/~wis/san_diego.htm*

Cabrillo National Monument, National Park Service: *http://www.nps.gov/cabr*

Salton Sea, State of California Web site: *http://www.saltonsea.ca.gov/*

San Diego Museum of Natural History: *http://www.sdnhm.org/*

California Geology: An Integrated View

The final two chapters of this book allow the reader to bring together the material presented in the previous chapters into a unified picture of California's geology. Chapter 18 summarizes important events in California's geologic history, presented in chronologic order, with emphasis on the role of plate tectonics and the important aspects of individual provinces. Finally, Chapter 19 discusses the role that geology plays in the lives of all Californians.

The Evolution of California Through Geologic Time

After looking at the highlights of California's provinces, we can see a unifying theme in California's geology: complexity caused by tectonics. The complexity has long been recognized by geologists attempting to decipher California's geologic history. Beginning with the gold rush, the search for mineral and petroleum resources spurred the preparation of accurate and detailed geologic maps in most of California. By the 1960s, the nature and approximate age of many of the rock formations, and the geometry of most of the faults and folds were fairly well established. Interpretations of California's geologic history, however, still contained a great many uncertainties.

During the past 30 years, the concept of plate tectonics has given geologists a framework for interpreting the complicated assortment of California rocks and faults. Advances in technology have enabled us to better identify and date geologic materials at the surface and to collect information about the Earth's interior. At the same time, societal concerns about seismic risks and other geologic hazards have led the public to provide the necessary resources to support geologic and geophysical investigations. Today, researchers are recognizing pieces of the tectonic puzzle with each new study. In turn, assembly of the pieces is shedding new light on California's tectonic setting during different periods in geologic time.

Because of the great complexity of California geology, our understanding of the evolution of California through geologic time is nowhere near complete. Nevertheless, it is useful to review California's history and synthesize the tectonic setting of all of the provinces at different points in geologic time.

CALIFORNIA'S EARLIEST HISTORY: BEFORE 600 MILLION YEARS AGO

Rocks older than about 600 million years are relatively rare in California. They are found only in southeastern California—in the southern Panamint and Nopah Ranges near Death Valley, in some of the ranges of the Mojave Desert, and in the San Gabriel and San Bernardino Mountains (Fig. 18-1; see also Chapter 16). Because these rocks are found in only a few places, and because they have been subjected to many geologic events since their formation, little is known about the earliest geologic environment of southeastern California. Based on their composition, these rocks are thought to be the remnants of very old terranes that were accreted early in the history of ancient continents that are preserved in today's North American Plate.

The oldest rocks in southeastern California, termed the ***crystalline basement***, consist of metasedimentary rocks, mainly gneiss, schist, and marble, and granitic plutonic rocks. The granitic rocks are about 1.4 to 1.7 billion years old, and, because they intruded into the metasedimentary rocks, those rocks are even older. At some localities, the oldest metamorphic rocks are as old as 2.4 billion years.

Fig. 18-1 The estimated edge of the Precambrian crystalline basement in the western United States, based on the distribution of rocks older than 600 million years old. The dashed line shows the presumed extent of buried crystalline basement. (Source: Stewart, J.H., and Suczek, C.A. 1977. *Cambrian and Late Pre-Cambrian Paleogeography and Tectonics in the Western U.S.* Los Angeles: Pacific Section, Society of Economic Paleontologists and Mineralogists.)

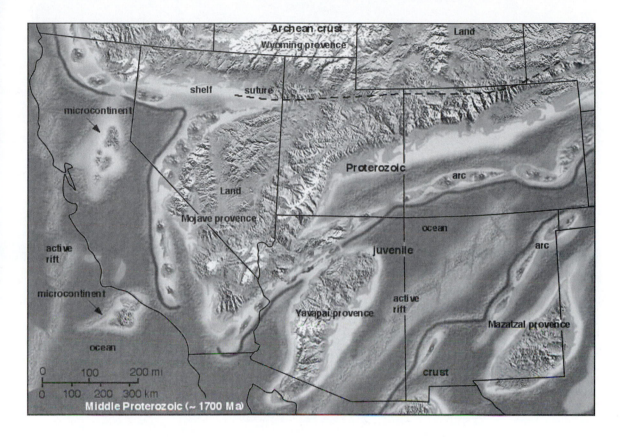

Fig. 18-2   A paleogeographic map showing the tectonic setting of the area that is now the southwestern United States about 1.7 billion years ago. Present state boundaries are shown for geographic reference. The reconstruction is based on the present distribution of different types of Middle Proterozoic rocks. (Source: Blakey, R. Northern Arizona University, *Paleogeography of the Southwestern U.S.: http://jan.ucc.nau.edu/~rcb7/ paleogeogwus.html.)*

Global reconstructions of the Earth at that time suggest that the ancient rocks of the southwestern United States were accreted to the most ancient core of North America (Fig. 18-2).

About 1.2 billion years ago, an ancient supercontinent formed by the accretion of smaller continental fragments, including those represented by southeastern California's most ancient rocks. Geologists have named this ancient continent ***Rodinia***, a word meaning "homeland" in Russian. The ancient core of North America formed the core of Rodinia. Based on the similarities of rock sequences and their ages, geologists believe that the future west coast of North America was next to the future Australia and Antarctica—a totally different configuration than today's (Figs. 18-3 and 18-4).

Fig. 18-3 A global reconstruction of the supercontinent Rodinia about 1 billion years ago. Today's Death Valley is shown by D.V. on the map. (Source: Karlstrom, K., and others. 1999. Refining Rodinia: Geologic evidence for the Australia–Western U.S. connection in the Proterozoic. *GSA Today* 9[10]).

About 900 million years ago, sediments were deposited on the crystalline basement rocks of the Rodinia supercontinent. Today these sediments are represented by the Kingston Peak Formation in the Death Valley region (see Chapter 7) and by more widespread formations in northeastern Washington, southern Idaho, and northwestern Utah. The Kingston Peak Formation contains coarse-grained gravels, and the size of the sediments indicates that they came from a landscape with considerable relief. Some of the cobbles in the formation appear to have been striated by glaciers, providing evidence that Rodinia was located in polar latitudes (see Chapter 8). In other regions, sedimentary rocks of this age are associated with volcanic rocks.

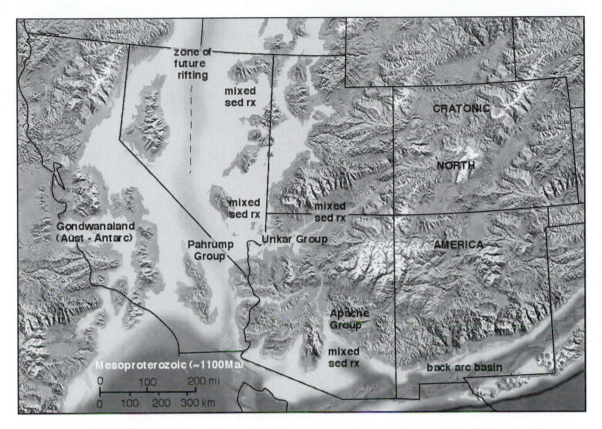

Fig. 18-4 A paleogeographic map showing the tectonic setting of the area that is now the southwestern United States about 1.1 billion years ago. Present state boundaries are shown for geographic reference. The reconstruction is based on the present distribution of different types of rocks that are about 1.1 billion years old. (Source: Blakey, R. Northern Arizona University, *Paleogeography of the Southwestern U.S.*: *http://jan.ucc.nau.edu/~rcb7/paleogeogwus.html.*)

LATE PROTEROZOIC AND PALEOZOIC CALIFORNIA: THE PASSIVE MARGIN

By about 700 to 800 million years ago, the supercontinent Rodinia began to break up, and rifts developed within the continent. Sediments began to accumulate along the western margin of North America in response to the sinking of the continental edge (Fig. 18-5).

Sedimentary rocks in several areas of southeastern California, including the San Bernardino Mountains (see Chapter 16) and Death Valley regions (see Chapter 7), and in other areas in Washington, Oregon, Utah, and Nevada appear to have formed in shallow marine waters during the initial subsidence. Quartz sandstone

Fig. 18-5 Thermal expansion, rifting, subsidence, and creation of a passive continental margin. (Source: Steward, J.H., and Suczek, C.A. 1977. *Cambrian and Late Pre-Cambrian Paleogeography and Tectonics in the Western U.S.* Los Angeles: Pacific Section, Society of Economic Paleontologists and Mineralogists.)

and associated shale, later metamorphosed to quartzite and phyllite, are the most common sedimentary rocks deposited, most dating to about 600 million years ago in late Proterozoic time.

By early Paleozoic time, a stable continental shelf had developed along the continental margin of western North America (Fig. 18-6; see Chapter 7). During the next 350 million years, thousands of meters of Paleozoic sedimentary rocks, mainly carbonates, accumulated on the carbonate platform. Geologists refer to this type of sedimentary environment, found along a tectonically quiet or passive margin (see Chapter 1), as a *miogeocline*. Today the rocks formed in the miogeocline can be seen in the ranges of the Basin and Range Province (see Figure 7-26) and the Mojave Desert, the San Bernardino Mountains (see Chapter 16), and the Peninsular Ranges (see Chapter 17). They are also preserved as metamorphic roof pendants in parts of the eastern High Sierra (see Chapter 8).

In general, the section of Paleozoic sedimentary rocks in southeastern California and adjacent parts of Nevada and Arizona increases in thickness from southeast to

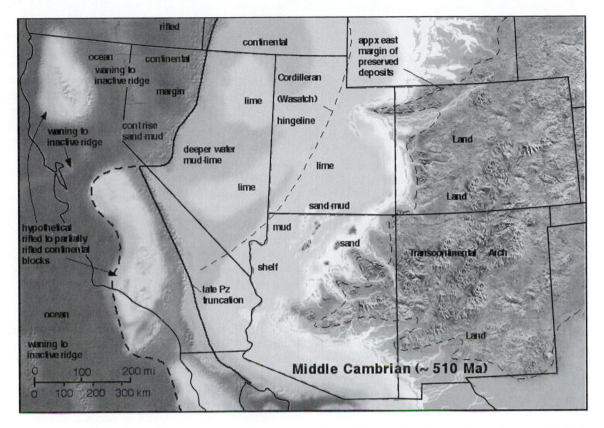

Fig. 18-6   A paleogeographic map showing the tectonic setting of the area that is now the southwestern United States during early Paleozoic time, about 510 million years ago. The area that is today's southeastern California was a passive continental margin. Present state boundaries are shown for geographic reference. The reconstruction is based on the type and present distribution of Cambrian rocks. (Source: Blakey, R. Northern Arizona University, *Paleogeography of the Southwestern U.S.: http://jan.ucc. nau.edu/~rcb7/paleogeogwus.html.*)

northwest. Assuming that the sediment accumulations decrease toward the ancient shoreline, geologists can reconstruct the continental margin (see Fig. 18-6). Based on the trends observed in Paleozoic sedimentary rocks, geologists believe that the continental margin faced toward the northwest in most areas.

Southeastern California's relatively quiet early tectonic setting was interrupted during the Devonian and Mississippian periods, when an episode of plate collision and mountain building was recorded by Paleozoic rocks in western Nevada and adjacent parts of California. These disturbances resulted in the accumulation of sediments shed from newly uplifted mountains (see Chapter 2) and caused preexisting sedimentary sequences to be folded and faulted. Clastic sedimentary rocks seen today in the northern Inyo Mountains and the Death Valley area record this episode of uplift, and the deformation is well preserved in Nevada, where geologists have named the ancient uplifted area the Antler Mountains.

Beyond the continental margin, other oceanic rocks were forming west of North America during Paleozoic time. Remnants of these rocks would be accreted to North America during Mesozoic time 100 to 200 million years later (see Figs. 8-25 and 9-3). Some of these rocks represent subduction-zone complexes, and others are remnants of oceanic volcanic terranes. During Mississippian and Permian time, a belt of island arcs lay west of the North American continent. Volcanoes from this chain erupted large volumes of andesite, and coral reefs ringed the volcanic islands. Today the ancient island arc system is preserved in the Paleozoic rocks of the northern Sierra Nevada and the eastern Klamath Mountains of California (see Chapter 9).

Although California was tectonically quiet during much of Paleozoic time, it was a time of tectonic upheaval along what is today the eastern edge of North America. The tectonic collisions that produced the complex rocks seen in the Appalachian Mountains occurred during Paleozoic time. By the Permian, about 250 million years ago, another supercontinent had been assembled. This great mass was named *Pangaea*, meaning "one Earth" in Greek.

MESOZOIC CALIFORNIA

Early in Mesozoic time, about 225 million years ago, the ancient supercontinent known as Pangaea began to break up. Between about 100 and 50 million years ago, North America moved westward, and a rift developed along the eastern edge of the continent. This rift would eventually evolve into the northern Atlantic Ocean, splitting the North American Plate from the western edge of the European continent. Along western North America, a north–south subduction zone developed, roughly in the same area as today's Sierra Nevada. A huge amount of oceanic lithosphere, equivalent to the entire Pacific basin, was subducted beneath western North America during Mesozoic time. Because of associated accretion, this was California's most important geologic building period.

During Mesozoic time, belts of oceanic rock were added to California by a series of collisions and accretions. The earliest rocks added to North America were Paleozoic sedimentary and volcanic rocks. These were accreted to North America during early Mesozoic time. Following each episode of accretion, the subduction zone shifted westward and accretion of a younger terrane began. The westernmost, youngest belt of accreted rocks in the Sierra Nevada is of middle to late Jurassic age, about 160 million years old (see Chapter 8 and Chapter 9).

Inland from the Mesozoic subduction zones, magma rose above the down-going plates, forming chains of andesitic volcanoes and granitic plutons beneath them. At its maximum development, the Mesozoic magmatic arc extended along western North America from Baja California, Mexico to British Columbia. In California, the arc was located approximately in the position of today's Sierra Nevada (see Fig. 12-12). Plutonic rocks representing the roots of the Mesozoic volcanic chain are found today throughout the Klamath Mountains, Sierra Nevada, Basin and Range, Mojave Desert, and Peninsular Ranges provinces. Mesozoic volcanic rocks, originally erupted along the arc, are also preserved in places in the same provinces. The greatest volume of

Fig. 18-7 Position of the subduction zone and the Farallon and Pacific Plates about 100 million years ago. (Source: *National Geographic.*)

magma was generated about 100 million years ago, during the subduction of the Farallon Plate.

By about 100 million years ago, the subduction zone had shifted westward to the approximate position of today's Coast Ranges (Fig 18-7). Pieces of the Farallon Plate were being accreted along western California as it was being subducted beneath North America (Fig. 18-8). Millions of years later, geologists would recognize these pieces as the Franciscan Assemblage of the Coast Ranges (see Chapter 12).

Fig. 18-8 A general cross section across California about 100 million years ago. (Source: Crouch, J., and Suppe, J. 1993. *Geological Society of America Bulletin* 105[11]. After Hamilton, W. 1989. In Geological Society of America Memoir 172.)

Between the trench and the Sierran volcanic arc, a marine forearc basin extended along most of California's length (Fig. 18-9; see also Fig. 18-8, Chapter 11, and Chapter 17). The Cretaceous Great Valley Sequence, including marine sedimentary rocks found today in the Transverse Ranges, accumulated in this linear basin (Fig. 18-10; see also Chapter 16). The sedimentary rocks of the Great Valley Sequence contain fragments of most of the immense volcanoes that once sat above the Sierra Nevada granitic rocks. The sedimentary rocks of the Great Valley Sequence accumulated on the Coast Range ophiolite, which geologists now recognize as representing the crust and upper mantle beneath the ocean floor (see Chapter 9).

The major tectonic events of the Mesozoic era in California had dramatic effects on the rocks that were already there. Folding and faulting accompanied the collision events of the Mesozoic, with the result that older rocks were folded and faulted during each episode. In many areas, Paleozoic and Mesozoic rocks were also metamorphosed during the intrusions of magma and episodic compressional events. Geologists attempting to understand California's pre-Mesozoic history have had to "see through" these later changes to determine the original nature of Paleozoic and older rocks.

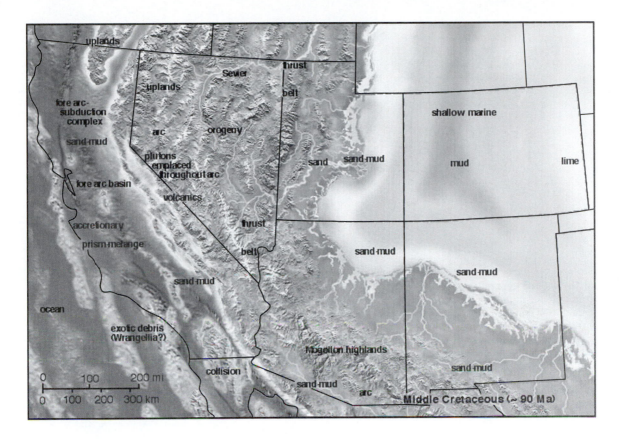

Fig. 18-9 A paleogeographic map showing the tectonic setting of the area that is now the southwestern United States during Mesozoic time, about 90 million years ago. Sediments of the Great Valley Sequence were deposited in the forearc basin that lay between the accretionary mélange of the future Franciscan complex and the plutons of the Sierran arc. Present state boundaries are shown for geographic reference. The reconstruction is based on the type and present distribution of Cretaceous rocks. (Source: Blakey, R. Northern Arizona University, *Paleogeography of the Southwestern U.S.: http://jan.ucc.nau.edu/~rcb7/paleogeogwus.html.*)

CENOZOIC CALIFORNIA

About 27 million years ago, the Pacific and North American Plates made direct contact for the first time when the closest part of Farallon Plate was completely consumed beneath North America (see Fig. 12-2). This event created the San Andreas transform boundary. North and south of the transform, subduction of the Farallon Plate continued. As increments of the Farallon Plate were consumed in the subduction zones, the transform margin lengthened. The development of the transform margin has been the dominant factor in California's later Cenozoic history. Geologic events throughout California, even in provinces not directly at the plate boundary, have been strongly influenced by the San Andreas system.

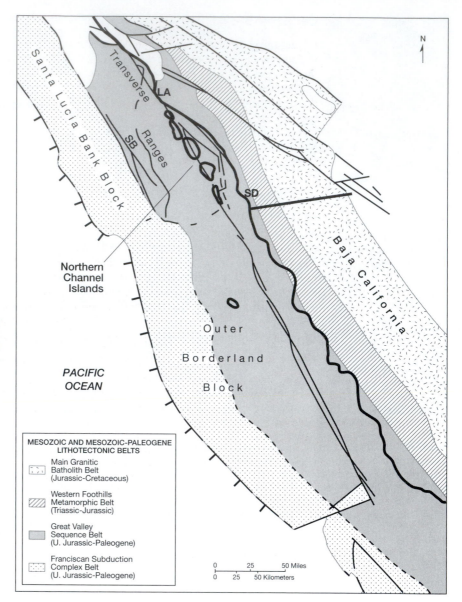

Fig. 18-10 The general tectonic blocks in southern California and Baja California, Mexico, during mid-Cenozoic time, prior to Miocene extension and rotation. Compare with Fig. 18-12 to see the changes in the configuration of the blocks. (Source: Atwater, T. 1998. Plate tectonic history of southern California with emphasis on the western Transverse Ranges and Santa Rosa Islands. In Weigand, P.W., ed. *Contributions to the Geology of the Northern Channel Islands, Southern California.* American Association of Petroleum Geologists, Pacific Section, MP 45.)

The development of the San Andreas transform had major impacts on the western edge of California. Large blocks have shifted by as much as 315 kilometers along the right-lateral faults of the San Andreas system. For example, rocks of the Salinian Block, which probably formed near the southern end of the Coast Ranges, moved northwestward along the San Andreas fault to its present position along the central California coast (see Fig. 12-5). Right-lateral movement along the transform boundary also brought crystalline basement rocks from southeastern California to their present positions in the San Gabriel Mountains.

Cenozoic Basins

During the early history of the San Andreas system, large volumes of sediment accumulated in marine basins along the continental margin. Geologists believe that the sediments were deposited in ***pull-apart basins***. The early boundary between the Pacific and North American Plates was oriented to create a transtensional tectonic regime in the area where the Transverse Ranges and Channel Islands are today (Fig. 18-10; see Fig. 1-17B). As a result, a number of fault-bounded basins formed and received sediments shed from nearby highlands (see Chapter 16). In the southern Great Valley, the San Joaquin basin was a very deep marine basin between 35 and 15 million years ago, and a shallow marine embayment persisted there until 2.5 to 3 million years ago.

Cenozoic Extension in the California Desert

In early Miocene time, about 22 to 17 million years ago, extension of the crust began along a roughly north–south zone that included parts of what is now the Mojave Desert and Basin and Range. Large crustal blocks were pulled apart along detachment faults (see Chapter 7). In some highly extended terranes, detachment faults completely removed the overlying rocks to expose the metamorphic rocks of the lower plate. Today these ***metamorphic core complexes*** can be seen in a belt that stretches from Canada to Mexico, including many that form the dome-shaped ranges of southeastern California.

Tectonic Rotation in the Transverse Ranges and Mojave Desert

Geologists have uncovered evidence that rocks in both the western Transverse Ranges (Fig. 18-11; see also Chapter 16) and in the Mojave Desert Province have been rotated in a clockwise direction during Miocene time. Geologists believe that the western Transverse Ranges rotated as a result of right-lateral motion along the edge of the North American Plate. As the early San Andreas fault developed, blocks of the North American Plate were added to the northwest-moving Pacific Plate; these blocks lay west of today's Transverse Ranges. The southern end of the western Transverse Ranges broke free of North America, but the eastern end remained attached to the continent, with the result that the block rotated clockwise (see Figs. 18-10 and 18-12).

Rotation of the western Transverse Ranges created a region of extension to the south, where today's Los Angeles basin and offshore islands are. Geologists believe that Miocene extension caused the uplift and unroofing of metamorphic rocks from lower levels of the crust, such as the Catalina schist (see Chapter 16).

Magmatic Activity

One major consequence of the development of the transform margin was the shutting off of magmatic activity beneath North America. In areas east of the growing transform margin, no oceanic plate was being subducted. As a result, no magma was generated. After about 27 million years ago, magmatic activity began to cease inland of the transform margin (see Chapter 12). Where remnants of the Farallon Plate and the East Pacific Rise still exist today, north of the Mendocino Triple Junction and south of Baja California, Mexico, magmatic activity persists on the North American Plate. East of the plate boundary in the Basin and Range Province, crustal extension is responsible for recent volcanic eruptions in Death Valley, along the eastern side of the Sierra Nevada, and in the Modoc Plateau (see Chapter 5).

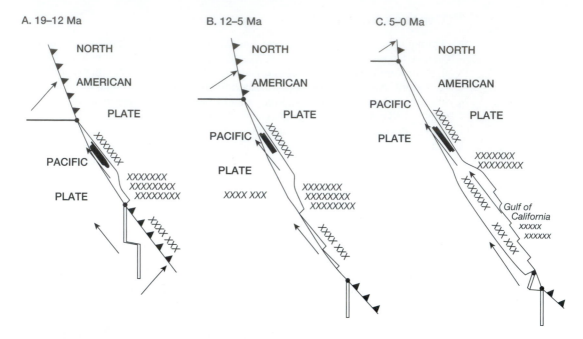

Fig. 18-11   Configuration of the transform boundary between the Pacific and North American Plates. **A** and **B**, Creating transtension and extension in the western Transverse Ranges. **C**, Creating transpression in the Transverse Ranges. (Source: Atwater, T. 1998. Plate tectonic history of southern California with emphasis on the western Transverse Ranges and Santa Rosa Islands. In Weigand, P.W., ed. *Contributions to the Geology of the Northern Channel Islands, Southern California.* American Association of Petroleum Geologists, Pacific Section, MP 45.)

The boundary between the Farallon and Pacific Plates, the East Pacific Rise, was an oceanic spreading ridge. As the Farallon Plate was being consumed by subduction, the East Pacific Rise approached closer to western North America. When the Farallon Plate was completely subducted, the East Pacific Rise itself encountered the subduction zone about 27 million years ago. Some geologists believe that this encounter might have triggered a period of volcanic activity close to the plate boundary. They hypothesize that the mid-Cenozoic volcanic rocks seen today in the Coast Ranges, near Point Conception, and in the Santa Monica Mountains, resulted from arrival of the East Pacific Rise at the western margin of the North American Plate.

California During the Past 5 Million Years

About 5 million years ago, the Gulf of California began to open, and the Baja California peninsula became part of the Pacific Plate. The transform boundary shifted inland in southern California, and the transtensional tectonic regime was replaced by transpression. The change in motion created greater convergence between the Pacific and North American Plates, including formation of the Big Bend segment of the San Andreas fault (see Chapter 14). The new configuration of the plate boundary resulted in a series of left-lateral faults along the southern edge of the western Transverse Ranges blocks and along the Garlock fault. The rate of right-lateral motion along the San Andreas system also increased.

Fig. 18-12 The present-day general tectonic blocks in southern California and Baja California, after Miocene extension and rotation. Compare with Fig. 18-10 to see the changes in the configuration of the blocks. The areas shown in cross-hatching are thought to have been created by extension since the time shown in Fig. 18-10. (Source: Atwater, T. 1998. Plate tectonic history of southern California with emphasis on the western Transverse Ranges and Santa Rosa Islands. In Weigand, P.W., ed. *Contributions to the Geology of the Northern Channel Islands, Southern California.* American Association of Petroleum Geologists, Pacific Section, MP 45.)

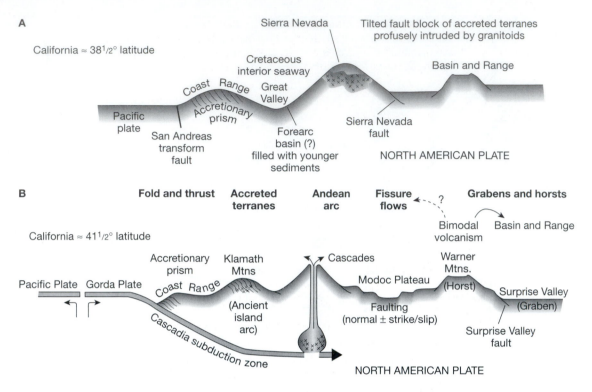

Fig. 18-13 Schematic cross sections across **A**, central and **B**, northern California, showing the tectonic settings of the two areas. (Source: Danielson, J., and May, A. 1992. Weed, CA: Shasta Valley College.)

During the past few million years, the transform boundary between the Pacific and North American Plates has continued to lengthen. Today it extends along most of California in the form of the San Andreas fault system (Fig. 18-13A). Right-lateral motion along the active faults of the San Andreas system is carrying the Pacific Plate northwest and the North American Plate to the southeast. North of Cape Mendocino, remnants of the Farallon Plate, known as the Gorda and Juan de Fuca Plates, are still being subducted beneath North America. As a result, a magmatic arc is still being generated north of Cape Mendocino, and the Cascades volcanic chain marks the surface above it (Fig. 18-13B; see Chapter 5).

Most of California's major mountain ranges were elevated about 4 to 5 million years ago in response to increased compression along the plate boundary. The present Sierra Nevada range rose in about the same position as the Mesozoic chain of volcanoes during the past 9 million years, but not to elevations as great as the Mesozoic volcanic arc. In coastal southern California, reverse faults and folds developed, forming deep basins as well as newly uplifted mountains. Beginning about 4 million years ago, basin subsidence in southern California increased as a result of increased compression. Along the coast, flights of marine terraces add to the evidence of ongoing uplift, particularly where compressional forces are strengthened near the Big Bend and the Mendocino Triple Junction. Finally, earthquakes on both reverse and transform faults serve as constant reminders that both compression and lateral plate motion are active in California today.

Further east, in the Mojave Desert and Basin and Range Provinces, crustal extension continues, as evidenced by active normal faults. Recent earthquakes with right-lateral displacement indicate that the Eastern California Shear Zone is a developing part of the transform boundary (see Chapter 14).

REFERENCES

ATWATER, T. 1998. Plate tectonic history of southern California with emphasis on the western Transverse Ranges and Santa Rosa Islands. In Weigand, P.W., ed. *Contributions to the Geology of the Northern Channel Islands, Southern California*. American Association of Petroleum Geologists, Pacific Section, MP 45.

ERNST, W.G. 1981. *The Geotectonic Development of California*. Upper Saddle River, NJ: Prentice Hall, 704 pp.

MCPHEE, J. 1994. *Assembling California*. New York: McGraw-Hill.

Web Sites and Media Resources

Atwater, T. Southern California: 20 MA to present, video animation: *http://www.geol.ucsb.edu/faculty/atwater*

Blakey, R.C. Paleogeography of the Southwestern U.S.: *http://jan.ucc.nau.edu/~rcb7/paleogeogwus.html*

Deep Time: PBS series, Evolution, WGBH Television, including paleogeographic reconstructions of Earth through geologic time: *http://www.pbs.org/wgbh/evolution/change/deeptime/index.html*

Paleomap Project, Scotese C.R.: *http://www.scotese.com/earth.htm*

19

Geology and California Citizens: Risks, Resources, and a Beautiful Place

GEOLOGIC HAZARDS

One of the most important benefits to understanding California's geologic history is an increased knowledge of geologic hazards. After decades of careful research, much of which was not directed at these practical goals, geologists have come to better understand the processes that create these hazards. As California's population increases, geology will play an even more important part in protecting the safety of its people and structures. A basic understanding of earthquakes, volcanic eruptions, floods, and landslides can benefit all Californians who seek to avoid becoming a victim of one of these hazards. With the exception of landslides, which are discussed here, these hazards and the processes that cause them have been discussed in previous chapters.

Earthquakes

As we have seen in Chapters 13 and 14, earthquakes are an unavoidable aspect of life in California. During the past 30 years, scientists have made considerable progress toward understanding the sources and mechanisms of earthquakes, but exact predictions are yet to come. The Alquist-Priolo Act prohibits new construction on active faults, and the Uniform Building Code is designed to make buildings earthquake-resistant. However, protection is far from complete. For example, patterns of ground shaking are not well identified in many areas, and seismologists are just beginning to assess the risks posed by buried faults. Unfortunately, public funding for earthquake studies is greatest immediately following a destructive event like the 1994 Northridge earthquake, but it then drops off sharply as strain energy continues to build on California's faults.

It is therefore important that every citizen be informed about the extent of active faulting in his or her area. Because earthquakes are possible at any time, particularly along the Cascadia subduction zone, the San Andreas fault system, and in the Transverse Ranges, all Californians must be prepared for earthquakes. Preparations should consider all the places where people spend their days and nights, including the home, the workplace, and schools. The earthquake scenario in Chapter 13 provides examples of the preparations that should be made in advance.

Volcanic Eruptions

The 1980 eruption of Mt. St. Helens in Washington (see Chapter 5) was an effective reminder of the possibility of eruptions from other Cascades volcanoes, including Mt. Shasta and Lassen Peak. Recent seismicity, ground bulging, and carbon dioxide emissions in the Mammoth area have also alerted scientists to the possibility of future eruptions there. As discussed in Chapter 5, geologists are in the process of evaluating volcanic hazards in the Pacific Northwest, including those posed by California's active centers. Continued monitoring of these volcanoes should be an important and effective part of mitigating the hazard posed by future eruptions.

Because of the location of California's active volcanoes, the most populated areas of the state are not at risk for experiencing catastrophic pyroclastic flows, debris flows, or blasts (Fig. 19-1). However, ash from major eruptions from the Lassen

Fig. 19-1 Map showing relative volcanic hazards in the western United States. (Source: U.S. Geological Survey. 1981. Professional Paper 1240-B.)

Legend

■ Highest hazard

▦ Moderately high hazard

▨ Moderate to low hazard

□ Minimal hazard

and Long Valley areas has blanketed much of California in the past (see Chapter 5). A future eruption from one of these areas could have destructive effects, particularly if east-blowing Santa Ana winds happened to be active during the eruption. Near the active volcanoes, hazards can be reduced by enactment of zoning ordinances based on the volcanic hazard maps, as well as by careful development of evacuation plans.

Floods

Although they might be unaware of the fact until a flood strikes, many Californians live on floodplains (Fig. 19-2; also see Chapter 10 and Chapter 11). Billions of dollars have been spent to reduce flood hazards by constructing dams, levees, and flood-control channels (Fig. 19-3), but it is not economically possible to eliminate all flood risk in most areas. Furthermore, flood elimination might not be desirable in areas where stream channels and riparian forests provide recreation for humans and habitat for other species. Many communities, including Napa, have recently adopted floodplain management plans that allow these areas to be preserved (Fig. 19-4).

The most effective way of reducing the risk of flooding is to avoid building on flood-prone areas. In addition to low-lying floodplains and coastal wetlands, other high-risk areas in California include desert washes, alluvial fans, and debris-flow fans at the mouth of steep canyons, where flash floods are a normal occurrence (see Chapter 6).

Fig. 19-2   Flooding in Napa, California, 1986. (Source: Napa Historical Society.)

Fig. 19-3 Flood-control channels, Los Angeles River near Downey, 1998. (Source: Gatley, D. Federal Emergency Management Agency.)

Fig. 19-4 Artist's rendering of the Napa River flood control project, showing the riparian corridor, the new First Street Bridge, and the proposed bypass channels in Napa. (Source: Napa Flood Control District.)

PLANNING AHEAD FOR A FLOOD

Water enthusiasts and anglers across the state enjoy Northern California's many beautiful rivers. But when heavy rains come, those same rivers have been known to escape their banks, endangering lives and property. In fact, northern California experienced heavy flooding in 1986 and as recently as 1995, 1996, and 1997. When heavy rains fall, the rule for being safe is simple: *plan ahead*.

Before the Storm

- Determine whether you are in a flood area by contacting your local insurance agency or planning office.
- Store supplies in a convenient location: first-aid kit and essential medicines, nonperishable foods, nonelectric can opener, radio, flashlights, batteries, and instructions on how and when to turn off your utilities.
- Decide where your family will go in case you must evacuate. Inform family and friends of the location.

When a Storm Warning Is Issued

- When flash flooding is *possible* in a designated area, a Flash Flood Watch is issued.
- When flash flooding is *reported* or is *imminent,* a Flash Flood Warning is issued.
- Store drinking water in closed containers.

- If advised by local authorities to leave your home, follow your planned route.
- Before leaving, turn off all electrical circuits at the fuse panel, gas service at the meter, and water at the main valve.

During the Storm

- Avoid areas prone to sudden flooding: dry creek beds, culverts, and river banks.
- A car can be swept away in as little as 12 inches of running water. If your car stalls, abandon it immediately and seek higher ground.
- Floodwater is often contaminated with sewage. Keep children away from floodwater and storm drains.

After the Storm

- Rely on utility company crews to turn on all services.
- Do not use fresh food or electrical and gas appliances that have come in contact with flood waters.
- Follow local instructions or purify water before drinking.
- Wash homes down with soap and water and disinfect them with a chlorine bleach solution of 1 tablespoon to 1 gallon of water.
- Contact your insurance agency immediately.

Source: Modified from KOVR 13 Northern California Online; © 1997 KOVR 13, *http://www.kovr.com/almanac/flood.htm.*

Landslides

Landslides are an all-too-familiar event to many of California's hillside residents. In many cases, landsliding damages roads, generating constant repair work and annoying traffic diversions. However, landslides in California have also dammed rivers, destroyed hundreds of homes, and killed several people during the past 30 years (see the box on southern California debris flows in Chapter 10).

Landsliding is the downhill *mass movement*, driven by gravity, of soil, rock, or a combination of the two known as *debris*. Slow mass movement, or *creep*, occurs on virtually all hillslopes, but in some areas, movement is rapid enough to send discrete masses of material down the hillsides. When this happens, a visible landslide scar is left in the area where debris was detached (Fig. 19-5). Landslide scars are typically bowl-shaped, with a steep slope at the head of the slide area. Downhill of the scar, the debris might come to rest on the lower slope, leaving a characteristic lumpy and disrupted mass. In most areas, geologists can recognize landslide-prone areas by mapping landslide scars and deposits using aerial photographs.

Geologists and engineers classify different types of landslide features, according to the depth of the material that fails, amount of water involved, rate of movement, and type of movement that occurs (Fig. 19-7). Landslide classification is important because the risks posed by different types of landslides are different (as discussed later). In addition, classification is important in deciding whether stabilization measures will be effective.

Fig. 19-5 Diagram showing features of a typical landslide. (Source: Highway Research Board. 1958. Special Report 29.)

A

B

Fig. 19-6 **A**, The St. Francis Dam, looking upstream. The photo was taken after completion in 1926, but before the dam failed in March 1928. **B**, The remains of the St. Francis Dam, taken shortly after it failed on March 12, 1928. (Source: Santa Clarita Valley Historical Society, Arthur Perkins.)

THE ST. FRANCIS DAM DISASTER

Just before midnight on March 12, 1928, the St. Francis Dam failed, sending a fast-moving wall of water rushing into San Francisquito Canyon (Fig. 19-6). About $5\frac{1}{2}$ hours later, the water, which was up to 10 stories deep, reached the Pacific Ocean near Ventura. That night, at least 470 people living in the canyon lost their lives, although some miraculously survived the flood. Thirteen-year-old Thelma Shaw left through the back door of her home, but her family went out the front. She was washed 9 miles downstream, deposited on the floodplain beneath a large tree, and rescued the next day. She never saw her family again. Of all California's disasters, only the 1906 earthquake and fire killed more people. Because many Mexican-American workers living downstream of the dam, in the Santa Clara River Valley near Santa Paula, were not counted as casualties, the death toll was probably even higher than reported. On the 75th anniversary of the disaster, *All Things Considered* broadcast several interviews, and a documentary film about the event is scheduled for completion in 2003 (see references at the end of the chapter).

The St. Francis Dam had been built between 1924 and 1926 by William Mulholland to store water brought to the area from the Owens Valley by the Los Angeles Aqueduct (see Chapter 10). Earlier that day, engineers had summoned Mulholland from the San Fernando Valley for an inspection because they were concerned about water seeping from the face of the earthen dam. Mulholland felt that the water was not a serious problem. He took responsibility for the disaster, but neither he nor his career ever recovered from the catastrophe.

The precise geological cause of the dam failure is still not completely understood, but the geology of the eastern side of the dam was partly to blame. In that area, the foliation of the Pelona schist is oriented parallel to the hillslope, which probably contributed to the weakening of the slopes. A fault contact between the schist and the Sespe formation might also have played a role. Old landslide scars in the area are evidence of predam slope failures. In the 1928 failure, the material holding the abutments and the foundation of the dam became saturated, causing the dam to float and fail.

Looking back, it is clear that geologists could have recognized that the rocks were susceptible to such an event. The St. Francis Dam disaster taught engineers the importance of geology. Since that time, geological inspections are required for dams, bridges, large buildings, and other structures throughout the world.

Of the many types of landslides, several are particularly common in California. *Rockfalls*, involving the free fall of loosened rock from cliffs, occur in the steep cliffs of the Yosemite area (Fig. 19-8) and in other very steep terrain. Rockfalls do not require water for their initiation, and they can occur without warning, posing a threat to unsuspecting hikers.

In July 1996, a massive slab of granitic rock suddenly broke loose from the cliffs above Yosemite Valley. The 120-meter-high chunk fell almost 800 meters to the valley floor near the Happy Isles area, where it shattered into debris that covered parts of the valley with rock dust up to 15 centimeters thick. One Yosemite hiker was killed and several were injured. The casualties would have been even greater if the rockfall had occurred during daylight hours rather than in the early evening. Rockfalls are a natural process of erosion on steep cliffs, including areas that are shaped by exfoliation (see Chapter 8). The sudden crashing of giant rock slabs is the result of slow, gradual weathering and loosening, and the exact moment of failure is unpredictable. Earthquakes can also trigger multiple rockfalls in mountainous areas.

FALL (ROCKFALL)

FALL—Mass travels most of the distance
in free fall, by leaps and bounds,
and rolling of bedrock or soil fragments.

SLIDE (ROCK SLUMP)

Sandstone
Shale

ROTATIONAL SLIDE—Movement involves turning
about a point (surface of rupture is concave upward).

SLIDE—movement of material by shear displacement along
one or more surfaces or within a relatively narrow zone

SLIDE ROCKSLIDE

TRANSLATIONAL SLIDE—Movement is predominantly along
planar or gently undulatory surfaces. Movement frequently is
structurally controlled by surfaces of weakness, such as
faults, joints, bedding planes, and variations in shear strength
between layers of bedded deposits, or by the contact
between firm bedrock and underlying detritus.

FLOW (EARTHFLOW)

Source area
Depositional
area

COMPLEX
(SLUMP-EARTHFLOW)

Sand
Clay
Slide
Earthflow

Fig. 19-7 Diagram showing different landslide types. (Source: U.S. Geological Survey. 1981. Professional Paper 1240-B; adapted from Varnes, D.)

Debris avalanches are relatively small, shallow slides that are also common on steep slopes held up by moderately resistant rocks such as sandstone. They commonly occur at the heads of steep canyons (Fig. 19-9). On slopes underlain by weak rocks, such as serpentinite or mudstone, landslides tend to be deeper and to cover a larger area. These types of landslides, called *slumps*, are especially common in the mélange units of the Franciscan Complex in the Coast Ranges (Fig. 19-10), although they occur in many other areas as well. At the downhill end of many slumps, the landslide debris loses its coherence, becoming an *earthflow*.

Debris flows are a particularly dangerous form of mass movement, as illustrated by the narrative in Chapter 10. Debris flows are created when rock and soil become completely saturated, sometimes becoming so mobile that they can attain speeds of hundreds of kilometers per hour. Debris flows form in steep stream channels, sometimes downhill of rockfalls or debris avalanches. In many mountainous areas of California, debris flows pose serious hazards to residents living in mountain canyons or on fans at the canyon mouths. Careful examination of these fans often reveals that they contain abundant materials carried by debris flows, sometimes even house-sized boulders (see Chapter 6).

Fig. 19-8 The July 1996 rockfall in Yosemite National Park, near Happy Isles. (Source: Harp, E. U.S. Geological Survey.)

The common triggering mechanisms for slope failures are present in many of California's hilly areas: steep slopes, weak rocks, severe storms and storm waves, earthquakes, and human activities. Because of active tectonics, steep slopes are abundant throughout the state. Tectonic activity has uplifted naturally weak materials, such as the unconsolidated sediments or marine sandstone and shale—materials that in most areas would be found beneath lowlands. In addition, shearing and faulting have weakened rocks along fault zones. As we have seen in Chapter 10, California regularly experiences severe winter storms, even though average rainfall is low during most years. Many landslides are generated during these wet winters. If streams are swollen by floodwaters, rivers could undermine the lower areas of unstable slopes, causing them to slide (Fig. 19-11). Along the coast, storm waves can undercut unstable sea cliffs, especially if high tides occur at the same time. In addition to weak rocks, steep slopes, and severe storms, California's hillslopes are also subjected to earthquakes. Ground shaking during earthquakes could trigger hundreds of rockfalls or shallow landslides, as was the case during the 1994 Northridge earthquake (see Chapter 13).

Landsliding is a natural process on California's hillslopes, and destruction by landsliding has predictably increased as hillsides have been developed. One of many notorious California landslides is the Portuguese Bend slide in the Palos Verdes Hills. In this area, portions of the marine terrace are sliding into the Pacific Ocean (Fig. 19-12). Between 1956 and 1959 alone, the Portuguese Bend slide caused more than $10 million in damages to homes, roads, and other structures.

Fig. 19-9 Debris avalanches near Ventura. (Source: Sarna-Wojcicki, A. U.S. Geological Survey.)

Fig. 19-10 Complex slump-earthflow, Redwood National Park, Humboldt County. (Source: Photo by author.)

Fig. 19-11   Aerial view of multiple landslides triggered by stream undercutting, Redwood Creek, Humboldt County. (Source: Photo by author.)

Fig. 19-12   Landslide along the southern California coast near Palos Verdes Hills. (Source: Bradley, W.)

In many counties, development has been restricted on landslide-prone hillslopes. In these areas, geologists have mapped vulnerable areas by recognition of landslide deposits and scars, and local agencies have enacted ordinances against developing landslide-prone zones. However, many unstable slopes have already been developed. In some of these areas, engineers have installed drains, built retaining walls, and planted vegetation to stabilize the hillslopes, with varying success. To avoid purchasing or building a home on an unstable hillslope, a prudent Californian seeking residence in any hilly area would be well advised to obtain a slope-stability evaluation from a geologist.

EARTH RESOURCES

California's geologic history has created significant Earth resources in addition to the geologic hazards already discussed. As a state with more than one automobile per citizen, California is highly dependent on petroleum, and most citizens are aware of this dependency. However, few people realize how important geologic materials are in meeting our everyday material needs. Minerals are necessary ingredients for everything from makeup to spark plugs, and rock materials are components of most modern building materials (see Table 19-1). In addition to providing these commodities, California's geologic history has also endowed the state with a rich and varied landscape, a resource to be enjoyed by all citizens.

The mineral industry has played an important role in California's history, beginning with the gold rush (see Chapter 8). Because of plate tectonics, California has a wide variety of rocks containing a wealth of mineral resources. In some areas, minerals bearing metallic elements like copper, gold, and tin have been profitably extracted by underground and surface mining. Nonmetallic mineral resources such as borate or gypsum have been mined from other regions.

We have investigated only a few of these resources, including evaporite minerals (Chapter 6), gold (Chapter 8), chromite (Chapter 9), mercury (Chapter 10), diatomite (Chapter 11), and lithium and gemstones (Chapter 17).

Mineral discoveries in other countries, combined with strict environmental regulations and the high cost of labor in the United States, have reduced the U.S. production of minerals during the past 30 years. Nevertheless, California continues to be a supplier of many mineral resources, despite the availability of many commodities from foreign suppliers. In 2001, the industry produced $3.7 billion of nonfuel mineral commodities.

Aggregates, Limestone, and Other Rock Materials

Although they are less glamorous than gold, copper, or gemstones, ordinary rock materials are very important resources in California. Sand, gravel, and crushed rock, known as *aggregates*, are essential as foundation materials for roads, commercial buildings, and residences (see Table 19-1). They are also essential for making concrete, which is a mixture of cement and aggregate. Cement is also made of geologic materials. Portland cement contains about 60 percent lime, which comes from carbonate sedimentary rocks (see Chapters 2 and 7) or crushed shells. The lime is mixed with silica and alumina, which come from clay, sand, or shale, and a small amount of iron oxide and gypsum. Because of California's geologic history, carbonate rocks are rare except in the sparsely populated desert area. As a result, relatively small bodies of carbonate rock in coastal California—for example, the Franciscan limestone block near Cupertino in Santa Clara County—are valuable resources.

TABLE 19-1 CALIFORNIA'S MINERAL PRODUCTION

Material	Type	Examples of mines	Uses	Production (in millions of dollars)[3]
Asbestos[1]	Silicate mineral	San Benito County	Refractory material	5.00
Borates[1]	Evaporite minerals	Boron, Inyo	See Chapter 6	462.00
Cement	Limestone	Permanente, Santa Clara	Construction, masonry	522.00
Clays	Silicate minerals	Ione, Sacramento	Ceramics, cement, cosmetics	32.00
Crushed rock	Commonly igneous	Numerous	Concrete, other construction	307.00
Diatomite[1]	Silica from diatoms	Lompoc, Santa Barbara	Filters (see Chapter 16)	90.00 (1989 value)
Dimension stone	Hard and durable rock	Fresno County	Building facades, counters	6.50
Gemstones	Tourmaline, beryl	Pala, San Diego	Jewelry	0.60
Gold[2]	Metal	McLaughlin, Lake, Mesquite, Imperial	Electronics and other (see Chapter 8)	354.00
Rare earths	Igneous rocks	Silver	NA	
Gypsum	Evaporite	Plaster City, Imperial	Wallboard, plaster	11.00 (1989 value)
Magnesium compounds	Sedimentary minerals	Bissel, Kern	Chemical industry	NA
Mercury[2]	Metallic mineral	McLaughlin, Napa, Lake, Yolo	Instrumentation, gold mining (see Chapter 12)	NA
Perlite[2], pumice[2], cinders	Volcanic rocks	Inyo, Modoc, Shasta counties	Absorbents, abrasives (see Chapter 5)	6.50
Rare earths[1]	Igneous rocks	Mountain Pass, San Bernardino	Televisions, superconductors (see Chapter 7)	24.00 (1989 value)
Sand and gravel	Alluvium metal	Numerous	Construction	548.00
Silver	Metal	Produced with gold	Photographic and other industries	2.00
Sulfur	Native element or sulfate		Fertilizer	51.0 (1989 value)
Tungsten[1]	Metal, associated with granitic rocks	Pine Creek, Inyo	Alloy	Zero in 1995; mine reopened in 1996
Zeolites	Silicates in altered volcanic rock	Barstow, San Bernardino	Ion exchange, absorbers and filters	NA

NA = Data not available; [1] California largest producer in U.S; [2] California second to fifth largest producer in U.S. [3] 1995 figures, unless otherwise noted.

Data sources: A. Tanner and S. Kohier-Antablin, 1995. California mineral industry survey, *California Geology*, p. 143; E.W. Tooker and D.J. Beeby, 1989. *Industrial Minerals in California: Economic Importance, Present Availability, and Future Development.* U.S. Geological Survey, Bulletin 1958, 127 pp.; W.H. Langer and V.M. Glanzmen, 1993. *Natural Aggregate: Building America's Future.* U.S. Geological Survey, Circular 1110, 39 pp.

Aggregates are extracted from the Earth by surface mining. Sand and gravel are mined from river beds (Fig. 19-13), and crushed rock is quarried from rock outcrops. Ironically, meeting the demand for aggregates in California has threatened their easy availability. Supplies of these materials are limited in developed areas. In addition, communities are unwilling to have active quarries next to residential and commercial developments. Developers must obtain these materials from more rural areas. Although the materials can be quarried at relatively low cost, transporting them to construction sites is very expensive. As California's population continues to expand, aggregates will become an increasingly valuable commodity.

Petroleum

As we saw in Chapter 16, petroleum is a major geologic resource in the San Joaquin Valley, the Transverse Ranges, and the offshore basins bordering the continent (see Table 16-1). At present, many of California's known oil fields are producing at low levels because of the lower cost of imported petroleum. However, as global supplies of oil decrease during the next several decades, production from California fields could increase, particularly from offshore platforms. Natural gas, which is plentiful in California's Cenozoic marine sedimentary rocks, could also be a significant source of revenue in the future.

Soils

As we saw in Chapter 11, fertile soils are one of California's most important resources. However, as California's population increases, residential areas are constantly encroaching onto the alluvial valley floors. If agriculture is to continue to be important in the state's economy, preservation of agricultural resources must be a priority, particularly in the Central Valley.

Water

Water is the natural resource in greatest demand, of greatest controversy, and with the most direct economic impact on our state. The material presented in Chapter 10

Fig. 19-13 Gravel mining operations. (Source: Langer, W. U.S. Geological Survey.)

serves only as an introduction to California's water, but it serves to highlight the major principles and issues.

CALIFORNIA'S LANDSCAPE

Because of its complex geology and varied climates, much of California's landscape is dramatically beautiful. The scenery varies from rocky deserts to alpine meadows, and from forested rivers to coastal sea cliffs. Despite being home to 33 million people, California has abundant space that remains open for its residents to enjoy. One important side effect of recent uplift along the San Andreas fault system was the creation of rugged mountains, natural open spaces for California's residents. With the exception of Sacramento, California's major urban areas are naturally confined, because rugged mountains enclose the urbanized valleys. The mountains have prevented limitless urban sprawl from overtaking the state. The creation of 15 national parks, monuments, and recreation areas, 273 state parks and beaches, and hundreds of local preserves guarantees that some of these areas will be left undeveloped for the enjoyment of future Californians. All California citizens owe a great debt to the foresight of early conservationists who set aside so many of the most beautiful areas for future enjoyment (Fig. 19-14).

Fig. 19-14 President Theodore Roosevelt and naturalist John Muir at Glacier Point in Yosemite, about 1906. (Source: Underwood and Underwood: Library of Congress, Prints and Photographs Division; reproduction number LC-USZ62-107389 DLC.)

The California landscape provides opportunities for a wide range of outdoor recreational activities. Equipped with even a small amount of knowledge about California geology, one can learn much from the rocks and landforms in different parts of the state while enjoying these activities. Investigations into California's geologic history provide learning for students of all ages and levels of expertise, particularly if they have an opportunity to visit geologically important localities in the field. To this end, it is my hope that those who read all or parts of this book are only at the beginning of their study of California geology.

REFERENCES

BAILEY, E., and HARDEN, D. 1974. *Mineral Resources in the San Francisco Bay Area: Present Availability and Future Demand.* U.S. Geological Survey, Miscellaneous Investigation Map I-909.

BLAIR, M., and others. 1979. *Seismic Safety and Land-Use planning: Selected Examples from California.* U.S. Geological Survey, Professional Paper 941-B, 82 pp.

HAYS, W.W. 1981. *Facing Geologic and Hydrologic Hazards: Earth-Science Considerations.* U.S. Geological Survey, Professional Paper 1240-B, 109 pp.

TYLER, M. 1995. *Look Before You Build: Geologic Studies for Safer Land Development in the San Francisco Bay Area.* U.S. Geological Survey, Circular 1130, 54 pp.

Web Sites and Media Resources

California Department of Resources: *http://www.consrv.ca.gov*

California State Parks, California Department of Parks and Recreation: *http://www.parks.ca.gov*

National Park Service, U.S. Department of Agriculture: *http://www.nps.gov*

San Francisquito Canyon and St. Francis Dam, Historic photographs: Santa Clarita Valley Historical Society, Santa Clarita Valley in Pictures: *http://www.scvhistory.com/scvhistory/scvhistory.htm*

"The St. Francis Dam Disaster: Anniversary of Second-Worst Disaster in California History." National Public Radio, *All Things Considered*, March 2003: *http:// discover.npr.org/features/feature.jhtml?wfId=1190341*

The St. Francis Dam Disaster, Documentary film by Wilkman Productions, scheduled for completion, 2003: *http://www.wilkman.com*

Glossary

A horizon The top layers of a soil where organic matter accumulates.

ablation The loss of mass from a glacier, mainly by melting.

accreted terrane A portion of a plate added to a larger block of crust along a convergent plate boundary.

accumulation area The part of a glacier where ice builds up.

acre-foot The amount of water needed to cover an acre of land to a depth of 1 foot; about 325,851 gallons.

active fault A fault capable of generating earthquakes today; *syn. capable fault.*

aftershock Smaller earthquake that follows the largest earthquake of a series in the same area.

aggregate Crushed rock, gravel, and sand used in construction.

alluvial fan A fan-shaped landform along a mountain front formed by the buildup of stream sediment and debris flow deposits.

alluvium Sediment deposited by a river.

alpine Pertaining to mountains.

alpine glaciers Streams of ice that flow down valleys in mountainous areas.

Alquist-Priolo Earthquake Fault Zoning Act 1972 California law that prohibits construction within 50 feet of an active fault. (Before 1994, *Special Studies Zone Act.*)

amplitude Height of a wave.

andesite Volcanic rock with intermediate silica content.

andesitic rock (lava) Volcanic rock (or lava) characteristically medium dark in color and containing 54 to 62 percent silica and moderate amounts of iron and magnesium; also called intermediate volcanic rock or lava.

anticline An upward-buckling fold.

appropriation rights The system of water law whereby water is allocated in the order of first historic use.

aqueduct An artificial channel constructed to carry water.

aquiclude Water-excluding layer. aquifer Water-carrying layer.

aquitard Water-retarding layer.

arroyo Canyon

artesian Condition in which groundwater is held under high pressure.

ash Fine-grained volcanic material erupted into the atmosphere.

asthenosphere The region of the mantle about 100 to 250 kilometers beneath Earth's surface where rocks are plastic and easily deformed.

attenuation The decrease in the amplitude and energy of seismic waves with distance from the epicenter.

axis The central hinge line of a fold.

B horizon The layers of a soil where weathering products accumulate.

bajada A broad area where alluvial fans overlap along the edge of a mountain range.

basalt Mafic volcanic rock.

basaltic *See mafic.*

base level The lowest point in a stream system.

basement The complex of older rocks, usually igneous and/or metamorphic, which are overlain by younger rocks and sediments in an area; *syn. crystalline basement.*

basement complex The oldest identifiable rocks in any region. May be of oceanic or continental origin.

basin A bowl-shaped depression.

bastnasite a yellow-to-brown mineral that is a source of rare earth elements.

batholith Very large body of plutonic igneous rock.

bay mud Dark colored clay and silt deposited in the San Francisco Bay estuary.

bed load The sediment that a stream moves along the bottom of its channel by rolling and bouncing.

bedding The layering in sedimentary rocks.

bimodal volcanism Eruptive pattern producing both high-silica and mafic rocks.

blind fault A fault that does not offset materials at the surface.

blueschist Foliated metamorphic rock containing diagnostic minerals that form under conditions of high pressure and low temperature in a subduction zone.

body wave Earthquake wave that travels through the earth's interior.

Brea Spanish for "tar." Origin of the name for Rancho La Brea tar pools, which trapped over a million specimens of plants and mammals during Pleistocene time.

breccia Coarse-grained rock composed of broken angular rock fragments.

calaverite A bronze-yellow massive mineral with metallic luster; a telluride of gold; first found in Calaveras County, California.

caldera A bowl-shaped depression formed by the collapse of a volcanic cone following a major summit eruption.

carbonate Mineral or rock containing CO_3.

carbonate platform A shallow marine shelf along the edge of a continent where most sediment is contributed by carbonate-secreting organisms like coral.

carbonatite Unusual plutonic rock composed mostly of carbonate minerals.

Cenozoic The geologic era between 66 million years ago and the present.

channel Pathway carved by a stream to carry runoff.

chemical weathering The chemical decomposition of minerals and rocks at the earth's surface.

chert Fine-gained sedimentary rock composed of silica, usually the skeletons of radiolaria.

chlorite Green silicate mineral with sheet structure, formed during low-grade metamorphism.

chromite A mineral containing chromium, iron, and oxygen.

cinder cone A small, steep-sided volcano built by the accumulation of mafic pyroclastic material ejected from a single vent.

cinders Volcanic fragments; small, and usually red, typically 5 to 2.5 cm across.

cinnabar Mercury sulfide, the most common mercury ore.

cirque Bowl-shaped depression at the head of a mountain valley, created by glacial erosion.

clastic Rock texture in which fragments are held together by a cementing mineral.

clay 1. Type of sheet silicate minerals containing water 2. Sediment composed of particles with diameters less than 2 microns.

composite cone A steep volcanic cone built by both lava flows and pyroclastic eruptions, also called stratovolcano.

composite volcano *See stratovolcano.*

composition The type and proportion of constituent elements or minerals in a rock.

compression Stress resulting from forces directed toward each other.

confined aquifer A water-bearing layer sandwiched between less permeable layers.

contact Boundary between two rock types.

contact metamorphism Changes in rocks adjacent to intruding magma.

convection The movement of liquid and gas in which hot, less dense materials rise and colder, denser materials sink.

converge Come together.

convergent boundary The line where two plates meet as they move toward each other.

core The inner part of the Earth, divided into the outer liquid and solid inner core, both of which are composed mostly of iron and nickel.

coseismic Taking place during an earthquake.

cradle Large box mounted on rockers, used to separate gold ore from sediment.

creep The very slow, generally continuous downslope movement of soil and debris under the influence of gravity.

crest The imaginary line running along the top of a mountain range.

cross beds Distinctive sedimentary layers, running criss-cross to each other, resulting from shifting currents or winds.

cross section A diagram showing a slice beneath the surface along a line and depicting the arrangement of rocks beneath the line.

cross-cutting relations Fundamental principle that states that a feature cutting through a rock must be younger than the rock it cuts.

crust The thin, outer layer of the Earth, composed mostly of silicate minerals; upper part of Earth's lithosphere.

crystalline The state of matter in which the atoms are arranged in a regular, repealing pattern.

crystalline basement *See basement.*

dacite Volcanic rock type with silica content intermediate between andesite and rhyolite.

Dam Safety Act Passed in 1973, required evaluations of the safety of existing dams in California, including preparation of evacuation plans and maps showing flood areas in case of dam failure.

debris Broken rock and soil material.

debris avalanche A very rapid, tumbling movement of rock and soil from a steep slope or a cliff.

debris basins Basins constructed to contain high volume of sediment washed from canyons during floods.

debris flow A water-laden movement of unconsolidated soil and rock material, including large blocks, down a channel.

deflation The removal of clay and dust from dry soil by strong winds.

deformation A general term for folding, faulting, and other changes in rocks, sediments, and the land surface that take place in response to stress.

deposition The settling of sediment.

desert pavement A veneer or armor of stones on the desert floor.

desert varnish Thin, dark coating on the surfaces of rocks formed by the gradual buildup of manganese and iron oxide.

detachment fault A large, gently dipping fault, along which a large crustal block moves over deeper rocks.

diatom Single celled plankton whose cell walls are composed of silica.

diatomite Siliceous rock composed of the remains of diatoms.

differential weathering Weathering that wears away rocks in an area at variable rates because of differences in the resistance of materials to erosion; *syn. differential erosion.*

dike A relatively narrow igneous body that cuts through the surrounding rocks.

dip The inclination of layered rocks, with 0 degrees being horizontal.

discharge The volume of water flowing in a stream per unit of time.

displacement The separation of formerly continuous features across a fault.

dissolved load Amount of material water carries in solution.

divergent boundary The line where two plates meet as they move apart from each other.

divide Imaginary line separating the drainage basins of two streams.

dome A hemisphere-shaped mountain.

drainage basin The area of land from which a given area collects runoff.

dredge To mechanically dig up the bed of a stream, lake, or ocean to extract sediment, deepen a channel or harbor, or mine placer deposits.

dredge field A pile of river sediment left after gold dredging operations.

dunite Ultramafic plutonic rock composed of the mineral olivine.

dynamic equilibrium A state of balance in which a system constantly adjust to disturbances to maintain the balance.

earthflow A detachment of soil and broken rock and its subsequent downslope movement at slow or moderate rates in a stream- or tongue-like form.

eclogite Metamorphic rock formed in the mantle, contains garnet.

emergent coast A coastline that has risen relative to sea level.

enclaves Included pieces of rock within a larger igneous body of contrasting composition.

epicenter The point on the earth's surface directly above the focus of an earthquake.

epoch Subdivision of a geologic period.

era Major division of the geologic time scale: Paleozoic, Mesozoic and Cenozoic.

erosion The removal of solid particles and ions in solution from an area at the surface.

erratic An out-of-place rock transported by a glacier and left behind when the ice melted.

estuary A semi-enclosed coastal body of water where fresh and salt water mix.

evaporation Change of state from liquid to gas (vapor).

evaporite Mineral formed from solution as a result of evaporation.

exfoliation The peeling off of the outside surfaces of massive rocks such as granite.

extension Stretching that occurs in reponse to stresses directed away from each other.

extinction The disappearance of a species from the Earth.

fault An abrupt break in rocks or sediments along which movement has occurred.

fault creep Gradual movement along a fault that takes place without earthquakes; *syn. tectonic creep.*

fault scarp A break in the ground surface created by vertical movement on a fault.

fault segment A section of an active fault that has the same earthquake characteristics and rupture history.

fault strand A single break within a fault zone.

fauna The assemblage of animals in a given habitat.

feeder dikes Linear conduits extending from a magma chamber to volcanic vents at the surface.

felsic *See silicic.*

Field Act Passed in 1933 after Long Beach earthquake. Requires state approval and inspection of school buildings to ensure seismic safety. Amended in 1967, but not all schools are yet in compliance.

fining-upward sequence A sedimentary deposit that is coarse-grained at the base and becomes gradually finer-grained towards the top.

fissure A linear crack.

flash flood A sudden flood of water through a stream channel.

floodplain The part of a stream valley adjacent to the channel that is covered during floods.

focus The point where an earthquake originates.

fold A bending of rocks or sediments without faulting.

fold-and-thrust belt An area of subparallel faults and folds indicating active compression.

foliation Sheet-like layering in metamorphic rocks that results from the growth of minerals in a preferred orientation.

foraminifera Single-celled protozoans whose shells are constructed from calcium carbonate.

fore-arc basin A marine depression between a volcanic island arc and an oceanic trench.

foreshock A minor tremor that precedes a larger earthquake.

formation A recognizable geologic unit that can be identified and mapped.

fossil The remains or traces of living organisms preserved in geologic materials.

fumarole A volcanic vent that emits only gas (fumes).

gabbro Mafic plutonic rock.

geobarometer A mineral whose presence indicates the pressure at which a rock was metamorphosed.

geomorphic province A region identified by its characteristic landscape, influenced by a combination of geology, climate, and topography.

geothermal Related to the heat of Earth's interior.

geothermometer A mineral whose presence indicates the temperature at which a rock was metamorphosed.

glacial outwash *See outwash.*

glacial polish Shiny rock surfaces created by the grinding of rock fragments contained within a moving glacier.

glacier A body of ice that moves over the land surface under its own weight.

glaucophane Blue silicate mineral of the amphibole family, found in blueschist.

gneiss High-grade metamorphic rock with characteristic light and dark banding.

graben A depressed segment of the Earth's crust bounded on at least two sides by faults; *see horst.*

graded bedding Sedimentary layer in which the particles decrease in size from bottom to top.

gradient The steepness of a hillslope or stream channel.

grain size Refers to the size of individual mineral crystals or particles within a rock or sediment deposit.

granite Silicic plutonic rock.

granitic rocks Igneous rocks with relatively high silica content, including granite, granodiorite, tonalite; *syn. granitoid.*

granitoid [granitic] A general term for intrusive igneous rocks that look similar to granite but may range in composition from quartz-diorite to granite. All granitic rocks are light colored; feldspar and quartz are visible in hand specimen.

gravel Rounded rock particles of larger size than sand.

greenstone A type of metamorphosed volcanic rock, usually basalt with a characteristic green color.

greywacke Type of sandstone containing abundant sand-sized rock fragments, typically gray or brown.

ground rupture Breaking of the surface along a fault during an earthquake.

groundwater Water stored underground in the spaces within rocks or sediments.

half-life The time required for one half of a given amount of a radioactive isotope to decay.

hanging valley A glacial valley whose mouth is much higher than the elevation of the valley it joins.

hardpan A soil layer formed by the buildup of carbonate, silica, or iron.

harmonic earthquakes Multiple, small to moderate earthquakes at regular time intervals, signaling the movement of magma rising toward the surface in an active volcanic area; *syn. harmonic tremors.*

hazard map A map portraying the relative risks from one or more geologic processes.

headwaters The highest-elevation areas of a stream system.

heap leaching A chemical process using cyanide to extract gold from quartz veins.

highly extended terranes Areas where extension of the Earth's crust has revealed mylonite and other metamorphic rocks from the lower crust.

Holocene The most recent geologic epoch; the past 10,000 years.

hornfels A high-temperature, low-pressure metamorphic rock of uniform grain size showing no foliation. Usually formed by contact metamorphism.

horst A block of the Earth's crust separated by faults from adjacent blocks; *see graben.*

Hospital Safety Act Passed in 1973 following San Fernando earthquake, strengthened construction standards for hospitals.

hot spot A stationary, localized zone of melting beneath the lithosphere.

humus Partially decayed plant and animal material; the organic part of soil.

hydraulic mining Washing of sediments from hillsides using high-pressure hoses to extract *placer* gold.

hydrocarbon compound A chemical compound composed of hydrogen and carbon.

hydrology The study of water.

hydrothermal Produced by hot fluids.

hydrothermal alteration The changing of minerals and crystallization of new minerals caused by the chemical action of groundwater, particularly at hot springs and geysers.

igneous Fire-formed; rocks crystallized from cooling magma.

inactive fault Fault that has not moved in millions of years. Unlikely to generate earthquakes.

incision Downcutting by a stream into its channel.

inclusion A fragment of older rock incorporated into a younger igneous rock; *syn. xenolith.*

index fossil A fossil that is very useful for geologic correlations.

infiltration The movement of water from the surface into soil or rock.

intensity A measure of ground shaking during an earthquake, based on types and extent of damage and human experience; *syn. Mercalli intensity.*

interior drainage A system of streams that converge in a closed basin and evaporate without reaching the sea.

intermittent stream or lake One that has water only periodically.

intrusive rock Igneous rock that solidified from magma below the earth's surface; also called plutonic rock.

invertebrate Lacking a spinal column.

inverted topography A landscape in which areas formerly in low-lying areas are now at the highest elevations, and vice versa.

isotope A form of a chemical element having the same number of protons, but a different number of neutrons in its nucleus, and hence a different atomic mass.

jadeite A green silicate mineral formed during high pressure metamorphism, found within the Franciscan complex.

landslide A general term for gravity-driven movement of rock and soil down a slope.

lanthanide elements Fifteen rare earth elements contained in bastnasite that have important commercial uses.

lapilli Fragments of volcanic rock formed when magma is ejected into the air by expanding gases. The sizes of the fragments range from sand- to cobble-size.

lava Molten silicate material (magma) at the Earth's surface.

lava tube Opening in volcanic rock formed when fluid lava flows through a recently cooled channel of lava, evacuating the center of the channel.

levee A naturally formed low ridge built along the side of a stream channel, or an artificial barrier constructed to increase the capacity of a channel.

limestone Sedimentary rock composed of calcium carbonate.

liquefaction The process by which ground shaking transforms wet, sandy sediment into an unstable, dense fluid during an earthquake.

lithification The process by which sediments become sedimentary rocks.

lithology Type of rock or sediment in a geologic unit.

lithosphere The Earth's rigid outer 100 kilometers, including the crust and the outer mantle.

load The dissolved and solid material transported by a stream.

local magnitude One commonly used measurement of earthquake magnitude; *see magnitude.*

lode Type of mineral deposit in which a valuable commodity is concentrated in a vein.

long tom A large version of the rocker used to separate placer gold from river sediments.

longshore current The flow of ocean water parallel to the coast within the surf zone.

longshore transport The movement of sand by the longshore current.

Love wave Type of surface earthquake wave.

mafic Term describing igneous rocks or minerals with high proportions of iron and magnesium and relatively low silica; *syn. basaltic.*

magma Molten silicate material.

magma differentiation Compositional changes in a magma caused by the early crystallization and isolation of early-formed minerals.

magnitude A measure of earthquake size determined from seismographs.

mantle The thickest of Earth's layers, between the crust and the core; composed of more dense silicate minerals than the crust.

marble Nonfoliated metamorphic rock formed from carbonate sedimentary rocks.

marine terrace A wave-cut platform that has been uplifted above the modem shoreline; *see wave-cut platform, shoreline angle.*

mariposite Unusual apple-green silicate mineral in the mica family; contains chromium.

marker bed An easily identified and unique geologic layer that can be used to correlate rock or sediment sequences from one area to another.

mass movement The downslope movement of rock, soil, and debris under the influence of gravity.

matrix The fine-grained material in a rock that contains larger particles.

maximum contaminant level (MCL) The highest level of a substance allowed by law in drinking water.

mechanical load Portion of material carried by a stream as solid particles.

Mediterranean climate A climate type characterized by warn, dry summers and mild, wet winters.

melange A mixture of rock types jumbled together by tectonic activity.

Mercalli scale *See intensity.*

Mesozoic The geologic era from about 235 million to about 66 million years ago.

metamorphic Rocks changed by the action of heat, pressure, and hot fluids.

metamorphic core complex Metamorphic rocks exposed in the lower plate of a detachment fault in a highly extended terrane.

metasedimentary Sedimentary rocks transformed by heat and pressure.

metavolcanic Volcanic rocks transformed by heat and pressure.

mima mounds Rolling hills and hollows, probably formed by weathering and burrowing animals.

mineral A naturally occurring, inorganic, crystalline solid composed of one or more chemical elements.

miogeocline A wedge of shallow marine sediment that accumulated on the continental shelf and slope of a passive margin; *see passive margin.*

moment magnitude Measure of earthquake size related to the force couple across the fault rupture surface; *see magnitude.*

moraine A ridge of sediment left by a glacier.

morro A round hill or point of land.

Mother Lode The major gold mining district in California's Sierran foothills; *see lode.*

mountain range Chain or group of mountains consisting of a number of individual peaks or mountains.

mud Fine-gained sediment composed of particles smaller than 1/16 mm.

mylonite Rock formed by intense grinding along the deeply buried parts of fault zones.

normal fault A fault that drops one side down relative to the other as a result of extension.

obsidian Volcanic glass with high silica content.

obsidian dome A round steep volcanic cone formed by the eruption of high-silica lava.

offset Displacement of a feature by an earthquake.

offshore bar A submerged ridge of sand.

ooze Fine-grained deep-sea sediment composed mostly of the remains of marine animals or plants.

ophiolite suite Assemblage of rocks thought to represent ocean-floor sediments and the crust and upper mantle beneath them; *syn. ophiolite.*

ore A concentration of a mineral or rock that can be extracted at a profit.

ore shoots Pockets of pure gold found within veins.

original horizontality Fundamental principle of sedimentary layers that states that sediments are deposited in flat-lying layers.

orogeny The process of mountain foundation, especially by folding of the Earth's crust.

orographic effect The effect of mountain ranges on precipitation.

outwash Sediment carried by a stream flowing from a glacier.

overdraft The pumping of groundwater in excess of recharge rates.

P wave The primary, fastest wave generated by an earthquake, a compressional body wave.

paleogeography The ancient environments and landscapes reconstructed, from the types and locations of rocks found in an area.

paleontology The study of fossils and the history of life.

paleoseismology The study of ancient earthquakes.

Paleozoic The geologic era from about 570 million to about 235 million years ago.

Pangaea The supercontinent that formed by the collision of all of the continents during late Paleozoic time.

panning Technique used to extract gold from river sediments by swirling the sediment in water using a round, shallow dish.

parent material The rock or sediment from which a soil formed.

pass A low point along the crest of a mountain range.

passive margin A continental edge far from active plate boundaries.

peak The highest point on a mountain; *syn. summit.*

peat A dark brown or black organic-rich material composed of partially decomposed wetland plants.

pegmatite Very coarse-grained igneous plutonic rock.

peridotite Ultramafic igneous rock composed of olivine together with plagioclase feldspar and pyroxene.

period Subdivision of a geologic era.

perlite A form of obsidian consisting of glassy globules, used as insulation or in plant growth media.

permeability The ability of a material to transmit fluid.

petroglyph A carving on a rock surface made by etching into the desert varnish.

petroleum Naturally occurring mixture of hydrocarbon compounds that includes natural gas, gasoline, and oil.

physical weathering Disintegration of rocks into smaller pieces through physical processes such as freezing and thawing, crystallization of salt, or the growth of roots.

pillow lava Distinctive pillow-shaped basalt that forms when lava solidifies under water.

placer A type of mineral deposit in which a valuable commodity is concentrated in river or beach sediments.

plate tectonics The concept that the Earth's lithosphere consists of a number of rigid, mobile pieces (plates) riding over the more plastic asthenosphere.

playa Dry lake bed in a desert basin.

Pleistocene The geologic time period from about 1.8 million until about 10,000 years ago; the Ice age.

pluton An identifiable body of igneous rock crystallized beneath the surface from a single body of magma.

plutonic Rocks crystallized from magma beneath the earth's surface*; syn. intrusive.*

polymerize To form a complex molecule by the chaining together of simple molecules.

porosity The percentage of the total volume of a rock or sediment that is empty spaces.

porphyritic Texture describing igneous rocks with some large, early-formed crystals in a mass of finer grained crystals or glass.

precursor A change in geologic conditions that precedes a geologic event such as an earthquake or volcanic eruption.

Proterozoic The geologic time period from about 2.5 billion to about 570 million years ago.

protolith The parent rock type of a metamorphic rock.

pull-apart basin A depression created by crustal extension.

pumice Lightweight volcanic glass containing abundant vesicles.

pyroclastic Fire-broken; fragmented volcanic material ejected during explosive eruptions.

pyroclastic flow Hot volcanic flow containing fragmental materials.

quartz Common silicate mineral composed of silicon and oxygen.

quartzite Metamorphosed quartz sandstone.

quicksilver Mercury.

radioactive decay Process through which unstable radioactive isotopes break down spontaneously to form more stable atoms, giving off radiation in the process.

radioactive isotopes Small fraction of an element that is unstable and breaks down spontaneously, giving off energy in the form of radiation.

radioactivity Spontaneous decay of an unstable isotope; *see isotope.*

radiolarian A single-celled marine organism whose tests are composed of silica; *pl. radiolaria.*

radiometric dating Age dating of Earth materials using the known half lives and decay rates of isotopes; *see half-life, isotope.*

rain shadow A dry area on the downwind side of a mountain range, created because the mountain range forms a barrier to moisture-carrying clouds.

range-front fault A young fault along the edge of an uplifting mountain range.

rare-earth elements Fifteen elements of the periodic table valued for their uses in technology industries.

Rayleigh wave Type of surface earthquake wave.

recharge The addition of water to groundwater from the surface.

recurrence interval The characteristic time period between repeated geologic events such as floods, earthquakes, or volcanic eruptions.

refraction The bending of waves as they travel from one medium to another and change velocity.

regional metamorphism Changes in rocks over a large scale in response to major magmatic and/or tectonic events.

relative age dating Determination of the ages of geologic formations, materials, or events relative to each other.

resonance Amplifying effect produced when the natural vibration frequency of a material or structure is matched by the frequency of seismic waves.

resurgent dome A volcanic mound built in the center of a caldera by eruptions that follow the caldera formation.

retreat The name given to the shrinking of a glacier, which happens when more snow melts in summer than accumulates in winter. The glacier does not actually move backwards.

reverse fault A fault along which one side is moved up over the other side as a result of compression.

rhyolite Silicic volcanic rock.

Richter magnitude (Richter scale) A commonly used measure of earthquake magnitude based on a logarithmic scale. Each integral step on the scale represents a tenfold increase in the extent of ground shaking, as recorded on a seismograph.

riparian Adjacent to a stream.

riparian rights A system of water laws whereby water is allocated to those owning land that borders a body of water

rip-up clasts Chunks of mud incorporated when turbidity currents rip up portions of underlying mud layers. Formed during the formation of a submarine fan.

roche moutonée An asymmetric bedrock dome created by flowing ice.

rock cycle A conceptual model used to illustrate the origins of and the relationships between earth materials.

rock fall Gravity-driven fall of loosened rock fragments from a steep cliff.

rocker A long box mounted on rockers and used to sort placer golds from river sediments; *syn. cradle.*

Rodinia The name (from the Russian for "homeland") given by geologists to an ancient PreCambrian supercontinent formed by the accretion of smaller continental fragments, including those represented by southeastern California's most ancient rocks.

roof pendant A block of metamorphic rock, originally at the top of a magma chamber, at the edge of a plutonic body.

rounding One measure of how well-traveled and tumbled grains of sediment are; refers to a lack of sharp edges.

runoff Water that flows over the land surface.

rupture surface The portion of a fault that slips during and earthquake.

S wave The secondary body wave generated in an earthquake, a shear wave.

safe yield The amount of groundwater that can be pumped without exceeding the recharge to the system.

sag pond A small depression in the land surface along an active fault zone.

salinization The buildup of soluble minerals in the lower part of a soil.

salts Minerals crystallized from supersaturated water.

salt-water intrusion The inflow of salt water into freshwater aquifers.

schist Strongly foliated metamorphic rock with visible platy minerals.

scoria Glassy volcanic rock low in silica that contains abundant vesicles.

sea stack A block of rock isolated from the shoreline by wave erosion.

sea-floor spreading The divergent motion and creation of oceanic crust caused by rising magma along divergent plate boundaries.

sediment Rock and mineral fragments derived from preexisting rocks.

sedimentary Rocks formed at the earth's surface from sediment, organic remains, and/or minerals precipitated from solution.

sedimentary basin A low-lying area where sediments accumulate.

seismic waves Vibrations resulting from released strain energy when rocks along fault planes break during earthquakes. There are four types: two surface waves: Love waves and Rayleigh waves; and two body waves: P(rimary) waves and S(econdary) waves.

seismograph Instrument for recording earthquake waves.

seismograph station Site where seismographs record distance between an earthquake's focus and the station. Measurement from three or more stations can be used to locate an earthquake's point of origin.

seismology The study of earthquakes.

seismometer Sensor used to measure earthquakes.

selenium A chemical element essential for living animals but toxic in large amounts.

septa Blocks of metamorphic rocks in the High Sierra that form screens or dividers at the edges of plutons.

serpentine (serpentinite) 1. A green, platy, hydrated silicate mineral containing magnesium. 2. Internal name for metamorphic rock composed of serpentine; *syn. serpentinite.*

shale Fine-grained sedimentary rock composed of mud-sized particles; *syn. mudstone.*

shear stress Forces causing two blocks of rock or other material to move past each other.

sheeted dikes Vertical basalt dikes carrying magma to the ocean floor; part of an ophiolite suite; *see dike.*

shelf The gently sloping part of the ocean floor below the low-tide line adjacent to the shoreline.

shield volcano A large, low, round volcano built mostly of basalt flows.

shoreline angle The line where a wavecut platform meets a sea cliff.

silica A term describing igneous rocks or minerals with high proportions of silica and relatively low iron and magnesium; *sin. folic.*

silica carbonate rock An unusual Franciscan rock, probably altered serpentinite, which is a common host rock for mercury ores.

silicate A mineral belonging to the group that contains the elements silicon and oxygen bonded together *in* a tetrahedron.

sinuous Snakelike, winding.

slate Fine-grained, foliated metamorphic rock that breaks into thin, regular layers.

slip rate The average rate of movement on an active fault.

slough A channel in a wetland area.

sluice box A long open box with a chain of baffles at the end designed to sort and concentrate *placer* gold from river sediment.

slump A type of deep landslide in which the upper part rotates as it moves downslope.

soil horizon A distinct soil layer whose properties differ from the layers above and below it.

soil profile The sequence of layers, extending from the surface to fresh rock or sediment, that result from the accumulation of organic material and the weathering of rock or sediment.

sorting A term used to describe how similar in size the particles in sediment or sedimentary rock are.

spatter cone Small volcanic cone formed when blobs of lava are thrown from a vent and pile up in a heap.

Special Studies Zone *See Alquist-Priolo Earthquake Fault Zoning Act.*

spit A sandy bar built by longshore currents from a promontory into a body of water.

stable carbonate platform The environment along the continental margin of western North America during the Paleozoic period, a broad continental shelf along a tectonically quiet margin, where carbonate-secreting organisms were dominant.

stage The height of water in a river at a given time.

stage recorder A device that continuously records a stream's water depth at a monitoring station or flood warning station.

stamp mill Device used to crush gold-bearing quartz.

strain energy The energy stored in rocks as they are subjected to stress.

strait A relatively narrow waterway or passage connecting two large bodies of water.

stratigraphic traps Type of geologic trap in which petroleum or natural gas accumulate because of variations in the permeability of reservoir rock.

stratigraphy The relationships and sequencing of layered rocks.

stratovolcano A large, steep-sided volcano built of both pyroclastic material and lava flows; *syn. composite volcano.*

stream channels Channels eroded by runoff from a stream's drainage basin, adjusting in size according to the volume of water carried, and forming a network that creates a system of drainage basins or watersheds.

stream terrace A remnant of a floodplain formed when a stream was flowing at a higher level.

stress The force per unit area acting on a rock; *syn. directed pressure.*

striations Small, parallel grooves created when a glacier flows over a bedrock surface.

strike The geographic orientation of an imaginary horizontal line drawn in the plane of an inclined rock layer.

strike-slip fault Fault with lateral motion; *syn. lateral fault*

stromatolite Carbonate mounds built by organisms such as algae.

structural trap Type of geologic trap in which faults or anticlines trap hydrocarbon compounds so that they accumulate in a pool of petroleum.

structurally controlled topography A landscape in which the mountains and valleys follow the trends of contacts and/or faults.

subduction The process by which an oceanic plate is driven beneath another plate into the mantle along a convergent boundary.

subduction complex The characteristic assemblage of rocks brought together along a subduction zone.

submarine canyon An underwater extension of a valley on land carved on the continental shelf.

submarine fan A fan-shaped feature on the ocean floor formed by deposits of turbidity currents.

submerged coast A coastal area that has been lowered relative to sea level.

subsidence Sinking.

summit Highest point of a mountain range.

superposition Fundamental principle of sedimentary layers that states that, in a sequence of undisturbed layers, the layers at the bottom are oldest.

surf zone The area along a coast where waves break.

surface waves Earthquake waves that travel on the earth's surface.

suspended sediment Solid particles carried within a body of flowing air or water

syncline A down-buckling fold.

table mountain River valley that becomes resistant to erosion when a volcanic eruption fills it with lava. As the surrounding area erodes, the sediments of the river channel beneath the rock form a flat-topped, river-shaped mountain.

Tahoe glaciers Extensive glaciers occupying the Sierra Nevada before the Tioga glaciation, probably between 130,000 and 200,000 years ago.

talc Silicate of magnesium common among metamorphic minerals, greasy, and extremely soft due to sheet structure. Sometimes known as soapstone.

tarn Lake formed in bottom of cirque after glacier ice has disappeared.

telluride Rare gold-bearing mineral ($AuTe_2$) found in Sierra Nevada foothills.

tephra Volcanic material thrown into the air during eruptions.

terrane A large block of crust with a distinct geologic character, originally part of the same crustal plate.

Tertiary gravels Deposits of rivers that flowed westward from a previous Sierra Nevada range during Tertiary time. Source of rich placer gold.

texture The shape, size, and orientation of particles in a rock, and the way they fit together.

thrust fault A type of reverse fault in which the fault surface is only slightly inclined; *see reverse fault.*

till Unsorted sediment transported by a glacier

Tioga glaciation Sierran glacial period during which Tiogan glaciers filled the valleys of the Sierra Nevada, between 20,000 and 100,000 years ago.

tonalite Intrusive igneous rock made of plagioclase feldspar, quartz, and amphibole or biotite. May be similar to diorite but contains considerable quartz and is not as dark, and chemically has less calcium, iron and magnesium.

topographic map A map portraying the elevation and shape of the land surface.

topography The form of the land surface, including the steepness.

tourmaline Rare mineral in pegmatite dike famous for pink, green and "watermelon" color.

transform boundary The line where two plates meet as they move parallel to each other

transpiration The process by which plants release water vapor into the atmosphere from their leaves.

transpression A combination of compression and transform motion.

transtension A combination of extension and transform motion.

trap A geologic barrier to the migration of petroleum.

trench (deep ocean) A very deep, long depression in the sea floor marking the surface of subduction zone.

tributary A stream that flows into a larger stream

trilobite Arthropods extinct since the end of Paleozoic time.

triple junction The point at which three tectonic plates meet.

tsunami An ocean wave generated when an earthquake displaces the sea floor

tufa Calcium carbonate formed along the edges of desert lakes where alkaline lake water mixes with springwater.

tuff Cemented (lithified) volcanic ash.

turbidite Rhythmically layered beds deposited by turbidity currents.

turbidity current A dense mixture of water and sediment that flows from the continental edge onto the deep ocean floor.

turtlebacks Form of detachment faulting from Cenozoic time found on eastern side of Death Valley.

U.S. Geological Survey Federal agency in the Department of Interior responsible for the nation's geologic programs, including mapping, hazard studies, resource inventories, and water safety oversight.

ultramafic Describes igneous rocks with very high iron and magnesium and low (about 40%) silica.

ultrametamorphic Rocks subjected to such a high degree of metamorphism that they have partly melted and are transitional to igneous rocks.

unconsolidated sediments (unconsolidated material) Nonlithified sediment that has no mineral cement or matrix binding its grains.

underplated Describes a situation in which one layer of rock is shoved beneath another.

Uniform Building Code (UBC) International standards for building construction adopted by California cities and counties.

Unreinforced Masonry (URM) Building Law 1986 California law mandating local governments to develop earthquake hazard mitigation programs for older brick buildings.

unsaturated zone Area of soil near the surface that is above the level of the water table and may therefore be dry during some periods.

vein An igneous intrusion filling a crack; may contain ore minerals.

vent The opening in a volcano where lava erupts.

vernal pool An ephemeral pool that forms in a topographic depression during the rainy seasons, giving life to various animals and plants.

vesicle A hole in a volcanic rock made by escaping gas.

viscosity The resistance to flow in a fluid; stickiness.

volcanic Rocks crystallized from magma at the earth's surface.

volcanic ash Instantly cooled high- or intermediate-silica lava less than 4 millimeters in diameter, produced by a pyroclastic eruption.

volcanic bomb A pyroclastic rock fragment that shows the effects of cooling in flight in its streamlined or "bread-crust" surface.

volcanic debris flows Mixtures of instantly melted snow and ice, as well as ash, produced by the eruption of an andesitic volcano. The saturated mixture can travel down the rivers on the flanks of a volcano at very high speed.

volcanic hazards maps Maps outlining areas likely to be impacted by future volcanic events. The maps are prepared based on studies of the geologic information from the area of known active volcanoes, and they can be used to define evacuation routes and plan future developments.

water table The depth below which rock and sediment are saturated with groundwater.

watershed *See drainage basin.*

wave-cut platform A nearly horizontal surface eroded on bedrock by waves along an uplifting coast.

weathering The physical disintegration and chemical alteration of rocks at the earth's surface.

xenolith *See inclusion.*

zone of saturation The area beneath the surface where all available space in a sediment or rock is filled with water.

Place Index

Subject Index

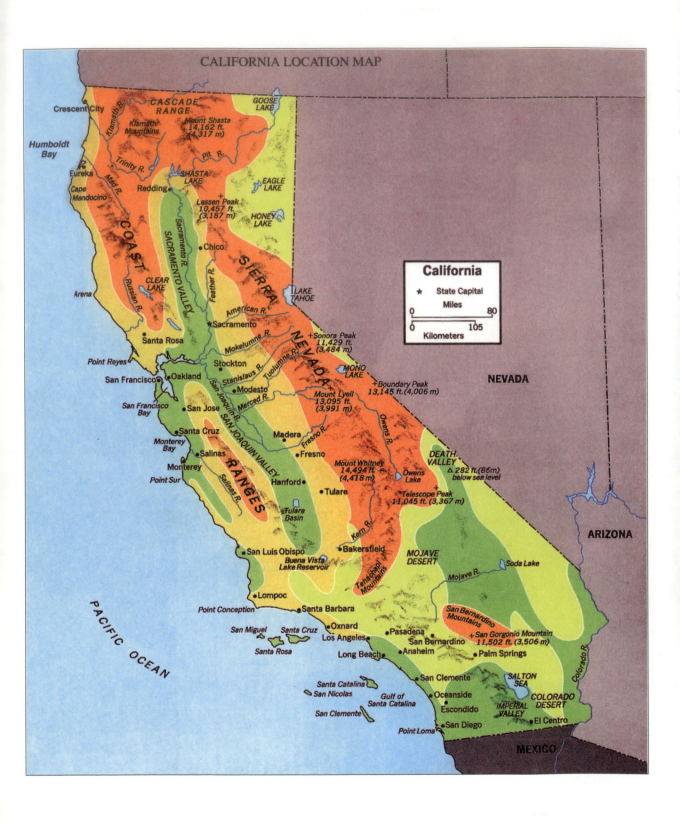